V. M. Popov

Hyperstability of Control Systems

With 8 Figures

Editura Academiei
București

Springer-Verlag
Berlin · Heidelberg · New York
1973

Vasile-Mihai Popov
Corresponding member of the Academy of the Socialist Republic of Romania

AMS Subject Classifications (1970)
Primary 93–02, 93C10, 93D15, 93D99

This is the revised version of
"HIPERSTABILITATEA SISTEMELOR AUTOMATE"

EDITURA ACADEMIEI REPUBLICII SOCIALISTE ROMÂNIA 1966
Str. Gutenberg 3 bis, Bucureşti

Translated by Radu Georgescu

ISBN 3-540-06373-0 Springer-Verlag Berlin Heidelberg New York
ISBN 0-387-06373-0 Springer-Verlag New York Heidelberg Berlin

Library of Congress Catalog Card Number 73-83000

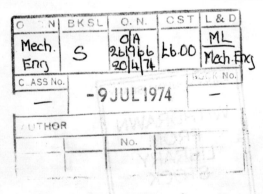

Die Grundlehren der mathematischen Wissenschaften

in Einzeldarstellungen
mit besonderer Berücksichtigung
der Anwendungsgebiete

Band 204

Herausgegeben von J. L. Doob A. Grothendieck E. Heinz
F. Hirzebruch E. Hopf W. Maak
S. MacLane W. Magnus J. K. Moser
M. M. Postnikov F. K. Schmidt
D. S. Scott K. Stein

Geschäftsführende Herausgeber B. Eckmann und B. L. van der Waerden

Die Grundlehren der mathematischen Wissenschaften

in Einzeldarstellungen
mit besonderer Berücksichtigung
der Anwendungsgebiete

Band 204

PREFACE

At the beginning of a book, we feel it necessary to give a brief account of our intentions even at the risk of oversimplifying the subject. The present work starts from the usual point of view of the control engineer who likes to have at his disposal a wide range of elements capable of being combined in various ways to form control systems as complex as desired, but who does not like to burden his creative imagination with instability problems. The "hyperstable blocks" introduced in the present work are characterized precisely by the property that they can be combined without loss of stability. They derive from the study of the concept of "hyperstability", introduced and developed by the author in previous papers where, however, it was treated fragmentarily.

The present work is the first comprehensive exposition of the subject and extends from general theory to applications. Being entirely devoted to this end, the work does not examine other related problems which are nevertheless the subject of extensive current research. The book does not even attempt to completely cover the previous results of the author in this field. The limitations of the book would have been more objectionable had there not existed such monographs as: A. Halanay, "The qualitative theory of differential equations", M. A. Aizerman and F. R. Gantmacher, "The absolute stability of control systems" and S. Lefschetz, "The stability of non-linear control systems", where complementary aspects of the problem of stability are amply discussed.

Intended as a self-contained unit, the work does not require from the reader more than a minimal background.

The author hopes that the book will offer a view of the many prospects for further developments as regards both the general theory and the applications and thus will incite the reader to personal contributions.

THE AUTHOR

Note

The present translation is based on a revised text of the work, prepared shortly after the publication of the Romanian original. The structure of the book and the main ideas remain the same, but the theorems take a more general form, since many simplifying assumptions have been removed.

More emphasis was placed on investigating the full implications of the frequency criteria — a leading theme for the whole work.

The book does not aim to reflect the particularly abundant literature published after 1966 on the problem of stability. We mention that the hyperstability of discrete systems has been recently and thoroughly investigated in a doctoral thesis by Sorin Gheţaru. We also mention the following recent monographs, related to our subject : E. B. Lee and L. Markus : "Foundations of optimal control theory", J. C. Hsu and A. U. Meyer : "Modern control principles and applications", D. D. Siljak : "Non-linear systems", R. W. Brockett : "Finite dimensional linear systems", J. L. Willems : "Stability theory of dynamical systems".

CONTENTS

Chapter 1. INTRODUCTION

Chapter 2. CLASSES OF EQUIVALENT SYSTEMS

Chapter 3. POSITIVE SYSTEMS

Chapter 5. APPLICATIONS

Appendix A. CONTROLLABILITY; OBSERVABILITY; NONDEGENERATION

INTRODUCTION

1. Stability as a property of a family of systems

The object of this monograph is the study of stability considered as a property not only of isolated systems but also of families of systems.

This point of view is most natural. In fact, the study of the stability of any isolated system leads immediately to conclusions concerning the stability of a family of systems.

Consider, for instance, the feedback system shown in Fig. 1.1, assuming that $\mu = 0$. Assume further that the input μ_1 and the output ν_1 of block B_1 are related by a linear differential equation with constant coefficients, and that the output ρ of block B_2 is related to the input ν of the same block by

$$\rho = \lambda \nu, \tag{1}$$

where λ is a constant. Assume also that the system admits a unique stationary solution, for which

$$\mu = \nu = \rho = 0. \tag{2}$$

Applying the methods described in any control textbook, one finds that if the stationary solution is asymptotically stable for a certain value λ_0 of λ, then this solution is also asymptotically stable for any value of λ contained in an interval of the form

$$\lambda_1 < \lambda < \lambda_2. \tag{3}$$

Thus if the system under investigation is asymptotically stable [1]) for $\lambda = \lambda_0$, then so also are all the neighbouring systems for which λ is contained in an interval of the form (3). Although,

---

[1]) If it is clear from the context which solution is being investigated for stability, we shall sometimes use the expression: "the stability of the system" which will mean "the stability of the solution" considered.

the magnitude of the interval (3) is unknown without further computations, one does, however, know that there exists a family of asymptotically stable systems characterized by (3).

One can extend the above result and find a considerably larger family of asymptotically stable systems if one resorts to the classical results of H. Poincaré and A. M. Liapunov on stability in the first approximation. Thus one finds that the stationary solution of the system under investigation is asymptotically stable even if the block B_2 is non-linear, the relation (1) being replaced by

$$\rho = \varphi(\nu), \tag{4}$$

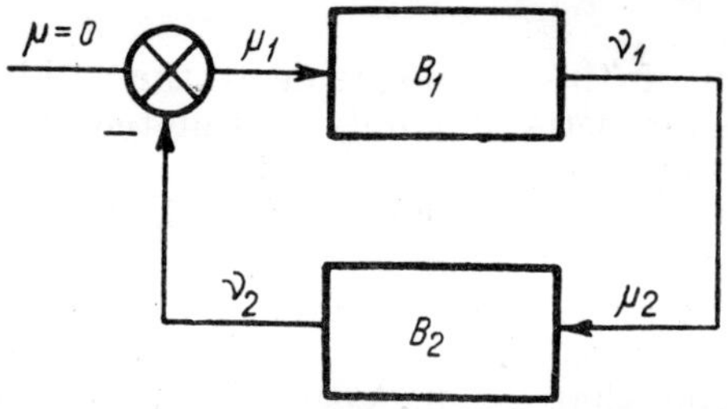

Fig. 1.1

where, for instance, φ is a continuously differentiable function such that $\varphi(0) = 0$ and

$$\lambda_1 < \frac{d\varphi(0)}{d\nu} < \lambda_2. \tag{5}$$

The above examples show only the simplest ways of obtaining information on the stability of a family of systems from a simple and unique criterion.

The extension of these results to more general cases constitutes a field of investigation of evident interest. This approach should lead to systematic methods of obtaining the maximum amount of information in stability problems from the least amount of data or computations. Having a comprehensive understanding of the behaviour of a whole family of different systems, one can more easily control the effects of the fortuitous changes occurring in such systems.

2. The families of systems considered in the problem of absolute stability

The families of systems in the foregoing examples are not precisely defined, since the numbers λ_1 and λ_2 in condition (3) have not been determined.

Liapunov's "second method" eliminates this drawback and leads to accurate conclusions regarding the stability of well-defined families of systems. This method made it natural to introduce the new notion of "absolute stability" whose origin can be traced to a work of A. I. Lur'e and V. N. Postnikov [1]. The results of the many investigations effected to date in the field of absolute stability have been presented in a number of monographs, from which we mention (in chronological order) A. I. Lur'e [1], A. M. Letov [1], A. Halanay

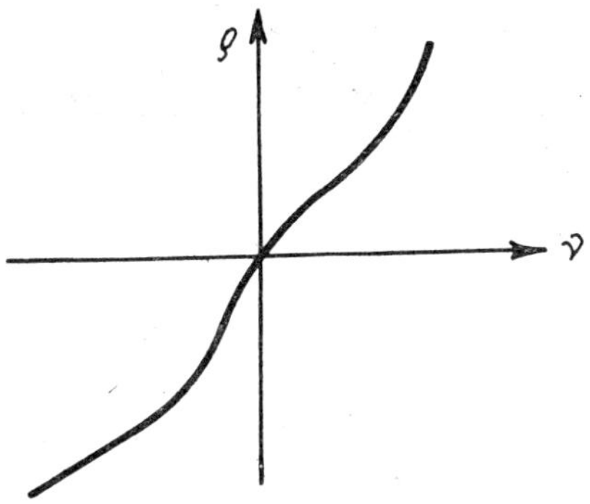

Fig. 1.2

[1], M. A. Aizerman and F. R. Gantmacher [1] and S. Lefschetz [1]. Without going into a detailed exposition of these results (see the final chapter of this book), we shall discuss here only the manner in which one defines the families of systems that are studied.

These systems are characterized by the fact that in Relation (4) — which describes the non-linear block B_2 (Fig. 1.1) — the function φ is continuous, vanishes for $\nu = 0$ and satisfies the inequality

$$\varphi(\nu)\nu > 0 \text{ for every } \nu \neq 0. \tag{6}$$

In other words, the graph of function φ is entirely contained in the quadrants I and III; it may have, for instance, a shape similar to that shown in Fig. 1.2.

The object of the study of absolute stability consists in finding a criterion which secures simultaneously the stability of all the systems characterized by Condition (6).

A direct generalization of Condition (6) is given by the inequalities

$$\lambda_1 \nu^2 < \varphi(\nu)\, \nu < \lambda_2 \nu^2 \text{ for every } \nu \neq 0, \tag{7}$$

which express the fact that the graph of the function φ is entirely located in the sector bounded by the lines $\rho = \lambda_1 \nu$ and $\rho = \lambda_2 \nu$ (see Fig. 1.3).

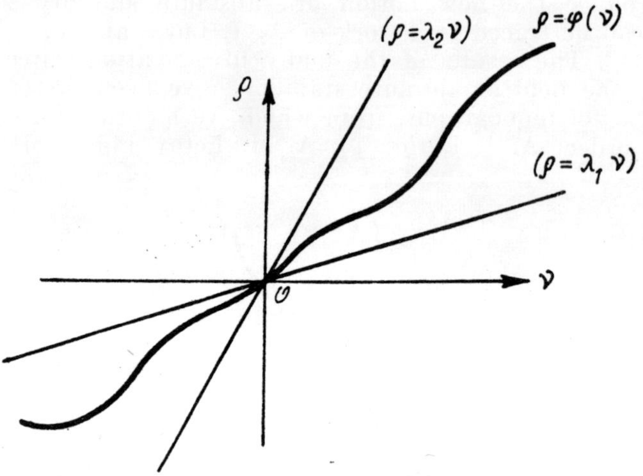

Fig. 1.3

The study of the families of systems characterized by the inequalities (7) received a strong impetus after M. A. Aizerman [1] stated the well-known problem which bears his name. Taking into account subsequent investigations, Aizerman's problem may be stated in the following way (which differs from the original statement) : Determine the most general conditions of the type (7) under which the system in Fig. 1.1 has the following stability property : every solution of the system tends to the stationary solution when time tends to infinity [1]).

The second method of Liapunov has been extensively used to obtain criteria for the stability of families of systems characterized by Condition (7).

3. Selecting the most natural families of systems

The choice of the family of systems for which we investigate the problem of stability is entirely up to us. Thus the first problem to be solved is to choose the studied family of system in the most "natural" way.

---

[1]) M. A. Aizerman [1] has conjectured that if all the linear systems which satisfy the inequality (7) for $\varphi(\nu) = \lambda\nu (\lambda = \text{const.})$ are stable, then the entire family characterized by Condition (7) has the same property. Further investigation (e.g. see V. A. Pliss [1]) has shown that this conjecture is not true in general, although is correct in many cases which are of practical interest.

To be more precise, we shall consider that a family of systems is more or less "natural" to the extent to which the following conditions are fulfilled:

A. *Large applicability*. The family of systems should be large enough to include as many practical systems as possible.

B. *Perfection of theoretical results*. The study of the family of systems should not leave any internal theoretical problems unsolved. In particular, the criteria for the simultaneous stability of all the systems of the family must be necessary and sufficient.

C. *Inclusion of all the known results*. The results obtained in the study of the families of systems should include as particular cases all the principal results known in the theory of the stability of control systems.

D. *Simplicity*. Both the characterization of the family of systems and the stability criteria obtained should be expressed in a simple form involving the least amount of calculation.

The above conditions, whose necessity is obvious, do not automatically determine the family of systems which is sought; however, they do considerably restrict the field of the possible selections.

4. Introducing new families of systems

In order to comply with requirements A, C and D we shall use in this monograph a modified form of the families of systems specified in Section 2. The important requirement B cannot be examined with the simple methods we use in the present Introduction; its fulfilment is in fact one of the main objectives of this book.

We have seen that in the problem of absolute stability, the non-linear block B_2 is described by a relation of the form (4), and this implies that, for a given value of the input there exists only one possible value of the output of the block. In most applications, the relation between the input and the output of an arbitrary block is best described by a dynamic relation, such as, for instance, a differential equation. It is therefore adequate (see requirement A), to include in the examined family also systems whose block B_2 is described by a differential equation or by some other dynamic relation different from Relation (4). Then, the input-output point $\nu(t)$, $\rho(t)$ will no longer be situated on a certain line (or characteristic) of the plane (ν, ρ).

The new family of blocks B_2 is to be described by certain relations which (see requirement C), must include the case of characteristics situated in a sector (see Fig. 1.3). For the sake of simplicity we shall consider the particular case in which the sector consists of the quadrants I and III (see Fig. 1.2). By a direct generalization of this condition we may require that the input-output point $(\nu(t), \rho(t))$ be situated, for every t, in the quadrants I and III, i.e.,

$$\rho(t)\,\nu(t) \geqslant 0 \text{ for every } t > 0. \tag{8}$$

Assume now that we have found a criterion securing the stability of the system for every dynamic block for which Condition (8) is satisfied. Intuitively (and this fact is justified rigorously in this monograph), it seems obvious that the property of stability should be maintained even if inequality (8) is violated on certain sufficiently small time intervals, but is satisfied "most of the time". This remark, together with Requirement A, suggests that Condition (8) need not be satisfied at every moment t, but only "in the mean". For simplicity (see requirement D) we consider the simplest mean and replace Condition (8) by the inequality

$$\int_0^t \rho(\tau)\,\nu(\tau)\,d\tau \leqslant 0 \text{ for every } t > 0. \tag{9}$$

Condition (9) is typical of the way in which we characterize families of systems in this monograph. In different parts of the work we shall also introduce families characterized by conditions which may be more general or more specialized than (9), but such conditions will always be strongly related to inequalities of the form (9).

5. The concept of hyperstability

The introduction of the families of systems characterized by conditions of the type (6) has led to the concept of absolute stability, which means stability for all the systems belonging to the family considered (see Section 2). Similarly, the introduction of the family of systems described by Condition (9) leads to a new concept of stability : stability for all the systems belonging to the new family of systems. To designate this concept, the author has introduced in [7] the term "hyperstability". Thus, by hyperstability we roughly mean the simultaneous stability of a collection of systems that satisfy a condition of the type (9).

In the particular case of the systems of the type shown in Fig. 1.1 where B_1 is given while B_2 is such that Condition (9) holds, the family of systems is completely defined as soon as B_1 is known. Therefore, we may consider that in this case the property of hyperstability is equivalent to the following property of block B_1: "Every system of the type shown in Fig. 1.1, consisting of our block B_1 associated with any other block B_2 which secures Condition (9), is stable". Instead of saying that the above property holds we shall use the shorter expression: "Block B_1 is hyperstable". Hyperstable blocks have remarkable properties which are investigated in Chapter 4 and used in most of the applications treated in the final chapter of the book (see also Appendix D).

There are, however, some applications which require a more general formulation of the problem of hyperstability. Observe that the family of systems of the type shown in Fig. 1.1, for which we have introduced the concept of hyperstability, is completely determined as soon as one knows (a) the integral which occurs in Condition (9) and (b) the law which describes the block B_1 and which is generally given by a differential equation. Thus, a differential equation and an integral contain all the information required for stating the problem of hyperstability. By a natural extension we can therefore state the problem of hyperstability for any set consisting of a differential equation and an integral, even if these are of much more general forms than those considered above. Such "objects" consisting of a differential equation and an integral will be the main object of study in this work.

6. Indications on the use of the monograph

The study of the systems on which the monograph is based begins with Chapter 2, where we examine various ways in which these systems can be described. These results are extensively used in the whole book.

Chapter 3 deals with the property of "positiveness". The results obtained in the study of this property, together with those obtained in Chapter 2, may also be applied in other problems, besides hyperstability.

The property of hyperstability is thoroughly investigated in Chapter 4 using the results of the previous chapters. Chapter 5 is devoted to applications.

The Appendices A, B and C contain some frequently used results which, strictly speaking, do not follow the logical line of Chapters 2—5, and which for this reason have been presented separately. The appendices may be read in the order preferred by the reader.

Appendix A includes results concerning controllability, observability and nondegeneration, properties which will be assumed to hold in some parts of the work. (The problem of hyperstability will, however, also be treated without resorting to these assumptions). The factorization of polynomial matrices — which arises only in connection with "multi-input" systems — is dealt with in Appendix B. Appendix C presents some properties of positive real functions. Appendix D includes a list of hyperstable blocks by means of which the reader may conceive new applications using the rules mentioned at the beginning of the appendix.

For the reader who is interested primarily in applications and for whom the general developments of the first chapters have a secondary interest, our manner of proceeding from the general to the particular will obviously cause some difficulty. From this point of view it may be advisable to begin with a cursory examination of Chapter 5 in order to obtain a rough idea of the applications and then take up reading of the rest of the monograph.

Each subject will be treated successively for various different types of systems (single-input, multi-input, discrete, non-linear, etc.). The exposition begins in each case with the simplest system and extends to more complicated ones.

In order to rapidly acquire an idea of the main results of the monograph it is recommended to restrict a first reading to the following paragraphs : §§ 1—4 of Chap. 2 ; § 8 of Chap. 3 ; §§ 13, 14, 15 and §§ 21, 22 and 25 of Chap. 4 ; §§ 25 and 27—30 of Chap. 5 ; §§ 31—33 of Appendix A and Appendices C and D.

In several paragraphs of Chap. 4 the proofs proceed from the simplified to the general case : in a first stage, the problem is solved under some additional conditions (e.g. complete controllability) ; the general proof is then completed in a second stage which can be omitted on a first reading. In all such cases we shall indicate the point where the reading may be interrupted so as to still obtain a complete proof of an important part of the theorem considered.

The multi-input systems have been treated in a much more succinct form under the assumption that the reader has already studied the single-input systems, and is thus able to carry out certain direct generalizations by himself. In a few cases the results have been stated briefly, with only some general indications concerning the proof.

Unless otherwise stated, small Greek letters will be used for scalars, small Latin letters for vectors and capital Latin letters for matrices.

The paragraphs are numbered starting from chapter 2 up to and including Appendix B. Each paragraph is treated as a self-contained unit and contains a body of results pertaining to a single main idea. In every paragraph the formulas are numbered starting from 1. Reference to a formula within a paragraph is made by simply quoting the number of the formula; reference to a formula from another paragraph is made by quoting also the number of the paragraph. Thus (m, § n) reads: formula (m) from paragraph n. The same rule applies also to the numbering and quotation of definitions, propositions and theorems stated in different paragraphs of the book.

A limited number of symbols maintain the same meaning throughout the work, while others change meaning from one paragraph to another. However, this should cause no confusion for the reader who follows the main idea of the paragraph he is reading. A list of the most common symbols used is given in Appendix E. The works quoted in the text can be found in Appendix F.

CHAPTER 2

CLASSES OF EQUIVALENT SYSTEMS

The main object of this chapter is the study of the different equivalent forms in which one can describe a system consisting of a differential equation and an integral. We will also show how one can pass from one form of description to another with the least amount of calculation.

Almost all the main ideas of this chapter are contained in §§ 2, 3 and 4 which treat single-input systems. The treatment of multi-input (§ 5) and discrete (§ 6) systems and of systems with time dependent coefficients (§ 7) follows the same scheme. Paragraph 1 includes a study of some special groups of transformations used throughout the chapter for the definition of the various classes of equivalent systems.

Another point treated in this chapter is the determination of certain quantities which are invariant under the mentioned groups of transformations (see § 3). Under the additional assumption that the systems investigated are completely controllable, one finds a "complete system of invariants", knowledge of which uniquely determines a class of equivalent systems (see § 4).

Various particular cases of the transformations studied in this chapter have been used previously by many authors for the purpose of simplifying the problems treated. A study of these transformations from the general point of view adopted in this chapter has been published in a previous paper of the author [16]. The main new points presented here are the introduction of the general transformations of § 1 and the determination of a complete system of invariants for completely controllable systems.

The reader who finds the first paragraph too general should take note that — besides indicating further possible generalizations not carried out in this book — this paragraph contains a general proof that the different classes of systems introduced in the rest of the chapter are equivalence classes. This may also be established directly by showing that the relations which define the classes are equivalence relations (i.e. are reflexive, symmetric and transitive).

§ 1. Equivalence classes for quadratic forms with relations between the variables

1. Transformations of quadratic forms with relations between the variables

Consider a quadratic form

$$\varphi(z) = z^* D z, \tag{1}$$

where z is a p-dimensional vector and D is a hermitian $p \times p$-matrix ($D^* = \bar{D}' = D$). Here and in the rest of the book (unless otherwise specified) all the vectors involved are finite sequences of complex numbers; the scalars and the entries of all the matrices are also complex numbers. Assume that z is required to satisfy the relation

$$Fz = 0, \tag{2}$$

where F is a $p \times p$-matrix. We assume that the matrix F is singular (otherwise (2) implies that $z = 0$ and the quadratic form (1) can take only the value zero). From (2) one obtains

$$z^*(T^*F + F^*T)\, z = 0, \tag{3}$$

where T is an arbitrary $p \times p$-matrix. Hence

$$\varphi(z) = z^*(D + T^*F + F^*T)\, z. \tag{4}$$

Effecting a change of variables

$$\tilde{z} = Sz, \tag{5}$$

where S is a $p \times p$-nonsingular-matrix ($\det S \neq 0$), the quadratic form (4) becomes

$$\varphi(S^{-1}\tilde{z}) = \tilde{z}^* \tilde{D} \tilde{z}, \tag{6}$$

where

$$\tilde{D} = (S^{-1})^* \, (D + T^*F + F^*T)\, S^{-1}. \tag{7}$$

Condition (2) becomes $FS^{-1}\tilde{z} = 0$. For convenience we multiply this equality on the left by the nonsingular matrix S, thus obtaining

$$\tilde{F}\tilde{z} = 0, \tag{8}$$

where

$$\tilde{F} = SFS^{-1}. \tag{9}$$

Consider the set $\mathcal{E}$ of the elements

$$W = (D\,;F), \quad (D^* = D). \tag{10}$$

In other words, any element of the set $\mathcal{E}$ consists of a pair of $p \times p$-matrices D and F, the matrix D being also hermitian. We shall adopt the following notation :

$$\text{“the equality } (\tilde{D}\,;\tilde{F}) = (D\,;F) \text{ means } \tilde{D} = D \text{ and } \tilde{F} = F\text{”}. \tag{11}$$

To any quadratic form (1) satisfying a relation of the form (2) there corresponds an element $(D\,;\,F)$ of the set $\mathcal{E}$.

Relations (7) and (9) define a transformation of an element $(D\,;\,F)$ of the set $\mathcal{E}$ into another element $(\tilde{D}\,;\,\tilde{F})$ of the same set. We consider the set of all the transformations of this form and also the set $\mathcal{G}$ of the corresponding “operators” of the form

$$\alpha = (S\,;\,T), \quad (\det\ S \neq 0). \tag{12}$$

In other words, every element of the set $\mathcal{G}$ consists of a pair of $p \times p$-matrices S and T, the matrix S being also nonsingular. Here too, we adopt the following convention :

$$\text{“the equality } (\tilde{S}\,;\tilde{T}) = (S\,;\,T) \text{ means } \tilde{S} = S \text{ and } \tilde{T} = T. \tag{13}$$

Using (11) we can write (7) and (9) in a more compact form

$$(\tilde{D}\,;\tilde{F}) = ((S^{-1})^*\,(D + T^*F + F^*T)\,S^{-1}\,;\,SFS^{-1}), \tag{14}$$

or in the symbolic form

$$(\tilde{D}\,;\tilde{F}) = (S\,;\,T) \perp (D\,;F), \tag{15}$$

where the sign $\perp$ has the meaning :

$$(S\,;\,T) \perp (D\,;F) = ((S^{-1})^*\,(D + T^*F + F^*T)\,S^{-1}\,;\,SFS^{-1}). \tag{16}$$

Relation (16) defines an "external law of composition" of the operator $(S\,;\,T)$ of the set $\mathcal{Q}$ and of the element $(D\,;\,F)$ of the set $\mathcal{E}$.

2. Successive transformations

If we apply a new transformation, of the same form as (14), by affixing the sign $\sim$ over all the symbols of (14), and if we eliminate $\tilde{D}$ and $\tilde{F}$ from the relations thus obtained, we can write the result in the form

$$(\tilde{\tilde{D}}\,;\,\tilde{\tilde{F}}) = (\check{S}\,;\,\check{T}) \perp (D\,;\,F), \tag{17}$$

where we have used Relation (16) (which gives the meaning of the sign $\perp$); the new operator $(\check{S}\,;\,\check{T})$ is defined by the relations

$$\check{S} = \tilde{S}S\,; \quad \check{T} = T + S^*\tilde{T}S, \tag{18}$$

or (see (13))

$$(\check{S}\,;\,\check{T}) = (\tilde{S}S\,;\,T + S^*\tilde{T}S). \tag{19}$$

Another way of writing the element $(\tilde{\tilde{D}}\,;\,\tilde{\tilde{F}})$ results from Relation (15) and from the relation obtained from it by affixing the distinctive sign $\sim$ over all the symbols; this leads to the equality

$$(\tilde{\tilde{D}}\,;\,\tilde{\tilde{F}}) = (\tilde{S}\,;\,\tilde{T}) \perp ((S\,;\,T) \perp (D\,;\,F)). \tag{20}$$

Relations (17), (19) and (20) suggest writing the operator $(\check{S}\,;\,\check{T})$ symbolically in the form

$$(\check{S}\,;\,\check{T}) = (\tilde{S}\,;\,\tilde{T}) \top (S\,;\,T), \tag{21}$$

where a new sign $\top$ has been introduced, whose meaning is as follows:

$$(\tilde{S}\,;\,\tilde{T}) \top (S\,;\,T) = (\tilde{S}S\,;\,T + S^*\tilde{T}S). \tag{22}$$

Relation (22) defines an "internal law of composition" in the set $\mathcal{Q}$ of the operators.

From Relations (20), (17) and (21) one obtains the equality

$$(\tilde{S}\,;\,\tilde{T}) \perp ((S\,;\,T) \perp (D\,;\,F)) = ((\tilde{S}\,;\,\tilde{T}) \top (S\,;\,T)) \perp (D\,;\,F), \tag{23}$$

which expresses the fact that the external law of composition defined by (17) is associative with respect to the internal law of composition (of the operators), defined by (22).

3. More about the group $\mathcal{G}$

The law (22) is defined for two arbitrary elements $(\tilde{S}\,;\,\tilde{T})$ and $(S\,;\,T)$ of the set $\mathcal{G}$ and the result obtained by applying the law (i.e., the element $(\tilde{S}S\,;\,T + S^*\tilde{T}S))$ belongs also to the set $\mathcal{G}$. Indeed, the set G is characterized — besides the fact that the matrices S and T have p rows and columns — , by the single condition that the matrix S be nonsingular. Therefore, we have only to note that the inequality $\det\,(\tilde{S}S) \neq 0$ is a consequence of the conditions $\det\,\tilde{S} \neq 0$ and $\det\,S \neq 0$.

We also remark that the internal law of composition (22) is associative, i.e., the relation

$$(\tilde{\tilde{S}}\,;\,\tilde{\tilde{T}}) \top ((\tilde{S}\,;\,\tilde{T}) \top (S\,;\,T)) = ((\tilde{\tilde{S}}\,;\,\tilde{\tilde{T}}) \top (\tilde{S}\,;\,\tilde{T})) \top (S\,;\,T), \qquad (24)$$

is valid for arbitrary elements $(\tilde{\tilde{S}}\,;\,\tilde{\tilde{T}})$, $(\tilde{S}\,;\,\tilde{T})$ and $(S\,;\,T)$ of the set $\mathcal{G}$. The set $\mathcal{G}$ contains a "neutral" element with respect to the law (22), namely the element

$$\varepsilon = (I_p\,;\,0), \qquad (25)$$

where I_p is the $p \times p$-identity matrix. This element satisfies the equalities

$$\varepsilon \top (S\,;\,T) = (S\,;\,T) \top \varepsilon = (S\,;\,T), \qquad (26)$$

for every element $(S\,;\,T)$ of the set $\mathcal{G}$.

Finally, for every element $(S\,;\,T)$ of $\mathcal{G}$, there exists in $\mathcal{G}$ a "symmetrical" (or "inverse") element

$$(S\,;\,T)^{-1} = (S^{-1}\,;\,(-S^{-1})^*\,TS^{-1}), \qquad (27)$$

which satisfies the relations

$$(S\,;\,T)^{-1} \top (S\,;\,T) = (S\,;\,T) \top (S\,;\,T)^{-1} = (I_p\,;\,0) = \varepsilon \qquad (28)$$

as one can easily see using (22).

The foregoing shows that $\mathcal{G}$ is a group under the law (22).

Moreover, the formula (23) shows that the law (16) is associative with respect to the law (22) of the group $\mathcal{G}$. We remark that the neutral element (25) of the group $\mathcal{G}$ is at the same time a neutral operator with respect to the law (16), since it satisfies the equality

$$(I_p\,;\,0) \perp (D\,;\,F) = (D\,;\,F), \qquad (29)$$

for every element $(D\,;\,F)$ of the set $\mathcal{E}$.

4. Partitioning of the set $\mathcal{E}$ into classes

We now introduce the following relation between the elements of the set $\mathcal{E}$:

An element $(\tilde{D};\tilde{F})$ of the set $\mathcal{E}$ is said to be in relation R_e to the element $(D;F)$ of the same set if there exists an operator $(S;T)$ which belongs to the group $\mathcal{G}$ and satisfies the relation

$$(\tilde{D};\tilde{F}) = (S;T) \perp (D;F). \tag{30}$$

It can be easily seen that the relation R_e is an equivalence relation, i.e., it is reflexive, symmetric and transitive. The reflexivity of R_e results from equality (29) which shows that any element $(D;F)$ is related through R_e to itself. The symmetry property of R_e results by observing that from (30) and Relations (23), (28) and (29) we have

$$\begin{aligned}(S;T)^{-1} \perp (\tilde{D};\tilde{F}) &= (S;T)^{-1} \perp ((S;T) \perp (D;F)) = \\ &= ((S;T)^{-1} \top (S;T)) \perp (D;F) = (D;F).\end{aligned} \tag{31}$$

Reading these equalities from right to left one sees that the element $(D;F)$ is related by R_e to the element $(\tilde{D};\tilde{F})$. Finally, the transitivity of R_e results by observing that, if beside Relation (30) we also consider the relation of the same form

$$(\tilde{\tilde{D}};\tilde{\tilde{F}}) = (\tilde{S};\tilde{T}) \perp (\tilde{D};\tilde{F}), \tag{32}$$

then we can write Relation (17); hence the element $(\tilde{\tilde{D}};\tilde{\tilde{F}})$ is related by R_e to the element $(D;F)$.

Since R_e is an equivalence relation the set $\mathcal{E}$ can be partitioned into classes [1]) according to the following:

Definition 1. *Two elements $(\tilde{D};\tilde{F})$ and $(D;F)$ of $\mathcal{E}$ are said to belong to the same class (with respect to the set $\mathcal{G}$) if there exists an operator $(S;T)$ in the set $\mathcal{G}$ such that one has relation (30) or the relations*

$$\tilde{D} = (S^{-1})^* (D + T^*F + F^*T)\, S^{-1}, \tag{33}$$

$$\tilde{F} = SFS^{-1}, \tag{34}$$

which — according to (16) — are equivalent to (30).

---

[1]) See N. Bourbaki [1], Chap. 1. § 7, Section 5 for a general theory which is particularly appropriate to our subject.

All the classes of equivalent systems introduced in the present chapter are defined by relations of the form (33), (34).

5. Other equivalence classes

In the sequel instead of the set $\mathcal{E}$ we consider certain subsets $\mathcal{E}'$, whose elements are of the form $(D;\ F)$ where, as before, D and F are $p \times p$-matrices and the matrix D is hermitian but the entries of these matrices satisfy some additional conditions. We also consider, instead of $\mathcal{G}$, subsets $\mathcal{G}'$ whose elements are of the form $(S;\ T)$ where, as before, S and T are $p \times p$-matrices and the matrix S is nonsingular but the elements of these matrices will be required to satisfy certain additional conditions. The subsets $\mathcal{E}'$ and $\mathcal{G}'$ will always have the following properties:

1° If two operators $(\tilde{S};\ \tilde{T})$ and $(S;\ T)$ belong to the subset $\mathcal{G}'$, then the operator $(\tilde{S}S;\ T+S^*\tilde{T}S)$, obtained by combining them according to the law (22), belongs to the subset $\mathcal{G}'$.

2° If an operator $(S;\ T)$ belongs to the subset $\mathcal{G}$, then the inverse operator $(S;\ T)^{-1}=(S^{-1};\ (-S^{-1})^*TS^{-1})$ (see (27)) belongs to the subset $\mathcal{G}'$.

3° If the element $(D;\ F)$ belongs to the subset $\mathcal{E}'$ and the operator $(S;\ T)$ to the subset $\mathcal{G}'$, then the element $((S^{-1})^*(D+T^*F+F^*T)S^{-1};\ SFS^{-1})$ (obtained by applying the law of composition (16) to the operator $(S;\ T)$ and to the element $(D;\ F)$), belongs to the subset $\mathcal{E}'$.

Under these conditions, all the arguments of the foregoing sections remain valid when we replace the sets $\mathcal{E}$ and $\mathcal{G}$ by the subsets $\mathcal{E}'$ and $\mathcal{G}'$, so that the following result may be stated:

Proposition 1. *If the subsets $\mathcal{E}'$ and $\mathcal{G}'$ have the properties specified above* [1]*), then the subset $\mathcal{E}'$ may be partitioned into*

[1]) More generally, if Conditions 1°—3° are suitably modified, Relations (35)—(36) still define certain equivalence classes in a subset $\mathcal{E}'$ of elements $(D;\ F)$, even if the subset $\mathcal{G}'$ of operators $(S;\ T)$ *depends* on the element $(D;F)$ of $\mathcal{E}'$ to which these operators are applied (i.e. if to every element $(D;\ F)\in\mathcal{E}'$ one may apply only the operators which satisfy the condition $(S;\ T)\in\mathcal{G}[D;F]$, where by $\mathcal{G}[D;\ F]$ we have denoted a subset of the set $\mathcal{G}$ depending on the element $(D;\ F)$. Re-examining the proofs given in Section 4, we see that this will take place when Conditions 1°—3° are replaced by the following ones:

1°′ If $(S;\ T)\in\mathcal{G}[D;F]$ and if $(\tilde{S};\ T)\in\mathcal{G}[\tilde{D};\tilde{F}]$ (where $(\tilde{D};\ \tilde{F})$ is given by (30)), then $(\tilde{S}S;\ \tilde{T}+S^*\tilde{T}S)\in\mathcal{G}[D;F]$.

2°′ If $(S;\ T)\in\mathcal{G}[D;\ F]$, then $(S^{-1};\ (-S^{-1})^*TS^{-1})\in\mathcal{G}[\tilde{D};\ \tilde{F}]$, where $(\tilde{D};\ \tilde{F})$ is given by (30).

3°′ If $(D;\ F)\in\mathcal{E}'$ and if $(S;\ T)\in\mathcal{G}[D;\ F]$, then

$$((S^{-1})^*(D+T^*F+F^*T)S^{-1};\ SFS^{-1})\in\mathcal{E}'.$$

classes as follows: two elements $(\tilde{D}; \tilde{F})$ and $(D; F)$ of the subset $\mathcal{E}'$ belong to the same class with respect to the subset $\mathcal{G}'$ if there exists an operator $(S; T)$ of the subset $\mathcal{G}'$ such that the two elements satisfy the relations:

$$\tilde{D} = (S^{-1})^* (D + T^*F + F^*T) S^{-1}, \tag{35}$$

$$\tilde{F} = SFS^{-1}. \tag{36}$$

§ 2. Classes of single-input systems

1. The system

Consider a system consisting of the differential equation

$$\frac{\mathrm{d}x}{\mathrm{d}t} = Ax + b\mu \tag{1}$$

and the integral

$$\eta(0, t_1) = [x^*Jx]_0^{t_1} + \int_0^{t_1} (\varkappa\mu^*\mu + \mu^*l^*x + x^*l\mu + x^*Mx)\,\mathrm{d}t. \tag{2}$$

In Eq. (1) A is a constant $n \times n$-matrix and b is a constant n-vector. In the same equation, μ (the "control function") is a scalar-valued function of t defined and piecewise continuous for $t \geqslant 0$ and x is a vector-valued function such that $x(t)$ is an n-vector representing the "state" of the system at time t.

In (2), J and M are hermitian constant $n \times n$-matrices, $\varkappa$ is a real scalar and l is an n-vector. The meaning of the first term of the right hand number of (2) is given by the equalities

$$[x^*Jx]_0^{t_1} = \int_0^{t_1} \frac{\mathrm{d}}{\mathrm{d}t}(x^*Jx)\,\mathrm{d}t = \int_0^{t_1} \left(\frac{\mathrm{d}x^*}{\mathrm{d}t} Jx + x^*J\frac{\mathrm{d}x}{\mathrm{d}t}\right)\mathrm{d}t. \tag{3}$$

To simplify the notations we always use x and μ instead of $x(t)$ and $\mu(t)$ under the integral sign and in equations like (1).

According to our general convention, we use the field of complex numbers. However $\varkappa$ and t are real [1]).

In order to simplify the developments which follow and use the general scheme of § 1, we shall write Relations (1), (2) in a more compact form. Introducing the vector-valued function z, with $2n + 1$ components, defined by the relation

$$z' = \left(\frac{dx'}{dt} \quad \mu \quad x' \right), \tag{4}$$

and using (3), we obtain a more compact expression of the integral $\eta(0, t)$ (see (2)) :

$$\eta(0, t_1) = \int_0^{t_1} z^* D z \, dt, \tag{5}$$

Here

$$D = \begin{pmatrix} \overbrace{0}^{n} & \overbrace{0}^{1} & \overbrace{J}^{n} \\ 0 & \varkappa & l^* \\ J & l & M \end{pmatrix} \begin{matrix} \}n \\ \}1 \\ \}n \end{matrix} \tag{6}$$

where we have also indicated the number of rows and columns of the blocks into which the matrix D is partitioned.

Using (4), Eq. (1) may be written as

$$Fz = 0, \tag{7}$$

where F is a matrix with $2n + 1$ rows and columns, given by

$$F = \begin{pmatrix} -I & b & A \\ 0 & 0 & 0 \\ 0 & 0 & 0 \end{pmatrix} \tag{8}$$

where I is the $n \times n$ identity matrix and every element of (8) represents a matrix, a vector or a scalar with the same number of dimensions as the corresponding element of the matrix D (6).

---

[1]) Therefore, the integral η (0, t) is always real.

Throughout this chapter the integration of (5) is not required and we may study only the integrand, i.e., the quadratic form $z^* D z$. This, together with Condition (7), constitutes a system of the type examined in § 1 and, therefore, we can apply the results of that paragraph.

To every system (5), (7) there corresponds an element

$$w = (D;\ F), \tag{9}$$

which can be written more explicitly by specifying the elements of the matrices D and F given by (6) and (8),

$$w = (J,\ \varkappa,\ l,\ M;\ A,\ b). \tag{10}$$

2. Transformations

As shown in § 1, the values of the quadratic form φ are not altered when we add a matrix such as $T^*F + F^*T$ to the matrix D of the quadratic form (see (3, § 1)).

We need consider only matrices T of the particular form

$$T = \begin{pmatrix} 0 & 0 & N \\ 0 & 0 & 0 \\ 0 & 0 & 0 \end{pmatrix}, \quad (N^* = N), \tag{11}$$

where N is a hermitian $n \times n$-matrix. Then (11) and (8) give

$$T^*F + F^*T = \begin{pmatrix} 0 & 0 & -N \\ 0 & 0 & b^*N \\ -N & Nb & NA + A^*N \end{pmatrix}. \tag{12}$$

Since N is a hermitian matrix, the expression above is of the same form as in (6).

We examine now the effect of a transformation of the form (5, § 1). We replace μ and x of (1) and (2) by $\tilde{\mu}$ and $\tilde{x}$, given by

$$\tilde{\mu} = \rho\mu - \rho q^*\ x, \tag{13}$$

$$\tilde{x} = Rx, \tag{14}$$

where ρ is a non-zero scalar constant, q is an n-vector and R is a nonsingular $n \times n$-matrix. Here again we use complex numbers.

The above transformation is less general than (5, § 1) since in (14) there is no term in μ. Therefore $\tilde{x}$ is differentiable and its derivative satisfies an equation of the form (1). The coefficient of x in (13) is written in a form which simplifies some of the subsequent formulas.

Introduce, in addition to z (see (4)), the similar expression

$$\tilde{z}' = \begin{pmatrix} \frac{d\tilde{x}'}{dt} & \tilde{\mu} & \tilde{x} \end{pmatrix}. \tag{15}$$

Owing to Relations (13) and (14), z and $\tilde{z}$ satisfy the relation

$$\tilde{z} = Sz, \tag{16}$$

where

$$S = \begin{pmatrix} R & 0 & 0 \\ 0 & \rho & -\rho q^* \\ 0 & 0 & R \end{pmatrix}. \tag{17}$$

From the conditions $\rho \neq 0$ and $\det R \neq 0$ specified above it follows that the matrix S is nonsingular.

The set $\mathcal{Q}'$, whose elements $(S\,;\,T)$ are given by the matrices (11) and (17), is a subset of the set $\mathcal{Q}$ introduced in § 1 (for $p = 2n + 1$). The subset $\mathcal{Q}'$ is characterized by the property that if the matrices S and T are partitioned as in (11) and (17), certain elements are equal to zero (as specified in (11) and (17)), and N is a hermitian matrix.

Likewise, the set $\mathcal{E}'$, whose elements $(D\,;\,F)$ are given by the matrices (6) and (8), is a subset of the set $\mathcal{E}$ of § 1. This subset is characterized by the property that if the matrices D and F are partitioned as in (6) and (8), certain elements are equal to zero or to $-I$, as specified in (6) and (8), and J is a hermitian matrix.

The set $\mathcal{E}'$ can be partitioned into classes by using Proposition 1, § 1. To that end we must show that the subsets $\mathcal{E}'$ and $\mathcal{Q}'$ satisfy Properties 1°—3°, Section 5, § 1. Property 1° follows from the remark that, if the matrices $\tilde{S}$ and $\tilde{T}$ are also given by the expressions (11) and (17) where all symbols bear the overscript $\sim$, then the matrices $\tilde{S}S$ and $T + S^*\tilde{T}S$ are expressed by

$$\tilde{S}S = \begin{bmatrix} \tilde{R}R & 0 & 0 \\ 0 & \tilde{\rho}\rho & -\tilde{\rho}\rho(q^* + \tilde{q}^*R) \\ 0 & 0 & \tilde{R}R \end{bmatrix}, \tag{18}$$

$$T + S^*\tilde{T}S = \begin{pmatrix} 0 & 0 & N + R^*\tilde{N}R \\ 0 & 0 & 0 \\ 0 & 0 & 0 \end{pmatrix} \tag{19}$$

and hence they have the form (11) and (17). Property 2° is verified by computing the matrix

$$S^{-1} = \begin{bmatrix} R^{-1} & 0 & 0 \\ 0 & \rho^{-1} & q^*R^{-1} \\ 0 & 0 & R^{-1} \end{bmatrix}, \tag{20}$$

which has obviously the form (17). As in the case of (19), one sees at once that the matrix $(-S^{-1})^* TS^{-1}$ is of the form (11). Thus, we find that Property 2° is satisfied. Finally, using Relations (20), (6), (12), (17) and (8) one can write the matrices $(S^{-1})^*(D + T^*F + F^*T)S^{-1}$ and SFS^{-1} in the form

$$(S^{-1})^* (D + T^*F + F^*T) S^{-1} = \begin{pmatrix} 0 & 0 & \tilde{J} \\ 0 & \tilde{\varkappa} & \tilde{l}^* \\ \tilde{J} & \tilde{l} & \tilde{M} \end{pmatrix}, \tag{21}$$

$$SFS^{-1} = \begin{pmatrix} -I & \tilde{b} & \tilde{A} \\ 0 & 0 & 0 \\ 0 & 0 & 0 \end{pmatrix} \tag{22}$$

where $\tilde{A}$, $\tilde{b}$, $\tilde{J}$, $\tilde{\varkappa}$, $\tilde{l}$ and M are expressed by

$$\tilde{A} = R(A + bq^*) R^{-1}, \tag{23}$$

$$\tilde{b} = \frac{1}{\rho} Rb, \tag{24}$$

$$\tilde{J} = (R^{-1})^* (J - N) R^{-1}, \tag{25}$$

$$\tilde{\varkappa} = \frac{1}{\rho^*\rho}\varkappa, \tag{26}$$

$$\tilde{l} = \frac{1}{\rho}(R^{-1})^* (\varkappa q + l + Nb), \tag{27}$$

$$\tilde{M} = (R^{-1})^*(\varkappa qq^* + ql^* + lq^* + M + N(A + bq^*) + (A + \\ + bq^*)^* N) R^{-1}. \tag{28}$$

Consequently, Property 3° is also satisfied. Using Proposition 1, § 1 one can establish the conditions under which a system of the form (1), (2) belongs to the same class as another system of the same form

$$\frac{d\tilde{x}}{dt} = \tilde{A}\tilde{x} + \tilde{b}\tilde{\mu} \tag{29}$$

$$\tilde{\eta}(0,t_1) = [\tilde{x}^*\tilde{J}\tilde{x}]_0^{t_1} + \int_0^{t_1} (\tilde{\varkappa}\tilde{\mu}^*\tilde{\mu} + \tilde{\mu}^*\tilde{l}^*\tilde{x} + \tilde{x}^*\tilde{l}\tilde{\mu} + \tilde{x}^*\tilde{M}\tilde{x})\, dt. \tag{30}$$

Definition 1. *System (1), (2) is said to belong to the same class as system (29), (30) (with respect to the set of matrices S, (17) and T (11)), if there exists a scalar $\rho \neq 0$, a vector q, a nonsingular matrix R and a hermitian matrix N such that the coefficients of the two systems satisfy the relations (23)—(28) (or in a more compact notation $\tilde{D} = (S^{-1})^*(D + T^*F + F^*T)S^{-1}$, $\tilde{F} = SFS^{-1}$).*

For simplicity we shall use the following shorter expression : "the systems belong to the same class", and add "with respect to the set of matrices S (17) and T (11)" only when confusion may arise.

If follows from Definition 1 that two systems belonging to the same class are related by a transformation of the type considered in Section 2 ; hence, we have the following :

Proposition 1. If the systems (1), (2) and (29), (30) belong to the same class in the sense of Definition (1), then for every pair of functions, μ and x which satisfy the differential Eq. (1), we have the following properties :

1° *Equation (29) is satisfied by the functions $\tilde{\mu}$ and $\tilde{x}$ given by the formulas (13) and (14)*

$$\tilde{\mu} = \rho\mu - \rho q^* x, \qquad \tilde{x} = Rx, \tag{31}$$

where ρ, q and R are the same as in Definition 1.

2° *The equality*

$$\tilde{z}^* \tilde{D} \tilde{z} = z^* Dz, \tag{32}$$

is satisfied, where the functions z and $\tilde{z}$ are obtained by introducing in (4) and (5) the functions μ, x, $\tilde{\mu}$ and $\tilde{x}$, considered above.

3° *The integrals $\eta(0, t_1)$ and $\tilde{\eta}(0, t_1)$ are equal for every positive number t_1 and for the functions μ, x, $\tilde{\mu}$ and $\tilde{x}$ considered above.*

The properties stated in Proposition 1 represent only another way of saying that the two systems are obtained from one another by a transformation of the type described in Section 2. Property 3° is an immediate and obvious consequence of Property 2°. Furthermore, the functions z and $\tilde{z}$ satisfy the relation $\tilde{z} = Sz$, where S is expressed by Relation (17) in which ρ, q and R are the same as in (23)—(28).

3. Some particular transformations

If in expression (11) one takes $N = 0$ and one requires that matrix S (17) be of the form

$$S = \begin{pmatrix} R & 0 & 0 \\ 0 & 1 & 0 \\ 0 & 0 & R \end{pmatrix}, \tag{33}$$

which happens if one introduces in (17) the restrictions

$$\rho = 1, \quad q = 0, \tag{34}$$

then one obtains a subset $\mathcal{G}'_s$ of the set $\mathcal{G}'$. One can easily see that $\mathcal{G}'_s$ is a subgroup of the group $\mathcal{G}'$. From (34), it follows that the transformations (13) and (14) modify only the state x and leave the control function μ unchanged. It follows at once that by applying these "state transformations" one obtains new classes of equivalent systems which, obviously, are obtained from the general classes described in Definition 1 by introducing in formulas (23)—(28) the restrictions : $\rho = 1$, $q = 0$ and $N = 0$.

Likewise, we obtain another subset of the set $\mathcal{G}'$ if $N = 0$ and if S has the form

$$S = \begin{pmatrix} I & 0 & 0 \\ 0 & \rho & -\rho q^* \\ 0 & 0 & I \end{pmatrix}, \tag{35}$$

(This happens if in (17) we take $R = I$ and $N = 0$). This defines another subgroup $\mathcal{G}'_c$ of the group $\mathcal{G}'$. Using the formulas (13) and (14) and the restriction $R = I$ one sees that these "control transformations" modify only the control function leaving unchanged the state x.

Finally, when the matrix S is equal to the identity matrix with $2n+1$ rows and columns, one obtains a new subgroup $\mathcal{G}'_n$ of $\mathcal{G}'$. Then both the state vector x and the control function μ remain unchanged and only the matrix D of the quadratic form z^*Dz is modified. The transformations thus obtained are precisely those described at the beginning of Section 2 and consist of the addition of a matrix of the form $T^*F + F^*T$ to the matrix D of the quadratic form.

By introducing the above subgroups we obtain at the same time new equivalence-classes which are easily deduced from the general classes described in Definition 1.

4. The polarized system and its properties

The results concerning the system (1), (2) can be immediately extended to the following more general system

$$\frac{\mathrm{d}x^\Delta}{\mathrm{d}t} = x^\Delta A^* + \mu^\Delta b^*, \qquad \frac{\mathrm{d}x}{\mathrm{d}t} = Ax + b\mu, \tag{36}$$

$$\eta_p(0, t_1) = \int_0^{t_1} z^\Delta Dz \,\mathrm{d}t, \tag{37}$$

where

$$z = \begin{pmatrix} \frac{\mathrm{d}x'}{\mathrm{d}t} & \mu & x' \end{pmatrix}, \quad z^\Delta = \begin{pmatrix} \frac{\mathrm{d}x^\Delta}{\mathrm{d}t} & \mu^\Delta & x^\Delta \end{pmatrix}. \tag{38}$$

In (36), (37) besides μ and x one finds the new variables μ^Δ and x^Δ which have the same significance and properties as μ and x except that $x^\Delta(t)$ is a row vector whereas $x(t)$ is a column vector. We shall say that the system (36), (37) is "polarized" and results from the polarization of the system (1), (2). We may say that the initial system is obtained from the polarized system if $\mu^\Delta = \mu^*$ and $x^\Delta = x^*$.

Notice that (36)—(37) is obtained from (1)—(2) by considering, besides Eq. (1), the conjugate-transpose equation $\mathrm{d}x^*/\mathrm{d}t = x^*A^* + \mu^*b^*$ and then replacing everywhere x^* by x^Δ and μ^* by μ^Δ. More generally, to "polarize" an arbitrary relation which is linear in x, μ and z, we introduce the conjugate-transpose relation and then replace x^* by x^Δ, μ^* by μ^Δ and z^* by z^Δ. This rule allows us to polarize all the developments of the foregoing sections. Thus, the operations performed in this paragraph are also valid for the polarized system. Since the polarized and the initial system have the same coefficients, the partition of the set of our systems into classes, given by Definition 1, extends immediately to polarized systems.

The introduction of the polarized system is useful owing to the following result, which is obtained from Proposition 1 by applying the polarization rule stated above:

Proposition 2. *Assume that the polarized system (36), (37) belongs to the same class as another system of the same form*

$$\frac{\mathrm{d}\tilde{x}^\Delta}{\mathrm{d}t} = \tilde{x}^\Delta \tilde{A}^* + \tilde{\mu}^\Delta \tilde{b}^*, \qquad \frac{\mathrm{d}\tilde{x}}{\mathrm{d}t} = \tilde{A}\tilde{x} + \tilde{b}\,\tilde{\mu}, \tag{39}$$

$$\eta_p(0, t_1) = \int^{t_1} \tilde{z}^\Delta \tilde{D} \tilde{z}\,\mathrm{d}t, \tag{40}$$

and hence the coefficients of the system (39), (40) and those of the system (36), (37) are connected by relations of the form (23)—(28), where $\rho \neq 0$, $\det R \neq 0$ and $N^* = N$. Then for every set of functions μ, x, μ^Δ and x^Δ which satisfy Eqs. (36), the following properties hold:

1°. *Equalities (39) are satisfied by the transformed functions $\tilde{\mu}$, $\tilde{x}$, $\tilde{\mu}^\Delta$ and $\tilde{x}^\Delta$, given by the relations (see (13) and (14) and the polarization rule):*

$$\tilde{\mu} = \rho\mu - \rho q^* x, \qquad \tilde{x} = Rx, \tag{41}$$

$$\tilde{\mu}^\Delta = \mu^\Delta \rho^* - x^\Delta q \rho^*, \qquad \tilde{x}^\Delta = \tilde{x}^\Delta R^*, \tag{42}$$

where ρ, q and R take the same values as in Relations (23)—(28) (which connect the coefficients of the two systems).

2°. *For every set of functions as in 1° one has*

$$\tilde{z}^\Delta \tilde{D} \tilde{z} = z^\Delta D z, \tag{43}$$

where z, z^Δ, $\tilde{z}$ and *$\tilde{z}^\Delta$ are given by (38) and by the expressions obtained from (38) after the introduction of the superscript $\sim$.*

3°. *For the functions introduced above, the integrals $\eta_p(0, t_1)$ and $\tilde{\eta}_p(0, t_1)$ are equal, for every positive number t_1.*

Just as in the case of Proposition 1, Property 3° is an immediate consequence of Property 2°.

The introduction of the polarized system simplifies considerably the calculations in the next paragraph.

§ 3. The characteristic polynomial of single-input systems

1. The characteristic function of single-input systems

Let us consider the system (36, § 2), (37, § 2) resulting from the polarization of the initial system (1, § 2), (2, § 2). If the control function μ has the form

$$\mu(t) = \mathrm{e}^{\sigma t}, \tag{1}$$

where σ is an arbitrary complex number, different from the eigenvalues of the matrix A, then the equation $\mathrm{d}x/\mathrm{d}t = Ax + b\mu$ (see (36, § 2)) has a particular solution of the form

$$x(t) = x_0(\sigma)\, \mathrm{e}^{\sigma t}. \tag{2}$$

By substituting (2) into the equation considered we uniquely determine the coefficient $x_0(\sigma)$:

$$x_0(\sigma) = (\sigma I - A)^{-1} b. \tag{3}$$

The condition that σ should be different from the eigenvalues of the matrix A is expressed as $\det(\sigma I - A) \neq 0$ and makes it possible to perform the operations specified in (3).

The vector $z(t)$, obtained by substituting (1) and (2) into (38, § 2), has the form

$$z(t) = z_0(\sigma)\, \mathrm{e}^{\sigma t}, \tag{4}$$

where

$$z_0(\sigma) = \begin{pmatrix} \sigma(\sigma I - A)^{-1} b \\ 1 \\ (\sigma I - A)^{-1} b \end{pmatrix}. \tag{5}$$

Treating similarly the equation $\mathrm{d}x^\Delta/\mathrm{d}t = x^\Delta A^* + \mu^\Delta b^*$ (see 36, § 2), one obtains that for the control function

$$\mu^\Delta(t) = \mathrm{e}^{\lambda t}, \tag{6}$$

where $\det(\lambda I - A) \neq 0$, the equation has the particular solution

$$x^{\Delta}(t) = e^{\lambda t} x_0^{\Delta}(\lambda), \tag{7}$$

where

$$x_0^{\Delta}(\lambda) = b^*(\lambda I - A^*)^{-1}. \tag{8}$$

Hence, the vector $z^{\Delta}(t)$ is expressed by

$$z^{\Delta}(t) = e^{\lambda t} z_0^{\Delta}(\lambda), \tag{9}$$

where [1])

$$z_0^{\Delta}(\lambda) = (\lambda b^*(\lambda I - A^*)^{-1} \quad 1 \quad b^*(\lambda I - A^*)^{-1}) = \bar{z}_0'(\lambda). \tag{10}$$

Substituting (4) and (6) into $z^{\Delta} D z$ (see (37, § 2)), we have

$$z^{\Delta}(t)\, D z(t) = \chi(\lambda, \sigma)\, e^{(\sigma+\lambda)t}, \tag{11}$$

where $\chi(\lambda, \sigma)$ is expressed by

$$\chi(\lambda, \sigma) = \bar{z}_0'(\lambda)\, D\, z_0(\sigma). \tag{12}$$

The function $(\lambda, \sigma) \mapsto \chi(\lambda, \sigma)$, (12), will be called the "characteristic" function" of the system (1, § 2), (2, § 2) or of the polarized system (36, § 2), (37, § 2).

Replacing in (12) the vectors $z_0(\sigma)$ and $\bar{z}_0'(\lambda)$ by their values (5) and (10), and the matrix D by its expression (6, § 2) we obtain the following explicit form of the characteristic function

$$\begin{aligned}\chi(\lambda, \sigma) = \varkappa + l^*(\sigma I - A)^{-1} b + b^*(\lambda I - A^*)^{-1} l + \\ + b^*(\lambda I - A^*)^{-1}(M + (\lambda + \sigma) J)(\sigma I - A)^{-1} b,\end{aligned} \tag{13}$$

for every ν and σ such that $\det(\nu I - A^*) \neq 0$ and $\det(\sigma I - A) \neq 0$ [2]). Clearly the characteristic function satisfies the equality [3]):

$$\chi(\lambda, \sigma) = \bar{\chi}(\sigma, \lambda) \tag{14}$$

---

[1]) We have $\bar{z}(\lambda) = \overline{z(\bar{\lambda})}$ and $\bar{z}_0'(\lambda) = \overset{*}{z}_0(\lambda) = \overline{z_0'(\bar{\lambda})}$ (see Appendix D).

[2]) In the following, whenever we encounter $(\sigma I - A)^{-1}$ (or $(\lambda I - A)^{-1}$) we will tacitly introduce the restriction $\det(\sigma I - A) \neq 0$ (or $\det(\lambda I - A^*) \neq 0$). However, all the relations expressed in terms of polynomials of ν and σ hold for every set of values of ν and σ.

[3]) We have $\bar{\chi}(\sigma, \nu) = \overline{\gamma(\bar{\sigma}, \bar{\nu})}$ (see Appendix D).

2. The characteristic polynomial and its properties

It follows from expression (5) that every element of z_0 is a quotient of two polynomials and that the denominator is equal to det $(\sigma I - A)$, while the degree of the numerator can not exceed the number n of rows and columns of the matrix A. Therefore, all the elements of the function $z_0(\sigma)$ det $(\sigma I - A)$ are polynomials whose degree cannot exceed the number n. A similar statement is also valid for det $(\lambda I - A^*)\hat{z}_0^{\wedge}(\lambda)$. Using these remarks and Relation (12) one can define the polynomial

$$\det(\lambda I - A^*)\,\chi(\lambda, \sigma)\det(\sigma I - A) = \sum_{j=0}^{n}\sum_{k=0}^{n} \varkappa_{jk}\lambda^j\sigma^k, \tag{15}$$

where $\varkappa_{jk}$ are constants. From (14) and (15) one obtains

$$\varkappa_{jk} = \overline{\varkappa}_{kj}, \qquad k, j = 1, 2, \ldots, n. \tag{16}$$

Let us introduce a positive normalizing factor ν, having the value

$$\nu = \begin{cases} \max |\varkappa_{jk}| & \text{if there exists a coefficient } \varkappa_{jk} \neq 0 \\ 1 & \text{if all the coefficients } \varkappa_{jk} \text{ are zero.} \end{cases} \tag{17}$$

Normalizing the polynomial (16) with the help of the factor ν, one obtains the polynomial

$$\pi(\lambda, \sigma) = \frac{1}{\nu}\det(\lambda I - A^*)\,\chi(\lambda, \sigma)\det(\sigma I - A). \tag{18}$$

which is called the "characteristic polynomial" of the system (1, §2), (2, §2) or of the polarized system (36, §2), (37, §2). (There is no danger of confusing the above polynomial with the characteristic polynomial of a matrix since the polynomial (18) depends on two complex variables, λ and σ.)

The characteristic polynomial, like the polynomial (15) from which it is derived, has the property

$$\pi(\lambda, \sigma) = \overline{\pi}(\sigma, \lambda). \tag{19}$$

Hence it can be written in the form

$$\pi(\lambda, \sigma) = \sum_{j=0}^{n}\sum_{k=0}^{n} \gamma_{jk}\lambda^j\sigma^k, \tag{20}$$

where (see 16))

$$\gamma_{jk} = \varkappa_{jk}/\nu, \qquad \gamma_{jk} = \overline{\gamma}_{kj}, \qquad k, j = 1, 2, \ldots, n. \tag{21}$$

Moreover, owing to the normalizing operation, we find that the coefficients γ_{jk} also have the property

$$\max |\gamma_{jk}| = \text{zero or } 1. \tag{22}$$

In most applications it is not necessary to know the exact value of the normalizing coefficient ν since it is automatically eliminated from the results.

3. Relations between the characteristic functions of systems belonging to the same class

Consider two polarized systems belonging to the same class (see Section 4, § 2) having the same form and the same properties as the systems (36, § 2), (37, § 2) and (39, § 2), (40, § 2) mentioned in the statement of Proposition 2, § 2. The functions μ, x, μ^{Δ} and x^{Δ} given by Relations (1), (2), (6) and (7) satisfy the conditions of Proposition 2, § 2. We compute the transformed functions given by Relations (41, § 2) and (42, § 2). The first of the Eqs. (41, § 2) yields

$$\tilde{\mu}(t) = \rho\,\delta(\sigma)\,e^{\sigma t}, \tag{23}$$

where

$$\delta(\sigma) = 1 - q^*(\sigma I - A)^{-1}b. \tag{24}$$

Using the second of the Relations (41, § 2) one can deduce the expression for $\tilde{x}(t)$. However, we shall not use that expression directly; we shall retain only the fact that it has the form $\tilde{x}(t) = \tilde{x}_0(\sigma)e^{\sigma t}$, where $\tilde{x}_0(\sigma)$ does not depend on t. Using also the fact that the functions $\tilde{\mu}$ and $\tilde{x}$ satisfy the equation $d\tilde{x}/dt = \tilde{A}\tilde{x} + \tilde{b}\mu$ (see Property 1° in Proposition 2, § 2), we can deduce the value of $\tilde{x}_0(\sigma)$ in the same way as we proceeded in the case of equality (2); since this value is uniquely determined, one obtains

$$\tilde{x}(t) = \tilde{x}_0(\sigma)\,\rho\,\delta(\sigma)\,e^{\sigma t}, \tag{25}$$

where $\tilde{x}_0(\sigma)$ is of the same form as in (3)*).

Using (23), (25) and (38, § 2) one obtains

$$\tilde{z}(t) = \tilde{z}_0(\sigma)\,\rho\,\delta(\sigma)\,e^{\sigma t}, \tag{26}$$

*) See footnote on page 43.

where $\tilde{z}_0(\sigma)$ is of the form (5) *). We likewise determine

$$\tilde{z}^{\Delta}(t) = e^{\lambda t}\bar{\delta}(\lambda)\rho^*\tilde{z}_0^{\Delta}(\lambda), \tag{27}$$

where $\tilde{z}_0^{\Delta}(\lambda)$ is of the form (10) *). Substituting the values (26), (27), (4) and (9) into the equality (43, § 2) written for $t = 0$, one obtains the relation *)

$$\chi(\lambda, \sigma) = \rho^*\rho\bar{\delta}(\lambda)\delta(\sigma)\tilde{\chi}(\lambda, \sigma). \tag{28}$$

Thus if two systems belong to the same class in the sense of Definition 1, § 2, their characteristic functions are connected by the relation (28), where $\delta(\sigma)$ is obtained from Relation (24) and the parameters ρ and q are the same as in the Relations (23, § 2)—(28, § 2) which are satisfied by the coefficients of the two systems.

4. Invariance of the characteristic polynomial under the transformations introduced in § 2

The equality (28) can be written in a more symmetric form if one uses the identity

$$1 - q^*(\sigma I - A)^{-1}b = \frac{\det(\sigma I - A - bq^*)}{\det(\sigma I - A)}, \tag{29}$$

which readily follows from the matrix identity

$$\begin{pmatrix} I & 0 \\ q^*(\sigma I - A)^{-1} & 1 \end{pmatrix}\begin{pmatrix} (\sigma I - A - bq^*) & -b \\ 0 & 1 \end{pmatrix} =$$

$$= \begin{pmatrix} (\sigma I - A) & -b \\ 0 & 1 - q^*(\sigma I - A)^{-1}b \end{pmatrix}\begin{pmatrix} I & 0 \\ q^* & 1 \end{pmatrix}. \tag{30}$$

whose verification is easy. By computing the determinants of the matrices in (30) one obtains the identity (29).

---

*) $\tilde{x}_0(\sigma)$, $\tilde{z}_0(\sigma)$, $\tilde{z}_0^{\Delta}(\lambda)$ and $\tilde{\chi}(\lambda, \sigma)$ are obtained from formulas (3), (6), (10) and (12) by putting the superscript $\sim$ over the symbols A, b, $z_0(\sigma)$, $z_0^{\Delta}(\lambda)$ and D.

Noting that the equality $\widetilde{A} = R(A + bq^*)R^{-1}$, (23, § 2) implies that

$$\det(\sigma I - A - bq^*) = \det(\sigma I - \widetilde{A}), \tag{31}$$

one obtains from (29) and (24)

$$\delta(\sigma) = \frac{\det(\sigma I - \widetilde{A})}{\det(\sigma I - A)}, \quad \delta(\lambda) = \frac{\det(\lambda I - \widetilde{A}^*)}{\det(\lambda I - A^*)}. \tag{32}$$

These relations, and the formula which gives the definition of the characteristic polynomial (18), allow one to deduce from (28) the relation

$$\nu\pi(\lambda, \sigma) = \rho^* \rho \widetilde{\nu} \widetilde{\pi}(\lambda, \sigma), \tag{33}$$

where ν and $\widetilde{\nu}$ are the normalizing factors of $\pi(\lambda, \sigma)$ and $\widetilde{\pi}(\lambda, \sigma)$, respectively. Relation (33) shows that $\pi(\lambda, \sigma)$ and $\widetilde{\pi}(\lambda, \sigma)$ are proportional. The constant of proportionality is positive as a consequence of the conditions $\nu > 0$, $\widetilde{\nu} > 0$ and $\rho \neq 0$ (see Definition 1, § 2 and Section 2 of this paragraph). Both polynomials can be written in the form (20) where the coefficients satisfy the Conditions (22). From the foregoing remarks it follows that the two polynomials are identical and we can state.

Proposition 1. *If two systems of the form (1, § 2), (2, § 2) belong to the same class in the sense of Definition 1, § 2, then the characteristic polynomials of the two systems are identical.*

Proposition 1 shows that to each class of systems there corresponds a uniquely determined characteristic polynomial. The characteristic functions of two systems belonging to the same class satisfy the relation

$$\widetilde{\chi}(\lambda, \sigma) = \frac{\det(\lambda I - A^*)\det(\sigma I - A)}{\rho^* \rho \det(\lambda I - \widetilde{A}^*)\det(\sigma I - \widetilde{A})}\, \chi(\lambda, \sigma), \tag{33 bis}$$

which results from (28) and (32). Thus, the characteristic function of an arbitrary system is equal to a positive constant multiplied by the characteristic polynomial of the class to which the system belongs, and divided by $\det(\lambda I - \widetilde{A}^*)\det(\sigma I - \widetilde{A})$.

The fact that the characteristic polynomial $\pi(\lambda, \sigma)$ remains unchanged if a transformation of the type investigated in § 2 is applied, recalls the property of the characteristic polynomial $\det(\sigma I - A)$ of a matrix A, that of being invariant under transformations of the type $\widetilde{A} = RAR^{-1}$. The connection between the characteristic polynomials of *systems* of the form considered in this chapter and the characteristic polynomials of certain *matrices* appears in a new light in Section 6 of § 8 where special forms of the positive systems are investigated.

From the invariance of the characteristic polynomial under the transformations given by formulas (23, § 2)—(28, § 2), it follows that all the coefficients of the characteristic polynomial written in the form (20) are invariants under these transformations. A natural problem which arises is whether or not the set of all these coefficients constitutes a "complete system of invariants". In other words, if one knows all these coefficients, can one determine uniquely the corresponding class of systems? This problem is examined in the next paragraph.

§ 4. Conditions under which all systems with the same characteristic polynomial belong to the same class

1. Some supplementary assumptions

As shown in the preceding paragraph (see Proposition 1, §3), to every class of systems there corresponds a uniquely determined characteristic polynomial; however, the inverse property is not true in general[1]). Indeed, consider two particular systems of the form (1, § 2), (2, § 2), whose coefficients have the values :

$$A = \begin{pmatrix} 1 & 0 \\ 0 & 2 \end{pmatrix}, \quad b = \begin{pmatrix} 1 \\ 0 \end{pmatrix}, \ \mathcal{J} = 0, \ \varkappa = 1, \ l = 0, \ M = 0, \tag{1}$$

$$\widetilde{A} = A, \ \widetilde{b} = \begin{pmatrix} 0 \\ 1 \end{pmatrix}, \ \widetilde{\mathcal{J}} = 0, \ \widetilde{\varkappa} = 1, \ \widetilde{l} = 0, \ \widetilde{M} = 0. \tag{2}$$

Using formula (13, § 3) we find that the characteristic functions of the two systems are both equal to 1. It follows that the polynomial $\det(\lambda I - A^*)\chi(\lambda, \sigma)\det(\sigma I - A)$ as well as the polynomial $\det(\lambda I - \widetilde{A}^*)\widetilde{\chi}(\lambda, \sigma)\det(\sigma I - \widetilde{A})$ — which occur in the relation (18, § 3) defining the characteristic polynomial — are both equal to $(\lambda - 1)(\lambda - 2)(\sigma - 1)(\sigma - 2)$. Since the normalizing factors, defined as in Section 2, § 3, are also equal, we conclude that the characteristic polynomials of the two systems are identical. However, the two systems do not belong to the same class since there would then exist a scalar $\rho \neq 0$,

[1]) Skipping the proof which follows is not detrimental to the understanding of the paragraph.

a vector q and a *nonsingular matrix* R such that (see (23, § 2), (24, § 2)):

$$R^{-1}\tilde{A} = (A + bq^*)\, R^{-1}, \; b = \rho R^{-1}\tilde{b}. \tag{3}$$

writing R^{-1} as

$$R^{-1} = \begin{pmatrix} \alpha & \beta \\ \gamma & \delta \end{pmatrix}. \tag{4}$$

and substituting this expression and the values (1) and (2) into the first of the equalities (3), one obtains the equality $\gamma = 2\gamma$ or $\gamma = 0$ [1]). This, together with the equality $\delta = 0$, which results by substituting Relations (1) and (2) into the second of the equalities (3), leads to the conclusion that the matrix R^{-1}, (4), is singular, which is a contradiction. Consequently, the property specified in the heading of this paragraph cannot be true unless some restrictive conditions are introduced. We shall show in the next sections that by introducing a natural restriction, namely assuming that the systems considered are completely controllable (see Appendix A), the specified property holds.

2. A one-to-one correspondence between the characteristic polynomials and certain particular systems

Consider systems which satisfy the relations

$$A = \operatorname{diag}\,(1, 2, \ldots, n), \;\; b' = (1\; 1 \ldots 1), \;\; M = 0, \tag{5}$$

$$\max\,(|\varkappa|, |l_j|, |(J)_{jk}|) = 0 \text{ or } 1, \tag{6}$$

where l_j are the components of the vector l, and $(J)_{jk}$ are the entries of the matrix J.

The characteristic polynomial of a system satisfying Relations (5) is given by the following relation, deduced from (13, § 3) and (18, § 3)

$$\pi(\lambda, \sigma) = \frac{1}{\nu}\prod_{k=1}^{n}(\lambda - k)(\sigma - k)\Bigg(\varkappa + \sum_{j=1}^{n}\frac{l_j^*}{\sigma - j} + \sum_{j=1}^{n}\frac{l_j}{\lambda - j} +$$

$$+ (\lambda + \sigma)\sum_{j=1}^{n}\sum_{k=1}^{n}\frac{(J)_{jk}}{(\lambda - j)(\sigma - k)}\Bigg), \tag{7}$$

where ν is the normalization factor as defined in Section 2, §3.

---

[1]) It is sufficient to write the equality for the elements located in the second row and the first column.

Conversely, consider an arbitrary polynomial of the form (20, § 3) which satisfies Conditions (19, § 3) and (22, § 3). We shall show that there exists one and only one system of the form (1, § 2), (2, § 2) whose characteristic polynomial is equal to the given polynomial and whose coefficients l, b and M satisfy Conditions (5) and (6).

To this end, note that from (7) one obtains the following relations

$$\frac{\varkappa}{\nu} = (-1)^n \lim_{\sigma \to \infty} \frac{\pi(-\sigma, \sigma)}{\sigma^{2n}}, \tag{8}$$

$$\frac{l_{k_0}}{\nu} = \frac{\pi(k_0, -k_0)}{\prod_{j=1}^{n} (k_0 - j) \prod_{j=1}^{n}{}' (-k_0 - j)}, \quad k_0 = 1,2, \ldots, n, \tag{9}$$

$$\frac{(J)_{k_0, k_2}}{\nu} = \frac{\pi(k_0, k_1)}{(k_0 + k_1) \prod_{j=1}^{n}{}' (k_0 - j)(k_1 - j)}, \quad k_0, k_1 = 1, 2, \ldots, n, \tag{10}$$

where the products of the form $\prod_{j=1}^{n}{}'$ (bearing the symbol "′") are obtained from the corresponding products (without this symbol) after deleting all the factors whose values are equal to zero.

Under our Assumptions (19, § 3) and $J^* = J$, the polynomials (7) and (20, § 3) are identical if their corresponding coefficients for all the terms in $\lambda^j \sigma^k$ with $j \leqslant k$ are equal. This gives a system of $(n+1)(n+2)/2$ linear equations with the same number of unknowns. (The unknowns are $\varkappa/\nu$, l_k/ν and $(J)_{ik}/\nu$ for $i \leqslant k$; recall that $J^* = J$.) This system has a unique solution if the corresponding homogeneous equation has no nontrivial solution. Now suppose this homogeneous equation does have a nontrivial solution. Then one can find $\varkappa/\nu$, l/ν and J/ν, not all zero, such that the corresponding polynomial (7) vanishes identically. This, however, contradicts (8)—(10) and proves that every polynomial (20, § 3), satisfying (19, § 3) can be written in the form (7) where $\varkappa/\nu$, l/ν and J/ν are uniquely determined. Choosing conveniently the positive constant ν one can always satisfy (6) — and this determines $\varkappa$, l and J uniquely. Finally, note that the polynomial (7) coincides with (20, § 3) and is the characteristic polynomial of the determined system (see also (21, § 3)).

3. A property of "completely controllable systems"

We shall show that if a system is completely controllable, — i.e., if the pair (A, b) is completely controllable — then there always exists a system which belongs to the same class and satisfies Conditions (5) and (6). Indeed, since the pair (A, b) is completely controllable, by Property 9° of Theorem 1, § 31 (Appendix A), we can find a vector q_0 with the property

$$\det (\sigma I - A - bq_0^*) = (\sigma - 1)(\sigma - 2)\ldots(\sigma - n). \qquad (11)$$

Therefore, the characteristic equation of the matrix $A + bq_0^*$ has the roots $1, 2, \ldots, n$ and hence this matrix can be written in the Jordan normal form given by the first equation of (5). It follows from Theorem 2, § 31 (Appendix A) that the pair $(A + bq_0^*, b)$ is completely controllable and therefore, by using Property 8° of Theorem 1, § 31, one can find a nonsingular matrix R_0 such that the pair

$$\widetilde{A} = R_0(A + bq_0^*) R_0^{-1}, \; \widetilde{b} = R_0 b \qquad (12)$$

has a form given by the first two equations of (5).

Besides the initial system with the coefficients $(J, \varkappa, l, M; A, b)$ we introduce another system : $(\widetilde{J}, \widetilde{\varkappa}, \widetilde{l}, \widetilde{M}; \widetilde{A}, \widetilde{b})$, belonging to the same class, and related to the first system by (23, § 2)—(28, § 2) where $\rho = 1$, $q = q_0$, $R = R_0$, $N = 0$. Since Eqs. (23, § 2) and (24, § 2) are identical to (12), it follows that the system $(\widetilde{J}, \widetilde{\varkappa}, \widetilde{l}, \widetilde{M}; \widetilde{A}, \widetilde{b})$ satisfies the first two conditions of (5). Apply again relations (23, § 2)—(28, § 2) by putting the superscript "$\sim$" over all the symbols and taking $\widetilde{R} = I$, $\widetilde{q} = 0$, $\widetilde{\rho} = 1$. Choose the hermitian matrix $\widetilde{N}$ so as to satisfy the equation

$$\widetilde{\widetilde{M}} = \widetilde{M} + \widetilde{N}\widetilde{A} + \widetilde{A}^*\widetilde{N} = 0. \qquad (13)$$

We have thus obtained a system $(\widetilde{\widetilde{J}}, \widetilde{\widetilde{\varkappa}}, \widetilde{\widetilde{l}}, \widetilde{\widetilde{M}}; \widetilde{\widetilde{A}}, \widetilde{\widetilde{b}})$, which belongs to the same class as the initial system and satisfies all the Conditions (5). The existence of a hermitian matrix $\widetilde{N}$ which satisfies Eq. (13) where $\widetilde{A}$ has the same form as in (5), results from a general theorem due to A. M. Liapunov [1]. Moreover, one can easily see that if the entries $(\widetilde{N})_{jk}$ of matrix $\widetilde{N}$ are given in terms of the entries $(\widetilde{M})_{jk}$ of matrix $\widetilde{M}$ by the relations

$$(\widetilde{N})_{jk} = -\frac{(\widetilde{M})_{jk}}{j+k}, \qquad j,\ k = 1,\ 2, \ldots, n,$$

then (13) is satisfied.

If, at the end of all these operations, the elements $\widetilde{\widetilde{J}}$, $\widetilde{\widetilde{\varkappa}}$ and $\widetilde{\widetilde{l}}$ are equal to zero, then (6) is satisfied and hence for these systems the proof is complete. If at least one of the elements mentioned is different from zero, we introduce the positive number

$$\rho_0 = \max\ (|\varkappa|,\ |l_j|,\ |(J)_{jk}|)$$

and again apply Eqs. (23, § 2)—(28, § 2), writing the superscript $\approx$ over all the symbols and using the values $\widetilde{\widetilde{\rho}} = \sqrt{\rho_0}$, $\widetilde{\widetilde{q}} = 0$, $\widetilde{\widetilde{R}} = I\sqrt{\rho_0}$, $\widetilde{\widetilde{N}} = 0$. The system obtained after these operations satisfies all the conditions required in Section 2.

4. Methods for bringing completely controllable systems to special forms

The foregoing sections lead to:

Proposition 1. *If two single-input, completely controllable systems of the form (1, § 2), (2, § 2) are of the same order and have the same characteristic polynomial, they belong to the same class (see Definition 1, § 2).*

The proof is almost immediate: as shown in Section 3, for each of the two systems considered (which shall be denoted by $\mathcal{S}$ and $\widetilde{\mathcal{S}}$) there exists a system ($\mathcal{S}_0$ and $\widetilde{\mathcal{S}}_0$, respectively) which belongs to the same class, has the form specified in Section 2 and which, according to Proposition 1, § 3, has the same characteristic polynomial as the initial system. Therefore, the systems $\mathcal{S}_0$ and $\widetilde{\mathcal{S}}_0$ have the same characteristic polynomial and from Section 2, it follows that they are identical. Consequently, the classes of the systems $\mathcal{S}$ and $\widetilde{\mathcal{S}}$ have a common element (given by the system $\mathcal{S}_0$ which is equal to $\widetilde{\mathcal{S}}_0$) and hence they are identical.

For certain applications it is convenient to state Proposition 1 in terms of the polynomial $\det(\lambda I - A^*)\,\chi(\lambda,\ \sigma)\,\det(\sigma I - A) = \nu\pi(\lambda,\ \sigma)$ which differs from the characteristic polynomial merely by the fact that it is not normalized (see Section 2, § 3). Thus we obtain:

Proposition 1′. *If two completely controllable systems of the same order, with the coefficients* $(J, \varkappa, l, M; A, b)$ *and* $(\tilde{J}, \tilde{\varkappa}, \tilde{l}, \tilde{M}; \tilde{A}, \tilde{b})$ *have the property that the polynomials* $\det(\lambda I - A^*)\chi(\lambda, \sigma)\det(\sigma I - A)$ *and* $\det(\lambda I - \tilde{A}^*)\chi(\lambda, \sigma)\det(\sigma I - \tilde{A})$ *are proportional and the constant of proportionality is positive, then there exists a transformation of the form (23, § 2)—(28, § 2) relating the coefficients of the two systems.*

Proposition 1′ supplies a convenient criterion for finding whether two completely controllable systems can be obtained from one another by means of a transformation of the type studied in § 2.

5. Properties of systems with the same $\pi(-\sigma, \sigma)$

In the next chapters $\chi(\lambda, \sigma)$ and $\pi(\lambda, \sigma)$ will be replaced by $\chi(-\sigma, \sigma)$ and $\pi(-\sigma, \sigma)$ (taking $\lambda = -\sigma$). For this reason it is interesting to see how the results presented in the foregoing sections are affected when instead of $\pi(\lambda, \sigma)$ one considers $\pi(-\sigma, \sigma)$. An examination of the problem in the particular case of systems of the form shown in Section 2 will again point out the form of the solution.

Consider two systems whose coefficients $(J_1, \varkappa_1, l_1, M_1; A_1, b_1)$ and $(J_2, \varkappa_2, l_2, M_2; A_2, b_2)$ have the form given in Section 2. Assume that the characteristic polynomials of the two systems satisfy the equality

$$\pi_1(-\sigma, \sigma) = \gamma_0\, \pi_2(-\sigma, \sigma), \quad (\gamma_0 > 0) \tag{14}$$

where γ_0 is a real and positive coefficient. From (14) and Relations (8) and (9) written successively for both systems, one obtains the following relations

$$\frac{\varkappa_1}{\nu_1} = \gamma_0 \frac{\varkappa_2}{\nu_2}, \quad \frac{l_1}{\nu_1} = \gamma_0 \frac{l_2}{\nu_2}. \tag{15}$$

(since the right-hand members of (8) and (9) depend only on the values of the polynomial $\pi(-\sigma, \sigma)$).

The expression (10) of matrix J does not involve $\pi(-\sigma, \sigma)$ but $\pi(\lambda, \sigma)$ for $\lambda \neq -\sigma$. Consequently, we can only state that there exists a hermitian $n \times n$-matrix X_0 such that

$$\frac{J_1}{\nu_1} = \gamma_0 \frac{J_2}{\nu_2} + X_0. \tag{16}$$

It follows from (15) and (16) that there exists a positive number $\gamma_1 = \frac{\nu_2}{\nu_1 \gamma_0}$ and a hermitian matrix $X_1 = -\nu_1 X_0$, such that

$$\varkappa_2 = \gamma_1 \varkappa_1, \; l_2 = \gamma_1 l_1, \qquad J_2 = \gamma_1 (J_1 + X_1). \tag{17}$$

These relations, together with the assumption that both systems are of the form given in Section 2, show that the systems are connected by relations of the form (23, § 2)—(28, § 2), where $q = 0$, $R = \frac{1}{\sqrt{\gamma_1}} I$, $\rho = \frac{1}{\sqrt{\gamma_1}}$, $N = 0$ [1]. We have reached the conclusion that there exists a hermitian matrix X_1 such that

"the system whose coefficients are $(J_1 + X_1, \varkappa_1, l_1, M_1; A_1, b_1)$ belongs to the same class as the system whose coefficients are

$$(J_2, \varkappa_2, l_2, M_2; A_2, b_2)\text{''}. \tag{18}$$

This result considerably simplifies the proof of

Proposition 2. *If two completely controllable systems of the same order, with the coefficients* $(\tilde{J}, \tilde{\varkappa}, \tilde{l}, \tilde{M}; \tilde{A}, \tilde{b})$ *and* $(J, \varkappa, l, M; A, b)$ *have the property that their characteristic polynomials* $\tilde{\pi}$ *and* π *satisfy the relation*

$$\tilde{\pi}(-\sigma, \sigma) = \gamma_0 \, \pi(-\sigma, \sigma), \; \gamma_0 > 0, \tag{19}$$

then there exists a hermitian matrix $\tilde{U}$ *such that*

"the system whose coefficients are $(\tilde{U}, \tilde{\varkappa}, \tilde{l}, \tilde{M}; \tilde{A}, \tilde{b})$ *and the system whose coefficients are* $(J, \varkappa, l, M; A, b)$, *belong to the same class".* (20)

Indeed, since the two systems are completely controllable one can find, as in Section 3, two other systems with the coefficients $(J_1, \varkappa_1, l_1, M_1; A_1, b_1)$ and $(J_2, \varkappa_2, l_2, M_2; A_2, b_2)$ such that the Conditions (5) and (6) are satisfied and

"the systems with the coefficients $(J_1, \varkappa_1, l_1, M_1; A_1, b_1)$ and the system with the coefficients $(\tilde{J}, \tilde{\varkappa}, \tilde{l}, \tilde{M}; \tilde{A}, \tilde{b})$ belong to the same class"; (21)

"the system with the coefficients $(J_2, \varkappa_2, l_2, M_2; A_2, b_2)$ and the system with the coefficients $(J, \varkappa, l, M; A, b)$ belong to the same class". (22)

---

[1]) Obviously, the coefficients $(J, \varkappa, l, M; A, b)$ and $(\tilde{J}, \tilde{\varkappa}, \tilde{l}, \tilde{M}; \tilde{A}, \tilde{b})$ from Relations (23, § 2)—(28, § 2) are replaced by $(J_1 + X_1, \varkappa_1, l_1, M_1; A_1, b_1)$ and $(J_2, \varkappa_2, l_2, M_2; A_2, b_2)$, respectively.

According to Proposition 1, § 3 the characteristic polynomials π_1 and π_2 of the new systems are equal to $\tilde{\pi}$ and π respectively and therefore, owing to equality (19), they satisfy (14). As shown above, Property (18) is fulfilled. Using (23, § 2)—(28, § 2) one sees that Property (21) leads to the following results :

"The system with coefficients $(J_1 + X_1, \varkappa_1, l_1, M_1; A_1, b_1)$ and the system with coefficients $(\tilde{J} + \tilde{X}, \tilde{\varkappa}, \tilde{l}, \tilde{M}; \tilde{A}, \tilde{b})$ belong to the same class". (23)

Here $\tilde{X}$ is the hermitian matrix given by relation $\tilde{X} = (R^{-1})^* X_1 R^{-1}$; and R is the matrix from the formulas obtained by expressing Condition (21) with the help of definition 1, § 2. Property (20) follows immediately from (23), (18) and (22) (moreover, the matrix U equals $\tilde{J} + \tilde{X}$).

With the help of Proposition 2 one can prove :

Proposition 3. *If two completely controllable systems of the same order, with the coefficients $(\tilde{J}, \tilde{\varkappa}, \tilde{l}, \tilde{M}; \tilde{A}, \tilde{b})$ and $(J, \varkappa, l, K; A, b)$ have the properties*

$$\tilde{A} = A, \tilde{b} = b, \tilde{\chi}(-\sigma, \sigma) = \chi(-\sigma, \sigma) \tag{24}$$

then there exist two hermitian matrices $\tilde{X}$ and N such that

$$\tilde{J} = J - \tilde{X} - N, \tilde{\varkappa} = \varkappa, \tilde{l} = l + Nb, \tilde{M} = M + NA + A^*N. \tag{25}$$

From (24) one obtains (19) and hence (20). Therefore, (23, § 2)—(28, § 2) are satisfied. From (23, § 2) and the first relation of (24), we get

$$\det(\sigma I - A) = \det(\sigma I - A - bq^*), \tag{26}$$

a relation which is obviously satisfied for $q = 0$. By virtue of Property 9° of Theorem 1, § 31 (appendix A), equation (26) has a unique solution, whence

$$q = 0. \tag{27}$$

Using this relation we may write (23, § 2) and (24, § 2) as

$$RA - AR = 0, \quad Rb = \rho b, \tag{28}$$

Obviously, the system (28) admits the solution $R = \rho I$. According to Property 5° of Theorem 1, § 31, (28) has a unique solution, hence $R = \rho I$. This relation, together with (27), allows one to write (25, § 2)—(28, § 2) (where $\widetilde{J}$ is replaced by $\widetilde{U} = \widetilde{J} + \widetilde{X}$) in the form

$$\begin{pmatrix} 0 & 0 & \widetilde{J} \\ 0 & \widetilde{\varkappa} & \widetilde{l}^* \\ \widetilde{J} & \widetilde{l} & \widetilde{M} \end{pmatrix} = \frac{1}{\rho^*\rho} \begin{pmatrix} 0 & 0 & J \\ 0 & \varkappa & l^* \\ J & l & M \end{pmatrix} + \frac{1}{\rho^*\rho} \begin{pmatrix} 0 & 0 & -\widetilde{X} \\ 0 & 0 & 0 \\ -\widetilde{X} & 0 & 0 \end{pmatrix} + \tag{29}$$

$$+ \frac{1}{\rho^*\rho} (T^*F + F^*T),$$

where the first two matrices have the form of matrix D (see (6, § 2)), and the matrices F and T are given by (8, § 2) and (11, § 2).

Multiplying (29) on the left by $z_0^\Delta(-\sigma)$ (see (10, § 3)) and on the right by $z_0(\sigma)$ (see (5, § 3)) and using the identities

$$F z_0(\sigma) = 0, \; z_0^\Delta(-\sigma) F^* = 0, \; z_0^\Delta(-\sigma) \begin{pmatrix} 0 & 0 & -\widetilde{X} \\ 0 & 0 & 0 \\ -\widetilde{X} & 0 & 0 \end{pmatrix} z_0(\sigma) = 0,$$

and formula (12, § 3), one obtains

$$\widetilde{\chi}(-\sigma, \sigma) = \frac{1}{\rho^*\rho} \chi(-\sigma, \sigma).$$

From this relation and the last equality of (24) one obtains — if $\chi(-\sigma, \sigma)$ is not identically zero — , the condition $\rho^*\rho = 1$. Substituting this condition into (29) and using the explicit expressions of the matrices F and T one obtains the Relations (25) which were to be proved.

We examine now the case $\chi(-\sigma, \sigma) \equiv 0$. First we prove

Proposition 4. *If the characteristic function of a completely controllable system has the property $\chi(-\sigma, \sigma) \equiv 0$, then the coefficients of the system satisfy the relations*

$$\varkappa = 0, \qquad l = Nb, \qquad M = NA + A^*N, \tag{30}$$

where N is a hermitian matrix.

For the proof that follows it is convenient to add the superscript "$\sim$" to all the symbols contained in Proposition 4 and therefore to consider a system $(\widetilde{J}, \widetilde{\varkappa}, \widetilde{l}, \widetilde{M}; \widetilde{A}, \widetilde{b})$ with the property $\tilde{\chi}(-\sigma, \sigma) \equiv 0$. Since the system is completely controllable, there exists a system which belongs to the same class and has the form given in Section 2, with coefficients $(J, \varkappa, l, M; A, b)$. The characteristic polynomial of this system is identical to that of the initial system (see Proposition 1, § 3) and hence, it vanishes for $\lambda = -\sigma$. Then, from (8)–(9) one obtains $\varkappa = 0$, $l = 0$ and from (5), $M = 0$. Since the two systems belong to the same class, their coefficients are related by (23, § 2)–(28, § 2). From $\varkappa = 0$ and from (26, §2) it follows that $\tilde{\varkappa} = 0$. From $l = 0$, (27, § 2) and (24, § 2) one obtains $\tilde{l} = \tilde{N}\tilde{b}$, where $\tilde{N} = (\tilde{R}^{-1})^* N R^{-1}$. From $M = 0$ and from (28, § 2) and (23, § 2) one obtains $\tilde{M} = \tilde{N}\tilde{A} + \tilde{A}^*\tilde{N}$. Thus we have obtained all the relations (30).

Using Proposition 4, we conclude the proof of Proposition 3 by simply remarking that when two systems satisfy Relations (24) and the condition $\chi(-\sigma, \sigma) \equiv 0$, then the Conditions (30) are satisfied for each system, whence we have also Relations (25).

§ 5. Equivalence classes for multi-input systems

1. Definition and properties of the classes of multi-input systems

Most of the results established in the foregoing chapters can be easily extended to multi-input systems of the form

$$\frac{dx}{dt} = Ax + Bu \tag{1}$$

$$\eta(0, t_1) = [x^*Jx]_0^{t_1} + \int_0^{t_1} (u^*Ku + u^*L^*x + x^*Lu + x^*Mx)\, dt, \tag{2}$$

$$(J^* = J, K^* = K, M^* = M).$$

In Eq. (1) x is, as before, a vector-valued function and $x(t)$ is an n-vector: the state of the system at time t; A is a constant $n \times n$-matrix. The scalar-valued control function μ of Eq. (1, § 2) is replaced here by u, which is a vector-valued function, defined for $t \geqslant 0$ and having m piecewise continuous components; accordingly, the vector b of equality (1, § 2) is replaced by the $n \times m$-matrix B. In (2) J and M are constant, hermitian $n \times n$-matrices; K is a hermitian $m \times m$-matrix and L is an $n \times m$-matrix. Here again we use the complex field (however t must be real). The first term of the right-hand member of expression (2) can also be written in the form (3, § 2).

We introduce the vector-valued function z, with $2n + m$ components, given by (cf. (4, § 2))

$$z' = \begin{pmatrix} \frac{dx'}{dt} & u' & x' \end{pmatrix}, \tag{3}$$

and write the integral (2) in the compact form (cf. (5, § 2))

$$\eta(0, t_1) = \int_0^{t_1} z^* D\, z\, dt, \tag{4}$$

where

$$D = \begin{pmatrix} \overbrace{0}^{n} & \overbrace{0}^{m} & \overbrace{J}^{n} \\ 0 & K & L^* \\ J & L & M \end{pmatrix} \begin{matrix} \}n \\ \}m \\ \}n \end{matrix} \tag{5}$$

Using z, (3), the differential equation (1) becomes $Fz = 0$, where F is a matrix with $2n + m$ rows and columns, expressed by

$$F = \begin{pmatrix} -I & B & A \\ 0 & 0 & 0 \\ 0 & 0 & 0 \end{pmatrix}. \tag{6}$$

In accordance with § 1, the values of the quadratic form z^*Dz are not altered when we add the matrix $T^*F + F^*T$ to the matrix D. We need only use matrices T of the form

$$T = \begin{pmatrix} 0 & 0 & N \\ 0 & 0 & 0 \\ 0 & 0 & 0 \end{pmatrix}, \quad (N^* = N) \tag{7}$$

where N is a hermitian $n \times n$-matrix.

Introduce the new variables $\tilde{u}$ and $\tilde{x}$, by relations similar to (13, § 2) and (14, § 2)

$$\tilde{u} = Pu - PQ^*x, \tag{8}$$

$$\tilde{x} = Rx, \tag{9}$$

where P is a nonsingular $m \times m$-matrix, R is a nonsingular $n \times n$-matrix and Q is an $n \times m$-matrix. Relations (8) and (9) show that $\tilde{z}$, defined by

$$\tilde{z}' = \begin{pmatrix} \frac{d\tilde{x}'}{dt} & \tilde{u}' & \tilde{x}' \end{pmatrix}, \tag{10}$$

is obtained from z, (3), by applying the transformation

$$\tilde{z} = Sz, \tag{11}$$

where

$$S = \begin{pmatrix} R & 0 & 0 \\ 0 & P & -PQ^* \\ 0 & 0 & R \end{pmatrix}. \tag{12}$$

Since P and R are nonsingular, S is also nonsingular.

As in the case of the single-input systems we may state the conditions under which the system (1), (2) belongs to the same class as another system of the same form

$$\frac{d\tilde{x}}{dt} = \tilde{A}\tilde{x} + \tilde{B}\tilde{u}, \tag{13}$$

$$\tilde{\eta}(0, t_1) = [\tilde{x}^*\tilde{J}\tilde{x}]_0^{t_1} + \int_0^{t_1} (\tilde{u}^*\tilde{K}\tilde{u} + \tilde{u}^*\tilde{L}^*\tilde{x} + \tilde{x}^*\tilde{L}\tilde{u} + \tilde{x}^*\tilde{M}\tilde{x})\,dt. \tag{14}$$

Definition 1. *System (1), (2) is said to belong to the same class as the system (13), (14) with respect to the group of operators (S; T) (see (7) and (12)) if the following relations hold:*

$$\tilde{A} = R(A + BQ^*)\,R^{-1}, \tag{15}$$

$$\tilde{B} = RBP^{-1}, \tag{16}$$

$$J = (R^{-1})^* (J - N)R^{-1}, \tag{17}$$

$$\widetilde{K} = (P^{-1})^* K P^{-1}, \tag{18}$$

$$\widetilde{L} = (R^{-1})^* (QK + L + NB)P^{-1}, \tag{19}$$

$$\widetilde{M} = (R^{-1})^* (QKQ^* + QL^* + LQ^* + M + N(A + BQ^*) +$$

$$+ (A + BQ^*)^* N)\, R^{-1}, \tag{20}$$

where P, Q, R *and* N *are some matrices with the properties* $\det P \neq 0$, $\det R \neq 0$ *and* $N^* = N$.

As in the case of the single-input systems we may consider besides (1), (2), the polarized system

$$\frac{\mathrm{d}x^\Delta}{\mathrm{d}t} = x^\Delta A^* + u^\Delta B^*, \ \frac{\mathrm{d}x}{\mathrm{d}t} = Ax + Bu, \ \eta_p(0, t_1) = \int_0^{t_1} z^\Delta D z \,\mathrm{d}t, \tag{21}$$

where

$$z^\Delta = \left(\frac{\mathrm{d}x^\Delta}{\mathrm{d}t} \quad u^\Delta \quad x^\Delta \right). \tag{22}$$

Definition 1 gives at the same time the conditions under which the polarized system (21) belongs to the same class as the polarized system

$$\frac{\mathrm{d}\tilde{x}^\Delta}{\mathrm{d}t} = \tilde{x}^\Delta \widetilde{A}^* + \tilde{u}^\Delta \widetilde{B}^*, \ \frac{\mathrm{d}\tilde{x}}{\mathrm{d}t} = \widetilde{A}\tilde{x} + \widetilde{B}\tilde{u}, \ \tilde{\eta}_p(0, t_1) = \int_0^{t_1} \tilde{z}^\Delta \widetilde{D} \tilde{z} \,\mathrm{d}t, \tag{23}$$

where

$$\tilde{z}^\Delta = \left(\frac{\mathrm{d}\tilde{x}^\Delta}{\mathrm{d}t} \quad \tilde{u}^\Delta \quad \tilde{x}^\Delta \right). \tag{24}$$

The properties of the polarized systems which belong to the same class are given by the following proposition:

Proposition 1. *If the elements* $(J, K, L, M; A, B)$ *and* $(\widetilde{J}, \widetilde{K}, \widetilde{L}, \widetilde{M}; \widetilde{A}, \widetilde{B})$, *formed with the coefficients of the polarized systems (21) and (23), satisfy the conditions of definition 1, then for every quadruple functions* u, x, u^Δ *and* x^Δ *satisfying* *the first two equations of (21), the following properties are satisfied.*

1°. *The first two equations of (23) are satisfied by transformed functions (see (8) and (9))*

$$\widetilde{u} = Pu - PQ^* x, \quad \widetilde{x} = Rx, \tag{25}$$

$$\widetilde{u}^\Delta = u^\Delta P^* - x^\Delta Q P^*, \quad \widetilde{x}^\Delta = x^\Delta R^*. \tag{26}$$

2° *One has*

$$\widetilde{z}^\Delta \widetilde{D} \widetilde{z} = z^\Delta D z, \tag{27}$$

where the functions z, z^Δ, $\widetilde{z}$ *and* $\widetilde{z}^\Delta$ *are defined by formulas (3), (10), (22) and (24).*

3° *For the functions introduced above the integrals* $\eta_p(0, t_1)$, *and* $\widetilde{\eta}_p(0, t_1)$ *are equal for any positive number* t_1.

In the particular case $u^\Delta(t) = u^*(t)$ and $x^\Delta(t) = x^*(t)$ we may say that the polarized systems reduce to the former systems of the form (1), (2), and from Proposition 1 one obtains a result which is a direct extension of Proposition 1, § 2.

2. The characteristic function

If the control function u is given by

$$u(t) = u_0 \, e^{\sigma t}, \tag{28}$$

where u_0 is a constant vector and the scalar σ satisfies the condition $\det(\sigma I - A) \neq 0$, then Eq. (1) has the particular solution

$$x(t) = x_0(\sigma) \, e^{\sigma t}, \quad \text{where} \quad x_0(\sigma) = (\sigma I - A)^{-1} B u_0. \tag{29}$$

Hence, the function z given by (3) has the form

$$z(t) = Z_0(\sigma) \, u_0 \, e^{\sigma t}, \tag{30}$$

where the matrix $Z_0(\sigma)$ is expressed by

$$Z_0'(\sigma) = ((\sigma(\sigma I - A)^{-1} B)' \;\; I_m \;\; ((\sigma I - A)^{-1} B)'), \tag{31}$$

where I_m is the $m \times m$-identity matrix.

Similarly, if the control function u^{Δ} is given by

$$u^{\Delta}(t) = u_0^{\Delta}\, e^{\lambda t}, \qquad (\det(\lambda I - A) \neq 0), \tag{32}$$

the first of Eqs. (21) has the particular solution

$$x^{\Delta}(t) = x_0^{\Delta}(\lambda)\, e^{\lambda t}, \quad \text{where} \quad x_0^{\Delta}(\lambda) = u_0^{\Delta} B^*(\lambda I - A^*)^{-1}, \tag{33}$$

and the vector $z^{\Delta}(t)$, (22), is given [1] by

$$z^{\Delta}(t) = u_0\, \overline{Z}_0'(\lambda)\, e^{\lambda t}, \tag{34}$$

where Z_0 has the same expression (31).

From (30) and (34) one obtains

$$z^{\Delta}(t)\, Dz(t) = u_0^{\Delta} H(\lambda, \sigma)\, u_0\, e^{(\lambda+\sigma)} \tag{35}$$

(cf. (11, § 3)). In our case the characteristic function is replaced by a matrix-valued characteristic function H expressed by

$$H(\lambda, \sigma) = \overline{Z}_0'(\lambda)\, DZ_0(\sigma). \tag{36}$$

The operations involved in (36) can be performed if the conditions $\det(\lambda I - A) \neq 0$ and $\det\,(\sigma I - A) \neq 0$ are satisfied (see footnote 2, § 3).

More explicitly, the characteristic function has the following form (obtained from (36) by using (31) and (5)) :

$$H(\lambda, \sigma) = K + L^*(\sigma I - A)^{-1}B + B^*(\lambda I - A^*)^{-1}L +$$

$$+ B^*(\lambda I - A^*)^{-1}\,(M + (\lambda + \sigma)\, J)\,(\sigma I - A)^{-1}B, \tag{37}$$

(cf. (13, § 3)). It follows that [2] (cf. (14, § 4))

$$H(\lambda, \sigma) = \overline{H}'(\sigma, \lambda). \tag{38}$$

We examine now the relation between the characteristic functions of the systems which belong to the same class. Consider the polarized systems (21) and (23) with the properties stated

---

[1]) Recall that $\overline{Z}_0'(\lambda) = \overline{Z_0'(\overline{\lambda})}$.

[2]) As before, the symbol $\overline{H}'(\sigma, \lambda)$ has the meaning $\overline{H}'(\sigma, \lambda) = \overline{H'(\overline{\sigma}, \overline{\lambda})}$.

in Proposition 1. The functions given by (28), (29), (32) and (33) satisfy the conditions of Proposition 1. Using the first of the relations (25) we find

$$\widetilde{u}(t) = PC(\sigma)\, u_0\, e^{\sigma t}, \tag{39}$$

where

$$C(\sigma) = I_m - Q^*(\sigma I - A)^{-1}B. \tag{40}$$

The second of the relations (25) leads to the conclusion that the function $\widetilde{x}$ has the form $\widetilde{x}(t) = \widetilde{x}_0 e^{\sigma t}$. From property 1° in Proposition 1, one sees that this function satisfies the second equation of (23). Substituting this solution of the function (39) into the second equation of (23) we uniquely determine the value of the coefficient $\widetilde{x}_0$ thus obtaining the expression $\widetilde{x}(t) = (\sigma I - \widetilde{A})^{-1} \widetilde{B} PC(\sigma) u_0 e^{\sigma t}$ [1]. Using this expression together with (39) we may write the corresponding $\widetilde{z}(t)$ given by (10), in the form

$$\widetilde{z}(t) = \widetilde{Z}_0(\sigma)\, PC(\sigma)\, u_0\, e^{\sigma t}, \tag{41}$$

where $\widetilde{Z}_0(\sigma)$ results from (31) by putting the superscript $\sim$ over A and B.

Similarly one finds that the functions

$$\widetilde{u}^{\Delta}(t) = e^{\lambda t} u_0^{\Delta} \overline{C}'(\lambda)\, P^*, \quad \widetilde{z}^{\Delta}(t) = e^{\lambda t} u_0^{\Delta} \overline{C}'(\lambda)\, P^* \overline{\widetilde{Z}}{}_0'(\lambda), \tag{42}$$

correspond to $u^{\Delta}(t) = e^{\lambda t}\, u_0^{\Delta}$.

According to Proposition 1 one obtains (27), i.e. (for $t = 0$)

$$u_0^{\Delta} H(\lambda, \sigma)\, u_0 = u_0^{\Delta} \overline{C}'(\lambda)\, P^* \widetilde{H}(\lambda, \sigma)\, PC(\sigma)\, u_0. \tag{43}$$

Since this equality is valid for every pair of vectors u_0^{Δ} and u_0, it follows that (cf. (28, § 3)):

$$H(\lambda, \sigma) = \overline{C}'(\lambda)\, P^* \widetilde{H}(\lambda, \sigma)\, PC(\sigma). \tag{44}$$

---

[1]) This equality, together with Relation (25), where we substitute (29), gives two different expressions $\widetilde{x}(t)$, namely

$$\widetilde{x}(t) = (\sigma I - \widetilde{A})^{-1} \widetilde{B} PC(\sigma)\, u_0 e^{\sigma t} = R(\sigma I - A)^{-1}\, B u_0 e^{\sigma t}.$$

Since the last equality holds for any vector u_0, we obtain the useful identity

$$(\sigma I - \widetilde{A})^{-1} \widetilde{B} PC(\sigma) = R(\sigma I - A)^{-1}\, B.$$

3. Properties of the determinants of $H(\lambda, \sigma)$ and $C(\sigma)$

We shall examine here some specific problems for multi-input systems. Relation (44) gives

$$\det H(\lambda, \sigma) = |\det P|^2 \det \bar{C}'(\lambda) \det C(\sigma) \det \widetilde{H}(\lambda, \sigma). \tag{45}$$

Note that $\det C(\sigma)$ can be written in a simple form by using the matrix identity (cf. (30, § 3)) :

$$\begin{pmatrix} I & 0 \\ Q^*(\sigma I - A)^{-1} & I_m \end{pmatrix} \begin{pmatrix} ((\sigma I - A) - BQ^*) & -B \\ 0 & I_m \end{pmatrix} = \\ = \begin{pmatrix} (\sigma I - A) & -B \\ 0 & I_m - Q^*(\sigma I - A)^{-1}B \end{pmatrix} \begin{pmatrix} I & 0 \\ Q^* & I_m \end{pmatrix}, \tag{46}$$

which implies that

$$\det(\sigma I - A - BQ^*) = \det(\sigma I - A) \det(I_m - Q^*(\sigma I - A)^{-1}B). \tag{47}$$

From (15) it follows that the left-hand member of (47) equals $\det(\sigma I - \widetilde{A})$. Using this remark and (40) one obtains from (47)

$$\det C(\sigma) = \frac{\det(\sigma I - \widetilde{A})}{\det(\sigma I - A)}, \quad \det \bar{C}'(\lambda) = \frac{\det(\lambda I - \widetilde{A}^*)}{\det(\lambda I - A^*)}, \tag{48}$$

(cf. (32, § 3)). Substituting these relations into (45) yields

$$\det(\lambda I - A^*) \det(\sigma I - A) \det H(\lambda, \sigma) = \\ = |\det P|^2 \det(\lambda I - \widetilde{A}^*) \det(\sigma I - \widetilde{A}) \det \widetilde{H}(\lambda, \sigma). \tag{49}$$

From (37) it follows that every entry of $H(\lambda, \sigma)$ is of the form

$$(H(\lambda, \sigma))_{jk} = \frac{\pi_{jk}(\lambda, \sigma)}{\det(\lambda I - A^*) \det(\sigma I - A)},$$

where π_{jk} are polynomials in λ and σ whose degrees in both σ and λ cannot exceed n. Hence

$$\det H(\lambda, \sigma) = \frac{\pi_0(\lambda, \sigma)}{(\det(\lambda I - A^*) \det(\sigma I - A))^m} \tag{50}$$

where π_0 is a polynomial whose degree cannot exceed mn. We shall show that the numerator and denominator of (50) have some common roots and hence (50) reduces to

$$\det H(\lambda, \sigma) = \frac{\pi_1(\lambda, \sigma)}{\det(\lambda I - A^*)\det(\sigma I - A)}, \tag{51}$$

where π_1 is a polynomial whose degree does not exceed n. Indeed, if $B = 0$ then $H(\lambda, \sigma) = K$ (see (37)) and we clearly can write (51). If $B \neq 0$, then by Proposition 3, 34 (Appendix A) there exists a nonsingular matrix R_0 such that the matrices $\widetilde{A} = R_0 A R_0^{-1}$ and $\widetilde{B} = R_0 B$ can be written as

$$\widetilde{A} = \begin{pmatrix} A_{11} & A_{12} \\ 0 & A_{22} \end{pmatrix}, \quad \widetilde{B} = \begin{pmatrix} B_1 \\ 0 \end{pmatrix}. \tag{52}$$

where the pair (A_{11}, B_1) is completely controllable. Then from $\widetilde{A} = R_0 A R_0^{-1}$, $\widetilde{B} = R_0 B$ and (52) one obtains

$$R_0(\sigma I - A)^{-1} B = (\sigma I - \widetilde{A})^{-1} \widetilde{B} = \begin{pmatrix} (\sigma I_1 - A_{11})^{-1} B_1 & 0 \\ 0 & 0 \end{pmatrix},$$

where I_1 is the identity-matrix with the same dimensions as A_{11}. If the matrices L, M and J are partitioned (in conformity with (52)) in the form

$$L^* R_0^{-1} = (L_1^* \quad L_2^*), \qquad (R_0^{-1})^* M R_0^{-1} = \begin{pmatrix} M_{11} & M_{12} \\ M_{21} & M_{22} \end{pmatrix},$$

$$(R_0^{-1})^* J R_0^{-1} = \begin{pmatrix} J_{11} & J_{12} \\ J_{21} & J_{22} \end{pmatrix},$$

one obtains from (37) the equality

$$H(\lambda, \sigma) = H_0(\lambda, \sigma),$$

where $H_0(\lambda, \sigma)$ is also defined by (37), after replacing J, K, L, M ; A, B ; I by J_{11}, K, L, M_{11} ; A_{11}, B_1 ; I_1. If we prove (51) for the latter system of matrices then one obtains the same relation for the initial system by simply multiplying the numerator and the denominator by $\det(\lambda I_2 - A_{22}^*)\det(\sigma I_2 - A_{22})$ (where I_2 is the identity-matrix with the same dimensions as A_{22}).

The foregoing considerations show that it is sufficient to prove Relation (51) under the assumption that the pair (A, B) is completely controllable. Then, by Property 9°, Theorem 1, § 34 (Appendix A), there exists a matrix Q_0 such that $\det(\sigma I - A)$ and $\det(\sigma I - A - BQ_0^*)$ have no common root. Hence, if we define a new system by means of Relations (15—20), where $P = I_m$, $Q = Q_0$, $R = I$ and $N = 0$, then $\det(\sigma I - A)$ and $\det(\sigma I - \widetilde{A})$ have no common root. For the system so defined one can write a relation of the form (50):

$$\det \widetilde{H}(\lambda, \sigma) = \frac{\widetilde{\pi}_0(\lambda, \sigma)}{(\det(\lambda I - \widetilde{A}^*) \det(\sigma I - \widetilde{A}))^m}. \tag{53}$$

Putting (50) and (53) into (49) and noting that in our case $P = I_m$ and hence $\det P = 1$, one obtains the equality

$$\begin{aligned}(\det(\lambda I - \widetilde{A}^*) \det(\sigma I - \widetilde{A}))^{m-1} \pi_0(\lambda, \sigma) &= \\ = (\det(\lambda I - A^*) \det(\sigma I - A))^{m-1} \widetilde{\pi}_0(\lambda, \sigma).\end{aligned} \tag{54}$$

It can be seen that since $\det(\sigma I - A)$ and $\det(\sigma I - \widetilde{A}$ have no common root, the polynomial π_0 has the form

$$\pi_0(\lambda, \sigma) = (\det(\lambda I - A^*) \det(\sigma I - A))^{m-1} \pi_2(\lambda, \sigma), \tag{55}$$

where π_2 is a new polynomial. Since the degree of the polynomial π_0 is no greater than nm and the degree of the polynomial $(\det(\lambda I - A^*) \det(\sigma I - A))^{m-1}$ is $n(m - 1)$, it follows that the degree of the polynomial π_2 cannot be greater than n. Putting (55) into (50) we obtain Relation (51) which was to be proved.

4. The characteristic polynomial and its invariance

Relation (51) implies the equality

$$\pi_1(\lambda, \sigma) = \det(\lambda I - A^*) \det(\sigma I - A) \det H(\lambda, \sigma),$$

which shows that the expression on the right-hand member is a polynomial. By normalizing this polynomial in the way shown in Section 2, § 3, one obtains a new polynomial

$$\pi(\lambda, \sigma) = \frac{1}{\nu} \det(\lambda I - A^*) \det(\sigma I - A) \det H(\lambda, \sigma), \tag{56}$$

which will be called the "characteristic polynomial" of the system considered or of the element $(J, K, L, M; A, B)$ formed with the coefficients of the system. From the properties of the polynomial π_1 stated in the previous section it follows that the degree of the characteristic polynomial (both in λ and in σ) cannot be greater than n. From (49) one sees that the characteristic polynomials of two systems belonging to the same class are identical.

Using Relation (38) one obtains

$$\pi(\lambda, \sigma) = \bar{\pi}(\sigma, \lambda), \tag{57}$$

which implies that the characteristic polynomial can be written in the form (20, § 3), where the coefficients satisfy the Relations (21, § 3)—(22, § 3).

5. Systems with a fixed differential equation

In the case of multi-input systems there no longer exists a one-to-one correspondence between the classes of equivalent systems and the characteristic polynomials. We have instead

Proposition 2. *If two completely controllable systems $\mathcal{S}$ and $\check{\mathcal{S}}$ with coefficients $\mathcal{S}$: $(J, K, L, M; A, B)$ and $\check{\mathcal{S}}$: $(\check{J}, \check{K}, \check{L}, \check{M}; \check{A}, \check{B})$ satisfy the conditions*

$$\check{A} = A, \quad \check{B} = B, \quad \check{H}(\lambda, \sigma) = H(\lambda, \sigma), \tag{58}$$

then the systems $\mathcal{S}$ and $\check{\mathcal{S}}$ belong to the same class in the sense of Definition 1. More precisely, there exists a hermitian matrix N such that

$$\check{J} = J - N, \quad \check{K} = K, \quad \check{L} = L + NB,$$
$$\check{M} = M + NA + A^*N. \tag{59}$$

If instead of Conditions (58) the systems satisfy the less restrictive condition

$$\check{A} = A, \quad \check{B} = B, \quad \check{H}(-\sigma, \sigma) = H(-\sigma, \sigma), \tag{60}$$

then there exist two $n \times n$ hermitian matrices X and N such that

$$\check{J} = J - X - N, \quad \check{K} = K, \quad \check{L} + NB, \quad \check{M} = M + NA + A^*N. \tag{61}$$

Proof. Let us first consider the completely controllable systems which satisfy the relations $M = 0$ and $A = A_0$, where

$$A_0 = \text{diag}\,(\alpha_1, \alpha_2, \ldots, \alpha_n), \tag{62}$$

where α_j are distinct, positive, arbitrary numbers. [1]

Matrix B^* can always be written in the form

$$(b_1\ b_2\ \ldots\ b_n), \tag{63}$$

where b_j are m-vectors. If we write the matrix L^* in the same form :

$$L^* = (l_1\ l_2\ \ldots\ l_n)$$

where the l_j are m-vectors, then the characteristic function of the system (see (37)), is given by

$$\begin{aligned} H(\lambda, \sigma) = K &+ \sum_{j=1}^{n} \left(\frac{l_j b_j^*}{\sigma - \alpha_j} + \frac{b_j l_j^*}{\lambda - \alpha_j} \right) + \\ &+ \sum_{j=1}^{n} \sum_{k=1}^{n} (\lambda + \sigma) \frac{b_j (J)_{jk} b_k^*}{(\lambda - \alpha_j)(\sigma - \alpha_k)} . \end{aligned} \tag{64}$$

Hence (cf. (8, § 4)—(10, § 4)),

$$K = \lim_{\sigma \to \infty} H(-\sigma, \sigma) \tag{65}$$

$$l_j b_j^* = [H(-\sigma, \sigma)(\sigma - \alpha_j)]_{\sigma = \alpha_j}, \qquad j = 1, 2, \ldots, n, \tag{66}$$

$$b_j (J)_{jk} b_k^* = \left[\frac{H(\lambda, \sigma)(\lambda - \alpha_j)(\sigma - \alpha_k)}{\lambda + \sigma} \right]_{\lambda = \alpha_j,\ \sigma = \alpha_k} \qquad j, k = 1, 2, \ldots, n, \tag{67}$$

To prove the first part of Proposition 2, let us consider the systems $\mathcal{S}$ and $\check{\mathcal{S}}$ which satisfy the conditions (58) stated in Proposition 2. Since the pair (A, B) is by hypothesis completely controllable, one can apply Property 9° of Theorem 1, § 34 and find a matrix Q_0 such that the matrix $A + BQ_0^*$ has the characteristic values $\alpha_1, \alpha_2, \ldots, \alpha_n$, satisfying the condi-

[1]) The values of the constants α_j are not fixed. Thus, the arguments which follow are valid, with minor changes, for other proofs.

tions required at the beginning of the proof. By virtue of Proposition 1, § 34 the pair $(A + BQ_0^*, B)$ will be completely controllable. Hence, we can apply to this pair Property 8° of Theorem 1, § 34 and determine a nonsingular matrix R_0, such that

$$\widetilde{A} = R_0(A + BQ_0^*) R_0^{-1} = \mathrm{diag}\,(\alpha_1, \alpha_2, \ldots, \alpha_n). \tag{68}$$

Using again Property 8° of Theorem 1, § 34 we conclude that if the matrix $\widetilde{B} = R_0 B$ is written in the form (63), then the vectors b_j satisfy the conditions

$$b_j \neq 0, \qquad j = 1, 2, \ldots, n. \tag{69}$$

Consider now another system

$$\widetilde{\mathcal{S}}: (\widetilde{J}, \widetilde{K}, \widetilde{L}, \widetilde{M};\, \widetilde{A}, \widetilde{B}), \tag{70}$$

with the property:

"The systems $\mathcal{S}$ and $\check{\mathcal{S}}$ belong to the same class" (71)

and whose coefficients are defined, — in terms of the coefficients of the systems $\mathcal{S}$ —, by the Relations (15)—(20) for the following particular values of the parameters:

$$P = I_m, \qquad Q = Q_0, \quad R = R_0, \qquad N = N_1, \tag{72}$$

where Q_0 and R_0 are the matrices which occur in (68), while N_1 is the hermitian solution of the equation

$$\widetilde{M} = 0, \tag{73}$$

where $\widetilde{M}$ is obtained by introducing the values (72) into (20). The fact that one can find a hermitian matrix N_1 which satisfies Eq. (73) follows by noting that this equation can be written in the form

$$M_1 + \widetilde{N}_1\widetilde{A} + \widetilde{A}^* \widetilde{N}_1 = 0, \tag{74}$$

where

$$M_1 = (R_0^{-1})^* (Q_0 K Q_0^* + Q_0 L^* + M) R_0^{-1};$$
$$\widetilde{N}_1 = (R_0^{-1})^* N_1 R_0^{-1}. \tag{75}$$

As in Section 3, § 4 (see (13, § 4)) one can show that there exists a hermitian matrix $\tilde{N}_1$ which satisfies Eq. (74). The matrix N_1 is then obtained using (75) which gives $N_1 = R_0^* \tilde{N}_1 R_0$. The matrix $\tilde{A}$ of the system $\tilde{\mathcal{S}}$ (70), given by Relation (15) for the values (72), is identical to the matrix A given by Relation (62). This, together with Condition (73), shows that the system $\check{\mathcal{S}}$, (70) satisfies Conditions (62). By formula (44) and the condition $P = I_m$, (see (72)), the characteristic function H of the system $\mathcal{S}$ is related to the characteristic function $\tilde{H}$ of the system $\tilde{\mathcal{S}}$, by

$$H(\lambda, \sigma) = \bar{C}'(\lambda)\, \tilde{H}(\lambda, \sigma)\, C(\sigma), \tag{76}$$

where $C(\sigma)$ is obtained by replacing in (40) Q by Q_0.

The same procedure can be applied to the system $\check{\mathcal{S}}$, thus defining a new system

$$\bar{\bar{\mathcal{S}}} = (\bar{\bar{J}}, \bar{\bar{K}}, \bar{\bar{L}}, \bar{\bar{M}}\,; \bar{\bar{A}}, \bar{\bar{B}}), \tag{77}$$

with the property

"The systems $\check{\mathcal{S}}$ and $\bar{\bar{\mathcal{S}}}$ belong to the same class", (78)

$\bar{\bar{\mathcal{S}}}$ being related to $\check{\mathcal{S}}$ by relations similar to those relating $\tilde{\mathcal{S}}$ to $\mathcal{S}$. Reexamining the procedure by which the system $\mathcal{S}$ was brought to the form $\mathcal{S}$ one sees that the parameters P, Q and R from the formulas of the type (15)—(20) which define the coefficients of the system $\bar{\bar{\mathcal{S}}}$ as functions of the coefficients of the system $\check{\mathcal{S}}$, have the same value as before, given by (72) (see the first two equalities of (58)). However, the matrix N will have, in general, a different value, $N = N_2$, such that the equality $\bar{\bar{M}} = 0$, similar to (73) is satisfied. Instead of (76) one obtains

$$\check{H}(\lambda, \sigma) = \bar{C}'(\lambda)\, \bar{\bar{H}}(\lambda, \sigma)\, C(\sigma), \tag{79}$$

where $C(\sigma)$ has the same expression as in (76).

We shall now show that the system $\bar{\bar{\mathcal{S}}}$ (77) is identical to the system $\check{\mathcal{S}}$ (70). From the first two equalities (58) and the fact that the parameters P, Q and R which occur in Relations (15) and (16) have in both cases the values given in (72), one obtains

$$\bar{\bar{A}} = \tilde{A}, \qquad \bar{\bar{B}} = \tilde{B}. \tag{80}$$

Moreover, the characteristic functions of the systems $\overline{\overline{\mathcal{S}}}$ and $\check{\mathcal{S}}$ are identical. To prove this, notice that the matrix $C(\sigma)$ can be singular only at a finite number of isolated points as shown by Relation (48). For all the points where the matrices $C(\sigma)$ and $C(\lambda)$ are nonsingular one can multiply Relations (76) and (79) on the left by $(\overline{C}'(\lambda))^{-1}$, and on the right by $(C(\sigma))^{-1}$, thus obtaining

$$\widetilde{H}(\lambda, \sigma) = (\overline{C}'(\lambda))^{-1} H(\lambda, \sigma) (C(\sigma))^{-1}, \tag{81}$$

$$\overline{\overline{H}}(\lambda, \sigma) = (\overline{C}'(\lambda))^{-1} \check{H}(\lambda, \sigma) (C(\sigma))^{-1} \tag{82}$$

whence, by using the last of the equalities (58) one sees that the matrices $\widetilde{H}(\lambda, \sigma)$ and $\overline{H}(\lambda, \sigma)$ are equal at all the points where the matrices $C(\lambda)$ and $C(\sigma)$ are nonsingular. Since all the elements of $\widetilde{H}$ and $\overline{H}$ can be written as quotients of polynomials, it follows from the foregoing that $\widetilde{H}$ and $\overline{H}$ are identical. Observe also that $\widetilde{\mathcal{S}}$ and $\overline{\overline{\mathcal{S}}}$ have the form presented at the beginning of the proof. By writing Relations (65)—(67) successively for $\mathcal{S}$ and $\overline{\overline{\mathcal{S}}}$ and taking into account (80) and (69) one sees that $\mathcal{S}$ and $\overline{\overline{\mathcal{S}}}$ are identical. This, together with (71) and (78) leads to the conclusion that the systems $\mathcal{S}$ and $\check{\mathcal{S}}$ belong to the same class. Denoting by $(J, K, L, M; A, B)$ the coefficients of the system $\widetilde{\mathcal{S}} = \overline{\overline{\mathcal{S}}}$, let us write the relations (17)—(20) for the pairs of systems $\widetilde{\mathcal{S}}$, $\mathcal{S}$ and $\widetilde{\mathcal{S}}$, $\check{\mathcal{S}}$. Since the parameters P, Q and R are the same in both cases, being given by formula (72), one obtains the following relations:

$$R^*\widetilde{J}R = J - N_1 = \check{J} - N_2, \tag{83}$$

$$P^*\widetilde{K}P = K = \check{K}, \tag{84}$$

$$R^*\widetilde{L}P = QK + L + N_1B = Q\check{K} + \check{L} + N_2\check{B}, \tag{85}$$

$$\begin{aligned} R^*\widetilde{M}R = QKQ^* + QL^* + LQ^* + M + N_1(A + BQ^*) + \\ + (A + BQ^*)^* N_1 = Q\check{K}Q^* + Q\check{L}^* + \check{L}Q^* + \check{M} + \\ + N_2(\check{A} + \check{B}Q^*) + (\check{A} + \check{B}Q^*)^*N_2. \end{aligned} \tag{86}$$

The above equalities imply (59), where $N = N_1 - N_2$ (the first two equalities of (59) follow immediately from (83) and (84); the third one results from (85) by using also the relations $\check{K} = K$ and $\check{B} = B$; the last one results from (86) by using also the relations $\check{K} = K$, $\check{L} = L + NB$, $\check{A} = A$ and $\check{B} = B$).

If the last relation of (58) is replaced by the last relation of (60), the matrices (81) and (82) are equal only for $\lambda = -\sigma$. Relations (65)—(67) written successively for $\mathcal{S}$ and $\bar{\bar{\mathcal{S}}}$ show that these systems differ only by the matrix J. Therefore, Relations (84)—(86) remain valid, and (83) is replaced by the equalities

$$R^*\tilde{J}R = J - N_1, \qquad R^*\bar{\bar{J}}R = \check{J} - N_2,$$

whence we get $\check{J} = R^*\bar{\bar{J}}R + N_2 - R^*\tilde{J}R + J - N_1$, that is, the first of the Relations (61) for $N = N_1 - N_2$ and $X = R^*(\tilde{J} - \bar{\bar{J}})R$. The other relations of (61) are obtained, as before, from (84)—(86).

§ 6. Equivalence classes for discrete systems

1. Definition of the classes of discrete systems

In the case of discrete multi-input systems the analogue of (1, § 5), (2, § 5) is given by

$$x_{k+1} = Ax_k + Bu_k, \qquad k = 0, 1, \dots, \tag{1}$$

$$\eta(0, k_1) = x^*_{k_1+1} J x_{k_1+1} - x^*_0 J x_0 + \sum_{k=0}^{k_1} (u^*_k K u_k +$$

$$+ u^*_k L^* x_k + x^*_k L u_k + x^*_k M x_k), \qquad k_1 = 0, 1, 2, \dots \tag{2}$$

In Relations (1) and (2), instead of the time functions u and x we have sequences which we shall again denote by u and x. Otherwise the symbols used above have the same meaning as in § 5. Introducing the $2n + m$-vector z_k as

$$z'_k = (x'_{k+1} \ u'_k \ x'_k), \qquad k = 0, 1, \dots, \tag{3}$$

one can write (2) in the compact form

$$\eta(0, k_1) = \sum_{k=0}^{k_1} z^*_k D z_k, \tag{4}$$

where

$$D = \begin{pmatrix} J & 0 & 0 \\ 0 & K & L^* \\ 0 & L & M-J \end{pmatrix}. \tag{5}$$

Using again the vector z_k (3) we can write (1) in the form $Fz_k = 0$, where the matrix F is the same as in § 5 (see (6, § 5)).

As shown in § 1 the quadratic form $z_k^* D z_k$ (with $Fz_k = 0$) is not altered if to the matrix D we add the matrix $T^* F + F^* T$. We need only consider here the particular case in which T is a matrix with $2n + m$ rows and columns of the form

$$T = \begin{pmatrix} \frac{N}{2} & \frac{NB}{2} & \frac{NA}{2} \\ 0 & 0 & 0 \\ 0 & 0 & 0 \end{pmatrix}, \tag{6}$$

where N is a hermitian $n \times n$-matrix. From (6) and (6, § 5) one obtains

$$T^*F + F^*T = \begin{pmatrix} -N & 0 & 0 \\ 0 & B^*NB & B^*NA \\ 0 & A^*NB & A^*NA \end{pmatrix} \tag{7}$$

and therefore $T^*F + F^*T$ has the same form as D (5). In fact, it is for this reason that we have considered the matrices T of the particular form (6).

We now apply to u and x the transformations (8, § 5), (9, § 5). Defining

$$\tilde{z}_k' = (\tilde{x}_{k+1} \quad \tilde{u}_k' \quad \tilde{x}_k') \tag{8}$$

we obtain the relation

$$\tilde{z}_k = S z_k, \tag{9}$$

where S has the same expression as in § 5 given by (12, § 5).

The elements $(D; F)$, where D has the form (5) and F the form (6, § 5) constitute a subset $\mathcal{E}'$ of the set $\mathcal{E}$ of § 1.

The elements $(S; T)$, where S has the form (12, § 5) and T is given by (6), constitute a subset $\mathcal{G}'$ of the set $\mathcal{G}$ of § 1.[1]) Proceeding as we did with the previous systems, we can establish the conditions under which (1), (2) belongs to the same class as another system of the same form:

$$\tilde{x}_{k+1} = \tilde{A}\tilde{x}_k + \tilde{B}\tilde{u}_k, \qquad k = 0, 1, \ldots \tag{10}$$

$$\tilde{\eta}(0, k_1) = \tilde{x}^*_{k_1+1}\tilde{J}\tilde{x}_{k_1+1} - \tilde{x}^*_0\tilde{J}\tilde{x}_0 + \sum_{k=0}^{k_1} (\tilde{u}^*_k\tilde{K}\tilde{u}_k + \tilde{u}^*_k\tilde{L}^*\tilde{x}_k + $$
$$+ \tilde{x}^*_k\tilde{L}\tilde{u}_k + \tilde{x}^*_k\tilde{M}\tilde{x}_k). \tag{11}$$

Definition 1. *The system (1), (2) is said to belong to the same class as the system (10), (11) (with respect to the set of matrices S, (12, § 5) and T, (6)), if the following relations are satisfied:*

$$\tilde{A} = R(A + BQ^*)\, R^{-1}, \tag{12}$$

$$\tilde{B} = RBP^{-1}, \tag{13}$$

$$\tilde{J} = (R^{-1})^*\, (J - N)\, R^{-1}, \tag{14}$$

$$\tilde{K} = (P^{-1})^*\, (K + B^*NB)\, P^{-1}, \tag{15}$$

$$\tilde{L} = (R^{-1})^*\, (L + QK + (A + BQ)^*\, NB)\, P^{-1}, \tag{16}$$

$$\tilde{M} = (R^{-1})^*\, (QKQ^* + QL^* + LQ^* + M +$$
$$+ (A + BQ^*)^*\, N(A + BQ^*) - N)\, R^{-1}. \tag{17}$$

Under the same conditions, the elements $(\tilde{J}, \tilde{K}, \tilde{L}, \tilde{M}; \tilde{A}, \tilde{B})$ and $(J, K, L, M; A, B)$ are said to belong to the same class.

Together with system (1), (2) we consider the polarized system

$$x^{\Delta}_{k+1} = x^{\Delta}_k A^* + u^{\Delta}_k B^*, \qquad x_{k+1} = Ax_k + Bu_k,$$
$$\eta_p(0, k_1) = \sum_{k=0}^{k_1} z^{\Delta}_k D z_k, \tag{18}$$

---

[1]) Since the matrix T, (6) depends on F, (6, § 5), we are in the case considered in footnote 1, page 29.

where u_k^Δ, x_k^Δ and z_k^Δ are new variables similar to u_k^*, x_k^* and z_k^*. If two systems of the form (1), (2) belong to the same class, the corresponding polarized systems are also said to belong to the same class. Assume that system (18) belongs to the same class as another system of the same form

$$\tilde{x}_{k+1}^\Delta = \tilde{x}_k^\Delta \tilde{A}^* + \tilde{u}_k^\Delta \tilde{B}^*, \quad \tilde{x}_{k+1} = \tilde{A}\tilde{x}_k + \tilde{B}\tilde{u}_k,$$

$$\tilde{\eta}_p(0, k_1) = \sum_{k=0}^{k_1} \tilde{z}_k^\Delta \tilde{D} \tilde{z}_k. \tag{19}$$

Let P, Q and R be the matrices in Relations (12)—(17) which express the fact that the two systems belong to the same class. Then we have

Proposition 1. *If systems (18) and (19) belong to the same class, then for every quadruple of sequences u, x, u^Δ and x^Δ which satisfy the first two equations of (18) the following properties are satisfied:*

1° *The first two equations of (19) are satisfied by the transformed sequences*

$$\tilde{u}_k = Pu_k - PQ^* x_k, \qquad \tilde{x}_k = Rx_k, \tag{20}$$

$$\tilde{u}_k^\Delta = u_k^\Delta P^* - x_k^\Delta Q P^*, \quad \tilde{x}_k^\Delta = x_k^\Delta R^*, \tag{21}$$

where P, Q, R have the same value as in the expressions (12)—(17) which relate the coefficients of the two systems.

2° *The following equality holds*

$$\tilde{z}_k^\Delta \tilde{D} \tilde{z}_k = z_k^\Delta D z_k, \tag{22}$$

where z_k, z_k^Δ, $\tilde{z}_k$ and $\tilde{z}_k^\Delta$ have expressions similar to (3).

3° *For the sequences introduced above, the expressions $\eta_p(0, k_1)$ and $\tilde{\eta}_p(0, k_1)$ are equal for all positive integers k_1.*

2. The characteristic function and the characteristic polynomial

If the control sequence is of the form

$$u_k = u_0 \sigma^k, \qquad k = 0, 1, 2, \ldots, \quad \text{then} \tag{23}$$

equation (1) has the particular solution

$$x_k = x_0(\sigma)\, \sigma^k, \text{ where } x_0(\sigma) = (\sigma I - A)^{-1} B u_0 \tag{24}$$

(provided, of course, that the condition $\det(\sigma I - A) \neq 0$ is satisfied). It follows that the corresponding sequence z (see(3)) has the form

$$z_k = Z_0(\sigma)\, u_0 \sigma^k, \tag{25}$$

where the matrix $Z_0(\sigma)$ is the same expression as in § 5 (see (31, § 5)).

Similarly, if the control sequence u^Δ has the form

$$u_k^\Delta = u_0^\Delta \lambda^k, \tag{26}$$

the first of the Eqs. (18) admits the particular solution

$$x_k^\Delta = x_0^\Delta(\lambda)\, \lambda^k, \text{ where } x_0^\Delta(\lambda) = u_0^\Delta B^*(\lambda I - A^*)^{-1} \tag{27}$$

(provided that $\det\,(I - A^*) \neq 0$) and the vectors $z_k^\Delta = (z_{k+1}^\Delta\ u_k^\Delta x_k^\Delta)$ have the form

$$z_0^\Delta = u_0^\Delta \bar{Z}_0'(\lambda)\, \lambda^k. \tag{28}$$

For the particular values (25) and (28) of z_k and z_k^Δ, the sum $\eta_p(0, k_1)$, given by (18), has the value

$$\eta_p(0, k_1) = \frac{(\lambda\sigma)^{k_1+1} - 1}{\lambda\sigma - 1}\, u_0^\Delta H(\lambda, \sigma)\, u_0, \tag{29}$$

which contains the "characteristic function" H, expressed by

$$H(\lambda, \sigma) = \bar{Z}_0'(\lambda)\, D Z_0(\sigma), \tag{30}$$

(cf. (36, § 5)). Written more explicitly, the characteristic function (deduced by replacing (5) and (31, § 5) in (30)) becomes

$$H(\lambda, \sigma) = K + L^*(\sigma I - A)^{-1}B + B^*(\lambda I - A^*)^{-1}L +$$

$$+ B^*(\lambda I - A^*)^{-1}\,(M + (\lambda\sigma - 1)\,J)\,(\sigma I - A)^{-1}B. \tag{31}$$

Proceeding as in Section (2, § 5), it can be shown that if systems (18) and (19) belong to the same class, then the characteristic function H of system (18) is related to the characteristic function $\tilde{H}$ of system (19) by (44, § 5), where $C(\sigma)$ is the same as in (40, § 5). All the developments of Section 3, § 5 are also valid for discrete systems. The characteristic polynomial of a discrete system is obtained by normalizing, as in Section (2, § 3), the expression

$$\det(\lambda I - A^*) \det(\sigma I - A) \det H(\lambda, \sigma).$$

Since the determinant of the characteristic function may be written as in (51, § 5) the above expression is a polynomial whose degree, both in σ and in λ, cannot exceed n. We thus obtain

$$\pi(\lambda, \sigma) = \frac{1}{\nu} \det(\lambda I - A^*) \det(\sigma I - A) \det H(\lambda, \sigma). \quad (32)$$

As in the previously examined cases the transformations given by the formulas (12)—(17) leave the characteristic polynomial invariant.

3. Relations between discrete systems with the same characteristic function

The analogue of Proposition 2, § 5 is given by

Proposition 2. *If two completely controllable systems $\mathcal{S}$ and $\check{\mathcal{S}}$ having the coefficients $\mathcal{S}$: $(J, K, L, M; A, B)$ and $\check{\mathcal{S}}$: $(\check{J}, \check{K}, \check{L}, \check{M}; \check{A}, \check{B})$ satisfy the conditions*

$$\check{A} = A, \quad \check{B} = B, \quad \check{H}(\lambda, \sigma) = H(\lambda, \sigma), \quad (33)$$

then $\mathcal{S}$ and $\check{\mathcal{S}}$ belong to the same class. More precisely, there exists a hermitian matrix N such that

$$\check{J} = J - N, \quad \check{K} = K + B^*NB, \quad \check{L} = L + A^*NB,$$

$$\check{M} = M + A^*NA - N. \quad (34)$$

If instead of conditions (33) the systems satisfy the conditions

$$\check{A} = A, \quad \check{B} = B, \quad \check{H}\left(\frac{1}{\sigma}, \sigma\right) = H\left(\frac{1}{\sigma}, \sigma\right), \quad (35)$$

then there exist two hermitian matrices X and N such that

$$\check{J} = J - X - N, \quad \check{K} = K + B^*NB, \quad \check{L} = L + A^*NB,$$

$$\check{M} = M + A^*NA - N. \quad (36$$

The proof follows the same lines as in the case of Proposition 2, § 5; consequently, we shall indicate here only the more important changes which have to be made. For discrete systems which satisfy (62, § 5), the matrix $H(\lambda, \sigma)$ differs from (64, § 5) only by the fact that the factor $(\lambda + \sigma)$ is to be replaced by $(\lambda\sigma - 1)$ in the double sum. Instead of Relations (65, § 5)—(67, § 5) one obtains (taking $\alpha_j > 1$)

$$l_j b_j^* = \left[H\left(\frac{1}{\sigma}, \sigma\right)(\sigma - \alpha_j) \right]_{\sigma=\alpha_j}, \qquad j = 1, 2, \ldots, n, \tag{37}$$

$$K - \sum_{j=1}^{n} \frac{b_j l_j^*}{\alpha_j} = \lim_{\sigma \to \infty} H\left(\frac{1}{\sigma}, \sigma\right), \tag{38}$$

$$b_j (J)_{jk} b_k^* = \left[\frac{H(\lambda, \sigma)(\lambda - \alpha_j)(\sigma - \alpha_k)}{\lambda\sigma - 1} \right]_{\lambda=\alpha_j,\ \sigma=\alpha_k},$$

$$j, k = 1, 2, \ldots, n. \tag{39}$$

Another change applies to Eq. (74, § 5) which is replaced by

$$M_1 + \widetilde{A}^* \widetilde{N}_1 \widetilde{A} - \widetilde{N}_1 = 0. \tag{40}$$

Since the matrix A has the form (68, § 5), Eq. (40) has a hermitian solution $\widetilde{N}_1$ whose elements $(\widetilde{N}_1)_{jk}$ are related to the elements $(M_1)_{jk}$ of the matrix M_1 by the expressions

$$(\widetilde{N}_1)_{jk} = -\frac{(M_1)_{jk}}{\alpha_j \alpha_k - 1}. \tag{41}$$

The proof can be completed as in Section 5, § 5.

§ 7. Equivalence classes for systems with time dependent coefficients

If all the coefficients of the system (1, § 5), (2, § 5) are given functions of time, the system takes the form

$$\frac{dx}{dt} = A(t)\, x + B(t)\, u, \tag{1}$$

$$\eta(t_0, t_1) = [x^* J(t)\, x]_{t_0}^{t_1} + \int_{t_0}^{t_1} (u^* K(t)\, u + u^* L^*(t)\, x +$$

$$+ x^* L(t)\, u + x^* M(t)\, x)\, dt, \tag{2}$$

where the symbols have the same meaning as in § 5. All the functions of time are assumed to be defined and piecewise continuous for $t \geqslant t_0$. The matrices $J(t)$, $K(t)$ and $M(t)$ are assumed to be hermitian for every $t \geqslant t_0$. Moreover, it is assumed that the matrix-valued function J is differentiable and that its derivative is piecewise continuous. The meaning of the first term in the right hand member of expression (2) is given by the equality

$$[x^* J(t)\, x]_{t_0}^{t_1} = \int_{t_0}^{t_1} \left(x^* J(t) \frac{\mathrm{d}x}{\mathrm{d}t} + \frac{\mathrm{d}x^*}{\mathrm{d}t} J(t)\, x + x^* \frac{\mathrm{d}\, J(t)}{\mathrm{d}t}\, x \right) \mathrm{d}t. \quad (3)$$

Introducing

$$z' = \left(\frac{\mathrm{d}x'}{\mathrm{d}t} \quad u' \quad x' \right), \quad (4)$$

Eq. (1) and the integral (2) can be written as

$$F(t)\, z = 0, \qquad \eta(t_0, t_1) = \int_{t_0}^{t_1} z^*(t)\, D(t)\, z(t)\, \mathrm{d}t, \quad (5)$$

where

$$F(t) = \begin{pmatrix} -I & B(t) & A(t) \\ 0 & 0 & 0 \\ 0 & 0 & 0 \end{pmatrix};$$

$$D(t) = \begin{pmatrix} 0 & 0 & J(t) \\ 0 & K(t) & L^*(t) \\ J(t) & L(t) & M(t) + \dfrac{\mathrm{d}J(t)}{\mathrm{d}t} \end{pmatrix}. \quad (6)$$

We also introduce the matrix

$$T(t) = \begin{pmatrix} 0 & 0 & N(t) \\ 0 & 0 & 0 \\ N(t) & 0 & \dfrac{\mathrm{d}N(t)}{\mathrm{d}t} \end{pmatrix}, \quad (7)$$

where $N(t)$ is defined and hermitian for all $t \geqslant t_0$, and the function N is differentiable and has a piecewise continuous derivative. In this case the matrix $T^*(t)\, F(t) + F^*(t)\, T(t)$ has the same form as the matrix $D(t)$ and one can write the equality

$$z^*(t)\,D(t)\,z(t) = z^*(t)\,(D(t) + T^*(t)\,F(t) + F^*(t)\,T(t))\,z(t). \quad (8)$$

We now pass from the variable $u(t)$ and $x(t)$ to the new variables $\widetilde{u}(t)$ and $\widetilde{x}(t)$ related to the former by the expressions (8, § 5), (9, § 5), where P, Q and R are now functions of t, defined for $t \geqslant t_0$, differentiable, having piecewise continuous derivatives and such that $P(t)$, $Q(t)$ and $R(t)$ have (respectively) the same properties as P, Q and R of §5 for every $t \geqslant t_0$. Then one can write

$$\widetilde{z}(t) = S(t)\,z(t), \quad (9)$$

where

$$S(t) = \begin{pmatrix} R(t) & 0 & \dfrac{\mathrm{d}R(t)}{\mathrm{d}t} \\ 0 & P(t) & -P(t)\,Q^*(t) \\ 0 & 0 & R(t) \end{pmatrix}. \quad (10)$$

For each $t \geqslant t_0$ the set of elements $(D(t)\,;\,F(t))$ is a subset $\mathscr{E}'$ of the set $\mathscr{E}$ of § 1, while the set of elements $(S(t)\,;\,T(t))$ is a subset $\mathscr{G}'$ of the set $\mathscr{G}$ of § 1. It can be easily seen that these subsets have the properties required in Proposition 1, § 1. Based on that proposition we can state the following definition which is a generalization of Definition 1, § 5:

Definition 1. *The elements*

$$(\widetilde{J}(t),\ \widetilde{K}(t),\ \widetilde{L}(t),\ \widetilde{M}(t)\,;\ \widetilde{A}(t),\ \widetilde{B}(t))$$

and

$$(J(t),\ K(t),\ L(t),\ M(t)\,;\ A(t),\ B(t)),$$

formed with the coefficients of two systems of the form (1), (2) belong to the same class (or the respective systems belong to the same class) if there exist $P(t)$, $Q(t)$, $R(t)$ and $N(t)$ with the properties stated above such that the following properties are satisfied

$$\widetilde{D}(t) = (S^{-1}(t))^*\,(D(t) + T^*(t)\,F(t) + F^*(t)\,T(t))\,S^{-1}(t), \quad (11)$$

$$\widetilde{F}(t) = S(t)\,F(t)\,S^{-1}(t), \quad (12)$$

or, more explicitly,

$$\tilde{A}(t) = R(t)\,(A(t) + B(t)\,Q^*(t))R^{-1}(t) + R(t)\frac{\mathrm{d}R^{-1}(t)}{\mathrm{d}t}, \tag{13}$$

$$\tilde{B}(t) = R(t)\; B(t)\; P^{-1}(t), \tag{14}$$

$$\tilde{J}(t) = (R^{-1}(t))^*\,(J(t) - N(t))R^{-1}(t), \tag{15}$$

$$\tilde{K}(t) = (P^{-1}(t))^*\; K(t)\; P^{-1}(t), \tag{16}$$

$$\tilde{L}(t) = (R^{-1}(t))^*\,(Q(t)\,K(t) + L(t) + N(t)\,B(t))\;P^{-1}(t), \tag{17}$$

$$\begin{aligned}\tilde{M}(t) = (R^{-1}(t))^*\,(Q(t)\;K(t)\;Q^*(t) + Q(t)\;L^*(t) + \\ + L(t)\,Q^*(t) + M(t) + N(t)\,(A(t) + B(t)\;Q^*(t)) + \\ + \left(A(t) + B(t)\;Q^*(t))^*\;N(t) + \frac{\mathrm{d}N(t)}{\mathrm{d}t}\right)\;R^{-1}(t).\end{aligned} \tag{18}$$

The properties of systems which belong to the same class are given by a generalization of Proposition 1, § 5 which can be easily established by the interested reader and which essentially expresses the fact that two systems belonging to the same class can be obtained from one another by a transformation defined by certain matrix-valued function T and S of the form (7) and (10).

POSITIVE SYSTEMS

In this chapter the so-called "positive systems" are defined and investigated. The positiveness property of single-input systems is treated in § 8. The same problem is then studied for multi-input (§ 9) and discrete (§ 10) systems, for systems with time dependent coefficients (§ 11) and for non-linear (§ 12) systems.

The concept of positiveness is abstracted from some recent papers — and especially, from the works of V. A. Yakubovich [2] and R. E. Kalman [5] on the solvability of the A. I. Lur'e equations — where, however, the positiveness property is amalgamated with other conditions. A study of the positiveness property in itself was carried out by the author in [16]. This chapter adds new results to the quoted work.

§ 8. Single-input positive systems

1. Definition of single-input positive systems

In this paragraph we shall examine the single-input systems described by the formulas (1, § 2), (2, § 2) and written in the compact form

$$\frac{dx}{dt} = Ax + b\mu, \qquad \eta(0, t_1) = \int_0^{t_1} z^* D z \, dt, \tag{1}$$

where

$$z = \begin{pmatrix} \dfrac{dx}{dt} \\ \mu \\ x \end{pmatrix}, \qquad D = \begin{pmatrix} 0 & 0 & J \\ 0 & \varkappa & l^* \\ J & l & M \end{pmatrix}, \tag{2}$$

the notations having the same meaning as in paragraph 2.

We shall say that system (1) is "positive" if there exist two real-valued functions, $y \mapsto \alpha(y)$ and $(y, \rho) \mapsto \beta(y, \rho)$, defined for every n-vector y (with complex components) and every complex scalar ρ, such that $\beta(y, \rho) \geqslant 0$ (for every y and every ρ) and the following property is true : For every pair of functions x, μ(x : continuous, μ : piecewise continuous), defined for $t \geqslant 0$ and satisfying the differential equation in (1), the corresponding value of η (see (1), (2)) satisfies, for every $t_1 \geqslant 0$, the equality

$$\eta(0, t_1) = \alpha(x(t_1)) - \alpha(x(0)) + \int_0^{t_1} \beta(x(t), \mu(t))\, dt.$$

This definition will be extended to other types of systems. For convenience, we abbreviate it as follows

Definition 1. *System (1) is said to be positive if* η *can be written in the form*

$$\eta(0, t_1) = [\alpha(x)]_0^{t_1} + \int_0^{t_1} \beta(x, \mu)\, dt \quad , \quad \text{where} \quad \beta(x, \mu) \geqslant 0, \qquad (3)$$

The exact meaning of the expression "can be written in the form (3)" is given above.

2. Theorem of positiveness for single-input systems

In the theorem which follows we make use of the characteristic function (12, § 3) given (for $\lambda = -\sigma$) by

$$\chi(-\sigma, \sigma) = \bar{z}_0'(-\sigma)\, D z_0(\sigma) \qquad (4)$$

(where $z_0(\sigma)$ has the expression (5, § 3)), as well as of the characteristic polynomial (18, § 3) which, for $\lambda = -\sigma$, becomes

$$\pi(-\sigma, \sigma) = \frac{1}{\nu} \det(-\sigma I - A^*) \det(\sigma I - A)\, \chi(-\sigma, \sigma), \quad (5)$$

where ν is a positive normalizing factor, chosen as in Section 2, § 3. We also make use of the matrix (12, § 2).

$$T^*F + F^*T = \begin{pmatrix} 0 & 0 & -N \\ 0 & 0 & (Nb)^* \\ -N & Nb & NA + A^*N \end{pmatrix}, \qquad (6)$$

where N is a hermitian matrix.

Theorem 1. *If the pair (A, b) is completely controllable (Appendix A) the following properties are equivalent:*

1° *System (1) is positive in the sense of Definition 1.*

2° *The inequality*

$$\chi(-\mathrm{i}\omega, \mathrm{i}\omega) \geqslant 0 \tag{7}$$

is satisfied for every real number ω which satisfies the condition $\det(\mathrm{i}\omega I - A) \neq 0$.

3° *There exists at least one polynomial ψ such that the polynomial (5) can be factorized in the form* [1])

$$\pi(-\sigma, \sigma) = \overline{\psi}(-\sigma)\, \psi(\sigma) \tag{8}$$

and for every polynomial ψ which satisfies Relation (8), there exists a scalar γ and an n-vector w such that the function (4) can be factorized as [2])

$$\chi(-\sigma, \sigma) = \overline{\varphi}(-\sigma)\, \varphi(\sigma), \tag{9}$$

where

$$\varphi(\sigma) = \gamma + w^*(\sigma I - A)^{-1} b, \tag{10}$$

and γ and w satisfy the relation

$$\gamma + w^*(\sigma I - A)^{-1} b = \sqrt{\nu}\, \frac{\psi(\sigma)}{\det(\sigma I - A)}. \tag{11}$$

Furthermore, there exists a hermitian matrix N such that after adding to the matrix D of (1) the matrix (6) [3]), *the integral $(0, t_1)$ can be written in the form*

$$\eta(0, t_1) = [x^*(J - N)\, x]_0^{t_1} + \int_0^{t_1} |\gamma\mu + w^* x|^2\, \mathrm{d}t. \tag{12}$$

If the system (1) has real coefficients and the polynomial ψ is chosen so as to have real coefficients, then γ, w and N can be chosen real.

---

[1]) We recall the notation $\overline{\psi}(\lambda) = \overline{\psi(\overline{\lambda})}$.

[2]) The factorizations of the form (9) have been used for the treatment of problems related to the stability of control systems by the author ([11, Lemma 1]) and by R. E. Kalman [5]; the papers were published practically simultaneously, but with different objectives in view. See also footnote 2, § 3.

[3]) It has been shown in Section 1, § 1 and in Section 2, § 2 that the equality $z^*Dz = z^*(D + T^*F + F^*T)z$ is satisfied.

4° *There exists a hermitian matrix N, a scalar γ and a vector w such that the matrices D, (2) and $T^*F + F^*T$, (6) satisfy the equality*

$$\begin{pmatrix} 0 & 0 & J \\ 0 & \varkappa & l^* \\ J & l & M \end{pmatrix} + \begin{pmatrix} 0 & 0 & -N \\ 0 & 0 & (Nb)^* \\ -N & Nb & NA + A^*N \end{pmatrix} = \begin{pmatrix} 0 & 0 & J-N \\ 0 & \gamma^*\gamma & \gamma^* w^* \\ J-N & \gamma w & ww^* \end{pmatrix}. \tag{13}$$

5° *There exists a hermitian matrix N such that the matrix*

$$\begin{pmatrix} \varkappa & (l + Nb)^* \\ l + Nb & M + NA + A^*N \end{pmatrix} \tag{14}$$

is positive semidefinite.

6° *There exists a hermitian matrix N such that after adding the matrix (6) to the matrix D, (2), the integral $\eta(0, t_1)$ takes the form*

$$\eta(0, t_1) = [x^*(J - N)\, x]_0^{t_1} + \int_0^{t_1} \varphi_0(\mu, x)\, dt, \tag{15}$$

where φ_0 is a positive semidefinite quadratic form in the variables μ and x.

7° $\chi(-\sigma, \sigma)$, *(4) admits a factorization of the form (9).*

8° $\pi(-\sigma, \sigma)$, *(5), admits a factorization of the form (8).*

9° $\pi(-\sigma, \sigma)$, *(5), satisfies the inequality*

$$\pi(-i\omega, i\omega) \geqslant 0, \tag{16}$$

for every real ω which satisfies the condition $\det (i\omega I - A) \neq 0$.

10° *The integral $\eta(0, t_1)$ can be written in the form (12).*

3. Remarks on the theorem of positiveness

Although the 10 properties stated in Theorem 1 are equivalent, they are used in different ways in applications. Properties 2° and 5°—10° are especially useful as criteria on the basis of which one can establish whether or not a given system is positive. However, from Theorem 1 it follows that any of the Properties 2°—10° constitutes a necessary and sufficient criterion for the positiveness of system 1 (assumed to be completely controllable). Property 3° obviously includes as particular cases the Properties 7°, 8° and 10° and should be used when we want to obtain the main consequences of the property of positiveness.

We also remark that under certain additional assumptions the condition that the matrix (14) be positive semidefinite (as required in Property 5°) can be written in simpler forms. From Property 5° it follows that $\varkappa \geqslant 0$. But if the strict inequality $\varkappa > 0$ is fulfilled, then the condition that the matrix (14) be positive semidefinite amounts to the requirement that the same property be satisfied by the matrix

$$M + NA + A^*N - \frac{(l + Nb)(l + Nb)^*}{\varkappa}$$

Conditions of this type are used in V. A. Yakubovich's papers on stability. From Property 5° it follows that the matrix $M + NA + A^*N$ must be positive semidefinite. Under the more restrictive assumption that this matrix be strictly positive definite, the condition that the matrix of the quadratic form (14) be positive definite reduces to a single scalar inequality

$$\varkappa - (l + Nb)^* (M + NA + A^*N)^{-1} (l + Nb) \geqslant 0,$$

a form employed by S. Lefschetz in his investigations on absolute stability.

Property 8° which is expressed exclusively in terms of the characteristic polynomial shows — taking into account Proposition 1, § 3 — that if a given system is (or is not) positive, then any system belonging to the same class as the system considered has the same property.

4. Proof of the theorem of positiveness

Theorem 1 will be proved according to the following scheme

$$1° \to 2° \to 3° \to 4° \to 5° \to 6° \to 1°, \tag{17}$$

$$3° \to 7° \to 8° \to 9° \to 2°, \tag{18}$$

$$3° \to 10° \to 6°. \tag{19}$$

where the numbers 1°—10° represent the respective properties stated in the theorem and the arrow signifies "implies". Indeed, from (17) it follows that Properties 1°—6° are equivalent. Using this conclusion one sees from (18) that Properties 7°—9°

are equivalent to the preceding ones. Finally, taking into account (19) one sees that Property 10° is also equivalent to the properties stated in Theorem 1.

Obviously, the order in which the implications (17)—(19) are proved is immaterial. Since most of the above implications are immediate it is convenient to start the proof with them. Thus, the implications from (19) are obvious since Property 10° also appears in the statement of Property 3°, while Property 6° is obviously satisfied if Property 10° holds. The implications in (18) are also easily proved. Property 7° is included in the statement of Property 3°. Using (5) and the inequality $\nu > 0$, we immediately see that Property 8° follows from Property 7°. Property 9° is an immediate consequence of Property 8° since the two factors of (8) are complex conjugates for $\sigma = i\omega$. Using (5) again we immediately obtain Property 2° from Property 9°.

Therefore, only the implications in (17) remain to be proved. We shall again begin with the most simple implications, namely we shall prove the chain of implications

$$3° \rightarrow 4° \rightarrow 5° \rightarrow 6° \rightarrow 1° \rightarrow 2° \rightarrow 3°, \tag{20}$$

which obviously is equivalent to (17). Property 4° is nothing more than the explicit formulation of a statement included in Property 3° namely that the integral $\eta(0, t_1)$ can be expressed in the form (12). From Property 4° it follows that the matrix (14) has the form

$$\begin{pmatrix} \gamma^*\gamma & \gamma^* w^* \\ \gamma w & w w^* \end{pmatrix} = \begin{pmatrix} \gamma^* \\ w \end{pmatrix} \begin{pmatrix} \gamma^* \\ w \end{pmatrix}^*$$

and hence is positive semidefinite. Thus Property 5° follows from Property 4°. If we add to matrix D of (1) the matrix (6), where N has the same value as in Property 5°, we can write the integral $\eta(0, t_1)$ in the form (15) (compare the expanded form (2, § 2) of the integral $\eta(0, t_1)$ with the expression (2) of the motrix D), where the quadratic form $\varphi_0(\mu, x)$ is expressed by

$$\varphi_0(\mu, x) = (\mu^* \; x^*) \begin{pmatrix} \varkappa & (l + Nb)^* \\ l + Nb & M + NA + A^*N \end{pmatrix} \begin{pmatrix} \mu \\ x \end{pmatrix}$$

and hence, according to Property 5°, is positive semidefinite, whence Property 6° follows. Property 1° (see Definition 1) results immediately from Property 6°.

To conclude the proof of Theorem 1 we have to prove the last two implications of (20) : 1° → 2° → 3°.

Proof of the implication 1° → 2°. We consider the polarized system (36, § 2), (37, § 2) and the particular control functions (1, § 3) and (6, § 3) for $\sigma = i\omega$ and $\lambda = -i\omega$. These functions satisfy the relations

$$\overline{\mu^{\Delta}(t)} = \mu(t) = e^{i\omega t}, \tag{21}$$

and the corresponding solutions (2, § 3) and (7, § 3) satisfy the conditions

$$((x^{\Delta}(t))^{*} = x(t) = x_0(i\omega)\, e^{i\omega t}. \tag{22}$$

Similarly, the functions (4, § 3) and (9, § 3) satisfy the relations

$$(z^{\Delta}(t))^{*} = z(t) = z_0(i\omega)\, e^{i\omega t}, \tag{23}$$

whence it follows that equality (11, § 3) can be written in the form

$$z^{*}(t)\, Dz(t) = \chi(-i\omega, i\omega). \tag{24}$$

Consider now system (1) for the function μ given by (21) and examine the solution x given by (22) to which there corresponds the function z given by (23). Then, the integrand from (1) satisfies Relation (24), whence

$$\eta(0, t_1) = \chi(-i\omega, i\omega)\, t_1. \tag{25}$$

We take the number t_1 to be equal to $\frac{2\pi}{|\omega|}$ (if $\omega \neq 0$) or to an arbitrary positive number if $\omega = 0$. Then from (22) one obtains $x(t_1) = x(0)$ and hence inequality (3) (which is included in property 1°) implies that $\eta(0, t_1) \geqslant 0$. This, together with (25) leads to (7).

Proof of the implication 2° → 3°

Let us introduce the polynomial

$$\pi_1(\sigma) = \pi(-\sigma, \sigma). \tag{26}$$

From (7) and (5) we have

$$\pi_1(i\omega) = \pi(-i\omega, i\omega) \geqslant 0, \tag{27}$$

and from (19, § 3) we obtain

$$\pi(-\sigma, \sigma) = \bar{\pi}(\sigma, -\sigma) \quad \text{or} \quad \pi_1(\sigma) = \bar{\pi}_1(-\sigma). \tag{28}$$

We will show that the polynomial can be factored into the form (8). This is obvious if the polynomial is identically zero or reduces to a constant (in the latter case from (27) and (28) it follows that the constant must be real and positive). Therefore, we assume that the polynomial π_1 does not reduce to a constant. Let σ_1 be a root of the polynomial π_1. By (28), from $\pi_1(\sigma_1) = 0$ it follows that $\overline{\pi}_1(-\sigma_1) = 0$ or $\pi_1(-\overline{\sigma}_1) = 0$. Hence, $\pi_1(\sigma)$ is divisible by both $(\sigma - \sigma_1)$ and $(\sigma + \overline{\sigma}_1)$. Therefore, (see also (27) in case $\sigma_1 = -\overline{\sigma}_1$) one can define a new polynomial

$$\pi_2(\sigma) = -\frac{\pi_1(\sigma)}{(\sigma - \sigma_1)(\sigma + \overline{\sigma}_1)},$$

which possesses the properties $\pi_2(i\omega) \geqslant 0$ and $\pi_2(\sigma) = -\overline{\pi}_2(-\sigma)$, similar to (27) and (28) (this can be readily seen using (27) and (28)). If the polynomial π_2 does not reduce to a constant we apply again the same procedure. In general we obtain a sequence of polynomials

$$\pi_{j+1}(\sigma) = -\frac{\pi_j(\sigma)}{(\sigma - \sigma_j)(\sigma + \overline{\sigma}_j)}, \tag{29}$$

which satisfy the relations (of the type (27) and (28)) $\pi_j(i\omega) \geqslant 0$ and $\pi_j(\sigma) = \overline{\pi}_j(-\sigma)$. Since the degree of the polynomial $\pi_{j+1}(\sigma)$ is smaller than the degree of the polynomial $\pi_j(\sigma)$ and the degree of the initial polynomial $\pi_1(\sigma)$ cannot be greater than $2n$ (see (20, § 3)), the procedure described above can be applied only $p \leqslant n$ times. The polynomial $\pi_{p+1}(\sigma)$ reduces to a constant which must be, as shown before, real and positive; hence there exists a number γ_0 such that

$$\pi_{p+1}(\sigma) = \overline{\gamma}_0\gamma_0. \tag{30}$$

From Relations (26), (29) (for $j = 1, 2, \ldots, p$) and (30) we obtain (8) where

$$\psi(\sigma) = \gamma_0(\sigma - \sigma_1)(\sigma - \sigma_2)\ldots(\sigma - \sigma_p), \quad p \leqslant n. \tag{31}$$

By permuting the order of some of the factors of (29) we obtain in general (when not all the factors are identical) a new factorization, different from the former.

Using (18, § 3), we deduce from (8) the factorization relation (9) where

$$\varphi(\sigma) = \sqrt{\nu}\,\frac{\psi(\sigma)}{\det(\sigma I - A)}. \tag{32}$$

From (31) it follows that the degree of the polynomial ψ cannot exceed n. Therefore we can define

$$\gamma = \lim_{\sigma\to\infty} \sqrt{\nu}\,\frac{\psi(\sigma)}{\det(\sigma I - A)}. \tag{33}$$

The degree of the polynomial

$$\tilde{\psi}(\sigma) = \sqrt{\nu}\,\psi(\sigma) - \gamma \det(\sigma I - A) \tag{34}$$

is smaller than n. Since the pair (A, b) is completely controllable we can apply Property 10° of Theorem 1, § 31 and deduce that there exists a vector w such that

$$w^*(\sigma I - A)^{-1} b = \frac{\tilde{\psi}(\sigma)}{\det(\sigma I - A)}. \tag{35}$$

Substituting (34) into (35) and using (32) we obtain formula (10). This, together with (32) gives (11).

Consider now besides the initial system (1) another system of the same form:

$$\frac{dx}{dt} = Ax + b\mu, \qquad \tilde{\eta}(0, t_1) = \int_0^{t_1} z^*\tilde{D}z\,dt \tag{36}$$

which differs from (1) only by the matrix $\tilde{D}$, having the form

$$\tilde{D} = \begin{pmatrix} 0 & 0 & \tilde{J} \\ 0 & \tilde{\varkappa} & \tilde{l}^* \\ \tilde{J} & \tilde{l} & \tilde{M} \end{pmatrix} = \begin{pmatrix} 0 & 0 & \tilde{J} \\ 0 & \gamma^*\gamma & \gamma^*w^* \\ \tilde{J} & \gamma^*w & ww^* \end{pmatrix} \tag{37}$$

or

$$\tilde{\varkappa} = \gamma^*\gamma, \qquad \tilde{l} = \gamma w, \qquad \tilde{M} = ww^*, \tag{38}$$

where the matrix $\tilde{J}$ is undetermined for the time being. Using Relations (38) and formula (13, § 3) one can easily see that the characteristic function $\chi(\lambda, \sigma)$ of system (36) for $\lambda = -\sigma$ is equal to the right-hand member of Relation (9), where $\varphi(\sigma)$ is given by (10). Therefore, the conditions of Proposition 3, § 4 are satisfied and hence, by that proposition, there exist two hermitian matrices $\tilde{X}$ and N such that Relations (25, § 4) are satisfied, that is, (see (38))

$$\tilde{J} = J - \tilde{X} - N, \quad \gamma^*\gamma = \varkappa, \quad \gamma w = l + Nb,$$

$$ww^* = M + NA + A^*N. \tag{39}$$

The last three relations of (39) show that for the matrix N given by Proposition 3, § 4 the matrix equality (13) is satisfied. This equality expresses the fact (which had to be proved) that the matrix $\tilde{D}$ of the integral (12) is obtained by adding to the matrix D of (1) a matrix of the form (6).

We still have to prove the last statement of Property 3°. If all the coefficients of system (1) are real, then the coefficients of $\pi(-\sigma, \sigma)$ are also real (see (18, § 3) and (13, § 3)). Therefore, those roots of $\pi(-\sigma, \sigma)$ which are not real can be grouped into pairs of complex conjugated roots. By permuting, if necessary, the order of the factors in the equality obtained by substituting (31) into (8), we can define a new polynomial ψ satisfying Relation (8) and which, together with any non-real root admits also the complex conjugate root with the same order of multiplicity. The coefficient γ_0 of (30) may always be taken real. Under these conditions the coefficients of the polynomial ψ are real. From (33) it follows that the scalar γ is real. The polynomial $\tilde{\psi}$, given by (34), also has real coefficients.

From Relation (35) written for an arbitrary real number σ_j we obtain

$$(\operatorname{Im} w)^* (\sigma_j I - A)^{-1} b = 0, \text{ where } \operatorname{Im} w = \frac{1}{2i}(w - \overline{w}). \tag{40}$$

Eq. (40) is valid, in particular, for n arbitrary real numbers σ_j, $j = 1, 2, \ldots, n$, distinct and different from the eigenvalues of the matrix A. By virtue of Property 13° of Theorem 1, § 31, the vectors $(\sigma_j I - A)^{-1}b$ are linearly independent. Therefore, Relations (40) lead (for $j = 1, 2, \ldots, n$) to the equality $\operatorname{Im} w = 0$ and hence the vector w is real.

Using these conclusions we deduce from the last two equalities of (39) the following relations:

$$(\operatorname{Im} N)\, b = 0, \tag{41}$$

$$(\operatorname{Im} N)\, A + A^* (\operatorname{Im} N) = 0, \text{ where } \operatorname{Im} N = \frac{1}{2i}(N - \overline{N}). \tag{42}$$

From (42) one also obtains the equations $(\operatorname{Im} N)A^k b = - A^* (\operatorname{Im} N) A^{k-1} b$, $k = 1, 2, \ldots, n$, whence, by using (41), one obtains successively the relations $(\operatorname{Im} N)A^k b = 0$, $k = 0, 1, 2, \ldots$, or:

$$(\operatorname{Im} N)\,(b \quad Ab \quad A^2 b \ldots A^{n-1} b) = 0. \tag{43}$$

By Property 4° of Theorem 1, § 31, the matrix which multiplies $\operatorname{Im} N$ in (43) is nonsingular, hence (43) yields $\operatorname{Im} N = 0$. Thus, N is real. We have obtained Property 3° and this concludes the proof of Theorem 1.

5. The Yakubovich-Kalman lemma

The equivalence of the Properties 2° and 4° of Theorem 1 has been proved by V. A. Yakubovich [2] and by R. E. Kalman [5] and represents an important result in the theory of stability [1]). This result can be also expressed as follows:

Lemma 1. *(Yakubovich-Kalman). If the pair (A, b) is completely controllable, the condition $\chi(-i\omega, i\omega) \geqslant 0$ (see Property 2° Theorem 1) is equivalent to the condition that the Lur'e equations* [2])

$$\varkappa = |\gamma|^2, \tag{44}$$

$$l + Nb = \gamma w, \tag{45}$$

$$M + NA + A^* N = w w^*, \qquad (N^* = N) \tag{46}$$

---

[1]) Stated as an open problem in the author's paper [8] the equivalence has been proved, for sufficiently general conditions, by V. A. Yakubovich [2]. Subsequently, Kalman has given a new proof [5] which includes some limit cases not treated in the initial work of V. A. Yakubovich. Kalman's proof is remarkable also because of the fact that it emphasizes the surprisingly simple connection between the solutions of the Lur'e equations and the coefficients of the relation of factorization (9)—(10).

[2]) The systems of equations of the form (31),(44)—(46) have been introduced and studied (in a scalar form and under some supplementary assumptions) by A. I. Lur'e [1] (see also the monograph of M. A. Aizerman and F. R. Gantmacher [1]).

have a solution (that is, there exist γ, w and a hermitian matrix N such that equations (44)—(46) are satisfied).

The Yakubovich-Kalman lemma is immediately obtained from Theorem 1 by observing that Eq. (13) results from Relations (44)—(46) and conversely. We also conclude that the solvability of the Lur'e equations is equivalent to any of the properties of Theorem 1. Furthermore it follows that if the Lure's equations are (or are not) solvable for a given system, then the same property holds for any other system which belongs to the same class as the first system in the sense of Definition 1, § 2.

6. Special forms for completely controllable single-input positive systems

We have seen that the integral $\eta(0, t_1)$ of a positive system can be expressed in the form (12). Moreover, with the help of the transformations studied in § 2, it is possible to bring the completely controllable positive system to even simpler forms.

Assume first that the condition $\varkappa \neq 0$ is satisfied. Then, by (13, § 3) and (18, § 3) the polynomial (5) is of degree $2n$. Hence, the polynomial ψ of (8) is of degree n and we can write

$$\psi(\sigma) = \gamma_0(\alpha_1 + \alpha_2\sigma + \alpha_3\sigma^2 + \dots + \alpha_n\sigma^{n-1} + \sigma^n), \quad (\gamma_0 \neq 0). \tag{47}$$

Introduce now

$$A_c = \begin{pmatrix} 0 & 1 & 0 & \dots & 1 \\ 0 & 0 & 1 & \dots & 0 \\ \cdot & \cdot & \cdot & \dots & \cdot \\ 0 & 0 & 0 & \dots & 1 \\ -\alpha_1 & -\alpha_2 & -\alpha_3 & \dots & -\alpha_n \end{pmatrix} \text{ and } b_c = \begin{pmatrix} 0 \\ 0 \\ \cdot \\ 0 \\ 1 \end{pmatrix}. \tag{48}$$

Clearly the pair (A_c, b_c) is completely controllable (see Property 7° of Theorem 1, § 31) and

$$\det(\sigma I - A_c) = \frac{\psi(\sigma)}{\gamma_0}. \tag{49}$$

Together with the initial system whose coefficients are $(J, \varkappa, l, M; A, b)$, consider a system of the same form having the coefficients $(0, 1, 0, 0; A_c, b_c)$. Using (49) and (8) one sees that the characteristic polynomial π_c of this system (see (13, § 3) and (18, § 3)) satisfies the relations

$$\pi_c(-\sigma, \sigma) = \frac{1}{\nu_c} \det(-\sigma I - A_c^*) \det(\sigma I - A_c) = \\ = \frac{1}{\nu_c |\gamma_0|^2} \bar{\psi}(-\sigma)\, \psi(\sigma) = \frac{1}{\nu_c |\gamma_0|^2} \pi(-\sigma, \sigma). \tag{50}$$

The conditions of Proposition 2, § 4 are satisfied and hence there exists a hermitian matrix $\tilde{U}$ such that the system ($\tilde{U}$, 1, 0, 0; A_c, b_c) belongs to the same class as the initial one. In other words there exists a transformation (13, § 2)—(14, § 2) which brings the initial system to the form

$$\frac{dx_c}{dt} = A_c x_c + b_c \mu_c, \qquad \eta_c(0, t_1) = [x_c^* \tilde{U} x_c]_0^{t_1} + \int_0^{t_1} |\mu_c|^2\, dt. \tag{51}$$

Notice the particularly simple form of the integral $\eta_c(0, t_1)$.

The foregoing operations can be performed for every factorization (8) and consequently we shall have as many systems (51) as there are distinct factorizations of the form (8) [1]).

We now examine the case $\varkappa = 0$, assuming first that $\pi(-\sigma, \sigma)$ is not identically zero. Therefore, the polynomial ψ of (8) is not identically zero and can be written in the form (47) where n is replaced by $q < n$. Together with system (1), we consider the system

$$\frac{dy_c}{dt} = A_{11} y_c + b_1 \xi_1, \tag{52}$$

$$\left.\begin{aligned} \frac{d\xi_j}{dt} &= \xi_{j+1}, \qquad j = 1, 2, \ldots, n-q-1, \\ \frac{d\xi_{n-q}}{dt} &= \mu_c, \end{aligned}\right\} \tag{53}$$

$$\eta_c(0, t_1) = [x_c^* \tilde{U} x_c]_0^{t_1} + \int_0^{t_1} |\xi_1|^2\, dt, \tag{54}$$

---

[1]) In the author's paper [16] a "canonical" system has been defined by adding the further condition that the polynomial ψ of (8) should not possess roots whose real part is positive. In this case, the system (51) is unique.

where $x_c' = (y_c' \xi_1 \xi_2 \ldots \xi_{n-q})$, $\widetilde{U}$ is a hermitian matrix, as yet undetermined and A_{11} and b_1 have the form (48) with n replaced by q. It follows that

$$\det(\sigma I_1 - A_{11}) = \frac{1}{\gamma_0} \psi(\sigma), \tag{55}$$

where I_1 is the $q \times q$-identity matrix. To prove that the above system is completely controllable and to compute its characteristic function, we need the expression for $x_c(\sigma) = (\sigma I - A_c)^{-1} b_c$, where A_c and b_c are the coefficients of systems (52), (53) written in the form $\mathrm{d}x_c/\mathrm{d}t = A_c x_c + b_c \mu_c$. Note that $x_c(\sigma)$ can be obtained by solving the system of equations $\sigma x_c(\sigma) = A_c x_c(\sigma) + b_c$. It follows that the components $y_c(\sigma)$ and $\xi_j(\sigma)$ of the vector $x_c(\sigma)$ written in the form

$$x_c'(\sigma) = ((\sigma I - A_c)^{-1} b_c)' = (y_c'(\sigma) \quad \xi_1(\sigma) \ldots \xi_{n-q}(\sigma)), \tag{56}$$

satisfy the system of algebraic equations obtained by replacing in (52) and (53) μ_c by 1, y_c and ξ_j by $y_c(\sigma)$ and $\xi_j(\sigma)$, and the derivatives $\mathrm{d}y_c/\mathrm{d}t$ and $\mathrm{d}\xi_j/\mathrm{d}t$ by $\sigma y_c(\sigma)$ and $\sigma \xi_j(\sigma)$ respectively. We thus find the relations

$$\xi_j(\sigma) = \frac{1}{\sigma^{n-q-j+1}}, \qquad j = 1, 2, \ldots, n-q, \tag{57}$$

$$y_c(\sigma) = (\sigma I_1 - A_{11})^{-1} b_1 \frac{1}{\sigma^{n-q}}. \tag{58}$$

We shall show that the system (52), (53) is completely controllable [1]). From Property 7° of Theorem 1, § 31, and from the fact that the pair (A_{11}, b_1) has the form (48), it follows that the pair (A_{11}, b_1) is completely controllable. Applying Property 10° of Theorem 1, § 31, one sees that there exists a q-vector q_0 such that

$$q_0^*(\sigma I_1 - A_{11})^{-1} b_1 = \frac{1}{\det(\sigma I_1 - A_{11})}$$

and then the expression $q_0^* y_c(\sigma) = 1/(\sigma^{n-q} \det(\sigma I_1 - A_{11}))$ (see (58)) is clearly irreducible. Therefore, by Property 11° of Theorem 1, § 31, (52)—(53) is completely controllable.

---

[1]) The complete controllability of the system (52), (53), can also be established by using Property 4° of Theorem 1, § 31.

To determine $\chi(-\sigma, \sigma)$ we note that it does not depend on J (see (13, § 3)) and hence, we may replace $\tilde{U}$ of (54) by zero. Then from (56), (5, § 3) and (10, § 3) it follows that (12, § 3) takes the form [1])

$$\chi_c(-\sigma, \sigma) = \bar{\xi}_1(-\sigma)\, \xi_1(\sigma), \tag{59}$$

where $\xi_1(\sigma)$ results from (57). Using (55) one also sees that the matrix A_c of the system (52), (53) satisfies the relation

$$\det(\sigma I - A_c) = \frac{1}{\gamma_0} \psi(\sigma)\, \sigma^{n-q}. \tag{60}$$

Using (59), (60) and (18, § 3) one obtains

$$\pi_c(-\sigma, \sigma) = \frac{1}{\nu_c} \det(-\sigma I - A_c^*)\, \bar{\xi}_1(-\sigma)\, \xi_1(\sigma) \det(\sigma I - A_c) =$$

$$= \frac{1}{\nu_c |\gamma_0|^2} \bar{\psi}(-\sigma)\, \psi(\sigma) = \frac{1}{\nu_c |\gamma_0|^2} \pi(-\sigma, \sigma),$$

the last equality resulting from (8).

Using Proposition 2, § 4 one finds that there exists a hermitian matrix $\tilde{U}$ such that the system (52)—(54) and the initial system belong to the same class.

It remains to examine the case $\pi(-\sigma, \sigma) \equiv 0$. Together with system (1) consider the following system

$$\begin{aligned} \frac{d\xi_j}{dt} &= \xi_{j+1}, \qquad j = 1, 2, \ldots, n-1, \\ \frac{d\xi_n}{dt} &= \mu_c, \end{aligned} \tag{61}$$

$$\eta(0, t_1) = [x_c^* \tilde{U} x_c]_0^{t_1}, \tag{62}$$

where $x_c' = (\xi_1 \xi_2 \ldots \xi_n)$. Using Property 7° of Theorem 1, § 31, we deduce that the system (61) is completely controllable. Its characteristic function χ_c satisfies the relation $\chi_c(-\sigma, \sigma) = 0$

---

[1]) Notice that in the polarized system (see 4, § 2) associated with system (52), (51) — for $\tilde{U} = 0$ — the integral $\eta_p(0, t_1)$ has the form $\int_0^{t_1} \xi_1^\Delta \xi_1\, dt$, whence by comparing with (37, § 2) we obtain the equality $z^\Delta D z = \xi_1^\Delta \xi_1$.

for any matrix $\widetilde{U}$ (see (13, § 3)). Applying again Proposition 2, § 4 we deduce that any system of the form (1) whose characteristic polynomial has the property $\pi(-\sigma, \sigma) \equiv 0$, can be brought to the form (61)—(62), with the help of a transformation of the type studied in § 2. In this case the usual integral term in the expression of $\eta(0, t_1)$ is absent (see (62)).

In chapter 4 we shall also use the following result:

Proposition 1. *Assume that system (1) is completely controllable and positive and that its characteristic polynomial can be written in the form*

$$\pi(\lambda, \sigma) = \overline{\varphi}(\lambda)\, \varphi(\sigma)\, \pi_1(\lambda, \sigma)\, (\lambda + \sigma), \tag{63}$$

where φ is a polynomial of degree $n_2 < n$, with the leading coefficient equal to 1, and π_1 is a polynomial in λ and σ with the property that there exists no constant σ_0 such that $\pi_1(\lambda, \sigma_0) \equiv 0$ for every λ. Then, by means of a transformation of the type studied in § 2, system (1) can be brought to the form

$$\frac{\mathrm{d}y}{\mathrm{d}t} = A_{11}y + b_1\mu(t), \quad (A_{11}, b_1) = \text{completely controllable pair} \tag{64}$$

$$\frac{\mathrm{d}z}{\mathrm{d}t} = \mathrm{d}c^*y + A_{22}z \tag{65}$$

$$\eta(0, t) = [y^*Uy]_0^{t_1}, \quad (U^* = U), \tag{66}$$

where y, b_1 and c are n_1-vectors, z and d are n_2-vectors ($n_1 + n_2 = n$), A_{11} and U are $n_1 \times n_1$-matrices and A_{22} is an $n_2 \times n_2$-matrix satisfying the condition

$$\det(\sigma I_2 - A_{22}) = \varphi(\sigma) \tag{67}$$

(where I_2 is the $n_2 \times n_2$-identity-matrix). Furthermore, for every pair of continuous functions y and μ which satisfy equation (64) the following statement is valid: if the identity $Uy(t) \equiv 0$ holds, then the identity $y(t) \equiv 0$ holds also. Moreover, the characteristic polynomial π_0 of system (64), (66) has the expression $\pi_0(\lambda, \sigma) = \pi_1(\lambda, \sigma)(\lambda + \sigma)$.

Proof. From (63) and (20, § 3) it follows that the degree of the polynomial π_0 cannot be greater than n_1. As in Section 2 of § 4 one can find a system of the form (64), (66), of order n_1, such that the pair (A_{11}, b_1) is completely controllable and its characteristic polynomial is equal to π_0. Moreover, one can find an $n_2 \times n_2$-matrix of the form (9, § 31), such that equality (67) is satisfied. This is accomplished by taking in (9, § 3) the numbers $\gamma_j (j = 1, 2, \ldots, n = n_2)$ equal to the coefficients of the polynomial φ written in the form $\varphi(\sigma) = \sigma^{n_2} - \sum_{j=1}^{n_2} \gamma_j^{\sigma^{j-1}}$. Then, it can be easily seen that the characteristic polynomial of the system (64)—(66) is given by the right-hand member (63) and that the order of the system is equal to $n_1 + n_2 = n$. To prove the first part of Proposition 1 one has only to apply Proposition 1, § 4; to that end it is sufficient to show that the vectors c and d (still undetermined) can be taken such that the system be completely controllable. Applying Property 10° of Theorem 1, § 31 to the completely controllable pair (A_{11}, b_1) one sees that there exists a vector c such that the expression $c^*(\sigma I_1 - A_{11})^{-1} b_{11}/\det\ (\sigma I_2 - A_{22})$ is irreducible (here I_1 is the $n \times n$-identity matrix). We choose the vector d of the same form as $\tilde{b}$ in (9, § 31). Since the matrix A_{22} has been taken as in (9, § 31), the pair (A_{22}, d) is completely controllable. Applying again Property 10° of Theorem 1, § 31 to the completely controllable pair (A_{22}, d), we find that there exists a vector r such that the expression $r^*(\sigma I_2 - A_{22})^{-1} d/\det\ (\sigma I_1 - A_{11})$ is irreducible. Consider the block obtained by introducing besides (64) and (65), the output $\nu = r^* z$. It can be easily seen that the transfer function of this block is

$$\gamma(\sigma) = r^*(\sigma I_2 - A_{22})^{-1} d c^*(\sigma I_1 - A_{11})^{-1} b_1. \tag{68}$$

From the way we chose the vectors c, d and r it follows that the transfer function (68) is irreducible. Therefore, by Definition 1, § 33 and Theorem 1, § 33, the system (64), (65) is completely controllable. We can now apply Proposition 1, § 4 and deduce that since system (64)—(66) and the initial system are both completely controllable, of the same order, and have the same characteristic polynomial, they belong to the same class.

To finish the proof of Proposition 1 we have to show that the identity $Uy(t) \equiv 0$ implies $y(t) \equiv 0$. Assume that the above statement is not true and hence there exists a continuous function μ and a continuous, non-identically zero function y, which satisfy (64) and $Uy(t) \equiv 0$. Then, from $Uy(t) \equiv 0$ one obtains by differentiation and using (64) the identity $UA_{11}y(t) + Ub_1\mu(t) \equiv 0$. If $Ub_1 = 0$, the obtained identity becomes $UA_{11}y(t) \equiv 0$, whence, by differentiation one obtains $UA_{11}^2 y(t) + U_{A_{11}} b_1 \mu(t) \equiv 0$. If $UA_{11}b_1 = 0$, can one again differentiate and one obtains a new identity. However, we cannot have the equalities $Ub_1 = 0$, $UA_{11}b_1 = 0, \ldots, UA_{11}^{n_1-1} b_1 = 0$

simultaneously, since in that case they could be written in the form

$$U(b_1 \quad A_{11}b_1 \ldots A_{11}^{n_1-1}b_1) = 0. \tag{69}$$

Since the pair (A_{11}, b_1) is completely controllable, the matrix which multiplies U in (69) is nonsingular (see Property 4° of Theorem 1, § 31). Hence (69) implies $U = 0$; but then the characteristic polynomial of system (64)—(66) is identically zero (see Relations (13, § 3) and (18, § 3)) and hence the conditions of Proposition 1 are not satisfied. From the foregoing it follows that after differentiating the identity $Uy(t) \equiv 0$ a sufficient number of times we always obtain an identity of the form $My(t) + v_0\mu(t) \equiv 0$, where the vector v_0 is different from zero. Multiplying the identity on the left by the vector $v_0^*/v_0^* v_0$ one obtains a relation of the form $\mu(t) = q^* y(t)$, where q is a constant vector. Taking this into account and considering Eq. (64), one finds that the function y satisfies the equation $dy/dt = (A_{11} + b_1q^*)\, y$. Hence, this equation admits a solution y which is not identically zero and which satisfies the condition $Uy(t) \equiv 0$, whence it follows that the pair $(U, A_{11} + b_1q^*)$ is not completely observable (see Definition 1, § 35). Using Property 1° of Theorem 1, § 35 (see Appendix A), we deduce that there exists a nonsingular matrix R such that the matrices $R(A_{11} + b_1q^*)R^{-1}$ and UR^{-1} can be written in the form

$$R(A_{11} + b_1q^*)\, R^{-1} = \begin{pmatrix} \overbrace{A_a}^{n_1-k} & \overbrace{0}^{k} \\ A_b & A_c \end{pmatrix} \begin{matrix} \}\ n_1 - k \\ \}\ k \end{matrix} \tag{70}$$

$$UR^{-1} = (U_0 \quad 0).$$

The last relation shows that the matrix $\tilde{U} = (R^{-1})^* UR^{-1}$ is of the form

$$\tilde{U} = (\overbrace{U_2}^{n_1-k} \quad \overbrace{0}^{k}).$$

But $\tilde{U}$ is a hermitian matrix, whence it follows that if its last k columns are zero, then its last k lines are also zero; hence the matrix $\tilde{U}$ can be written in the form

$$\tilde{U} = \begin{pmatrix} \overbrace{U_2}^{n_1-k} & \overbrace{0}^{k} \\ 0 & 0 \end{pmatrix} \begin{matrix} \}\ n_1 - k \\ \}\ k \end{matrix} \tag{71}$$

Consider now the system consisting of Eqs. (64) and (66) and apply the transformation $\tilde{\mu} = \mu - q^*y$, $\tilde{y} = Ry$ (i.e. the same transformation as in § 2 for $N = 0$ and $\rho = 1$). Then the transformed matrices $\tilde{A}_{11}$ and $\tilde{U}$ take the forms (70) and (71) (see (23, § 2) and (25, § 2)). It can be easily seen that the characteristic polynomial of the transformed system contains det $(\sigma I_c - A_c)$ (see (13, § 3) and (18, § 3); I_c is the identity matrix with the same dimensions as A_c). Since the transformations studied in § 2 leave the characteristic polynomial invariant (see Proposition 1, § 3) the characteristic polynomial of the system considered is identically zero for any constant $\sigma = \sigma_0$ which satisfies the equation det $(\sigma_0 I_c - A_c) = 0$. This completes the proof, since the foregoing conclusion contradicts the conditions of Proposition 1.

The proof also yields the following proposition (the statement of which is obtained from Proposition 1 when the polynomial φ reduces to a constant).

Proposition 2. *Assume that system (1) is completely controllable and positive and that its characteristic polynomial is of the form* $\pi(\lambda, \sigma) = \pi_1(\lambda, \sigma)(\lambda + \sigma)$, *where* π_1 *is a polynomial in* λ *and* σ *with the property that there exists no constant* σ_0 *such that* $\pi_1(\lambda, \sigma_0) \equiv 0$. *Then, by means of a transformation of the type studied in § 2, the system can be brought to the form*

$$\frac{dx}{dt} = A_c x + b_c \mu, \quad (A_c, b_c) = \textit{completely controllable pair} \tag{72}$$

$$\eta(0, t_1) = [x^* U x]_0^{t_1}. \tag{73}$$

Moreover, for all continuous functions μ *and* x *which satisfy equation (64), the following statement is true: The identity* $Ux(t) \equiv 0$ *implies the identity* $x(t) \equiv 0$.

§ 9. Multi-input positive systems

1. The theorem of positiveness for multi-input systems

We reexamine the problems treated in § 8 replacing system (1, § 8) by the multi-input system (1, § 5), (2, § 5), which can be written in the condensed form

$$\frac{dx}{dt} = Ax + Bu, \quad \eta(0, t_1) = \int_0^{t_1} z^* D z \, dt, \tag{1}$$

with the same notation as in § 5. We adopt a definition similar to the Definition 1, § 8.

Definition 1. *The system (1) is positive if the integral* $\eta(0, t_1)$ *can be written in the form*

$$\eta(0, t_1) = [\alpha(x)]_0^{t_1} + \int_0^{t_1} \beta(x, u)\, dt, \text{ where } \beta(x, u) \geqslant 0. \quad (2)$$

The generalization of Theorem 1, § 8 to multi-input systems is given by

Theorem 1. *If the pair (A, B) is completely controllable, then the following properties are equivalent:*

1° *The system (1) is positive in the sense of Definition 1.*

2° *The matrix* $H(-i\omega, i\omega)$ *is positive semidefinite for every real number* ω *satisfying the condition* $\det(i\omega I - A) \neq 0$ *(see (36, § 5) and (31, § 5)).*

3° *There exists at least one polynomial* ψ *such that the characteristic polynomial* π*(56, § 5) can be factored as*

$$\pi(-\sigma, \sigma) = \overline{\psi}(-\sigma)\, \psi(\sigma). \quad (3)$$

Moreover, for every polynomial ψ *satisfying Relation (3) there exist an* $m \times m$*-matrix* V*, and an* $n \times n$*-matrix* W*, such that*

$$H(-\sigma, \sigma) = \overline{F}_0'(-\sigma)\, F_0(\sigma), \quad (4)$$

where

$$F_0(\sigma) = V + W^*(\sigma I - A)^{-1} B, \quad (5)$$

$$\det F_0(\sigma) = \sqrt{\nu}\, \frac{\psi(\sigma)}{\det(\sigma I - A)}. \quad (6)$$

Furthermore, there exists an $n \times n$*-hermitian matrix such that the equality*

$$\begin{pmatrix} 0 & 0 & J \\ 0 & K & L^* \\ J & L & M \end{pmatrix} + \begin{pmatrix} 0 & 0 & -N \\ 0 & 0 & B^*N \\ -N & NB & NA + A^*N \end{pmatrix} =$$

$$= \begin{pmatrix} 0 & 0 & J-N \\ 0 & V^*V & V^*W^* \\ J-N & WV & WW^* \end{pmatrix} \quad (7)$$

is satisfied and η *can be written as*

$$\eta(0, t_1) = [x^*(J - N)\, x]_0^{t_1} + \int_0^{t_1} \| Vu + W^*x \|^2\, dt. \quad (8)$$

If the coefficients of system (1) are real and if one chooses in (3) a polynomial with real coefficients, then one can choose matrices V, W and N with real entries.

4° *There exists a hermitian matrix N and two matrices V and W such that equality (7) is satisfied.*

5° *There exists a hermitian matrix N such that the matrix*

$$\begin{pmatrix} K & (L+NB)^* \\ L+NB & M+NA+A^*N \end{pmatrix} \tag{9}$$

is positive semidefinite.

Remarks. To shorten the exposition we do not present the generalizations of Properties 6°—10° of Theorem 1, § 8. We only remark that Properties 6°, 7° and 10° of Theorem 1, § 8 can be directly and easily extended to the present case. However, Properties 8° and 9° of Theorem 1, § 8 do not hold in the same form. If the matrix $H(-\,i\omega,\, i\omega)$ of Property 2° is not only positive semidefinite, but even positive definite, Property 2° can be stated in the same way as Property 9° of Theorem 1, § 8, namely : "$\pi(-\,\sigma, \sigma)$ satisfies the inequality $\pi(-\,i\omega,\, i\omega) \geqslant 0$". The last condition can be replaced by Relation (3).

2. Proof of the theorem

We shall prove Theorem 1 according to the scheme

$$3° \to 4° \to 5° \to 1° \to 2° \to 3°, \tag{10}$$

the first implications of which are almost immediate. Thus, Property 4° is explicitly contained in the statement of Property 3°. Property 5° is a consequence of Property 4°, since from (7) it follows that the matrix (9) has the form $(V\,W^*)^*\,(V\,W^*)$ and hence is positive semidefinite. Property 1° results from 5° since, by taking into account the equality $z^*Dz = z^*\,(D + T^*F + F^*T)z$, where F and T have the forms (6, § 5) and (7, § 5), η can be written as

$$\eta(0, t_1) = [x^*(J - N)\,x]_0^{t_1} +$$

$$+\int_0^{t_1} \begin{pmatrix} u \\ x \end{pmatrix}^* \begin{pmatrix} K & (L+NB)^* \\ L+NB & M+NA+A^*N \end{pmatrix} \begin{pmatrix} u \\ x \end{pmatrix} \mathrm{d}t, \tag{11}$$

hence the property from Definition 1 is satisfied since the matrix (9) is positive semidefinite. It remains to prove the implications 1° → 2° → 3°.

Proof of the implication 1° → 2°. Consider the polarized system (21, § 5) and the particular control functions (28, § 5) and (32, § 5) for $\sigma = i\omega$, $\lambda = -i\omega$, and $u_0^\Delta = u_0^*$. These functions satisfy the relations

$$(u^\Delta(t))^* = u(t) = u_0 \, e^{i\omega t}, \tag{12}$$

and the corresponding solutions (29, § 5) and (33, § 5) satisfy conditions of the form (22, § 8). The corresponding functions $z^\Delta(t)$ and $z(t)$ satisfy relations of the form (23, § 8). Hence, the equality (35, § 5) can be written as

$$z^*(t) \, Dz(t) = u_0^* H(-i\omega, i\omega) \, u_0. \tag{13}$$

Consider now system (1) for the function u given by (12) and examine the solution x of the form (22, § 8) and the corresponding function z given by (23, § 8). The integrand in the expression of η (see (1)) satisfies (13), whence

$$\eta(0, t_1) = u_0^* H(-i\omega, i\omega) \, u_0 t_1. \tag{14}$$

Taking t_1 equal to $2\pi/|\omega|$ (if $\omega \neq 0$) or to an arbitrary positive number (if $\omega = 0$) one obtains from (22, § 8) the equality $\omega(t_1) = x(0)$. Therefore, (2) reduces to $\eta(0, t_1) \geqslant 0$. This and (14) imply $u_0^* H(-i\omega, i\omega) u_0 \geqslant 0$. Since u_0 is arbitrary, one obtains Property 2°.

Proof of the implication 2° → 3°. Since the matrix $H(-i\omega, i\omega)$ is positive semidefinite, we have the inequality $\det H(-i\omega, i\omega) \geqslant 0$. Using expression (56, § 5) one obtains the inequality $\pi(-i\omega, i\omega) \geqslant 0$ and hence, as in the case of the single-input systems, the polynomial π is factorizable as in (3). From the expression (37, § 5) it follows that all the entries of

$$Y(\sigma) = H(-\sigma, \sigma) \det(-\sigma I - A^\star) \det(\sigma I - A) \tag{15}$$

are polynomials whose degrees cannot exceed $2n$. From Property 2° it follows that the matrix $Y(i\omega)$ is positive semidefinite for any real [1]). Furthermore, from (38, § 5) it follows that

$$Y(\sigma) = \overline{Y}'(-\sigma).$$

[1]) It follows directly that the matrix $Y(i\omega)$ is positive semidefinite for every real ω which satisfies the condition $\det(i\omega I - A) \neq 0$. Since the condition $\det(i\omega I - A) = 0$ can be satisfied at most at n isolated points ω_j and since Y is a matrix polynomial, it follows, by continuity, that $Y(i\omega)$ is also positive semidefinite at the points ω_j.

Note that from (15) and (56, § 5) it follows that

$$\det Y(\sigma) = \nu\pi(-\sigma, \sigma)\,(\det(-\sigma I - A^*)\det(\sigma I - A))^{m-1} \tag{16}$$

and hence, for every polynomial ψ which satisfies (3), the polynomial

$$\psi_1(\sigma) = \sqrt{\nu}\,\psi(\sigma)\,(\det(\sigma I - A))^{m-1} \tag{17}$$

satisfies

$$\det Y(\sigma) = \bar{\psi}_1(-\sigma)\,\psi_1(\sigma). \tag{18}$$

By Theorem 1, § 39, for every polynomial ψ_1 which satisfies (18), there exists a matrix polynomial C which satisfies the relations

$$Y(\sigma) = \bar{C}'(-\sigma)\,C(\sigma), \tag{19}$$

$$\det C(\sigma) = \psi_1(\sigma). \tag{20}$$

From (15) and (19) one obtains (4) where

$$F_0(\sigma) = \frac{C(\sigma)}{\det(\sigma I - A)}\cdot \tag{21}$$

This and Relations (20) and (17) give (6).

We now show that $F_0(\sigma)$ can be expressed as in (5). To this end we temporarily introduce a few additional hypotheses: Assume that $\pi(-\sigma, \sigma) \not\equiv 0$ and that $M = 0$, while A is a diagonal matrix of the form

$$A = \operatorname{diag}(\alpha_1, \alpha_2, \ldots, \alpha_n), \tag{22}$$

where the numbers α_j are real, distinct positive and satisfy the conditions

$$\pi(-\alpha_j, \alpha_j) \neq 0, \quad j = 1, 2, \ldots, n. \tag{23}$$

In this case one can use formulas (63, § 5)—(67, § 5).

From (19) it follows that the diagonal entries $(Y(i\omega))_{jj}$ of matrix $Y(-i\omega, i\omega)$ are related to the entries $(C(i\omega))_{kj}$ of matrix $C(i\omega)$ by the equality

$$(Y(i\omega))_{jj} = \sum_{k=0}^{n} |(C(i\omega))_{kj}|^2 \quad (\omega \text{ real}) . \tag{24}$$

Let p_j be the highest degree of $(C(i\omega))_{kj}$, for $k = 1, 2, \ldots, n$. Then from (24) one obtains

$$\lim_{\omega \to \infty} \frac{Y(i\omega)_{jj}}{\omega^{2p_j}} = \gamma_j ,$$

for some γ_j. We have seen, however, that the degrees $(Y(i\omega))_{jj}$ cannot exceed $2n$. Thus, $p_j \leqslant n$. In other words the elements of the matrix polynomial C cannot be of a degree higher than n.

Hence, one can define the matrix (see (21))

$$V = \lim_{\sigma \to \infty} F_0(\sigma) . \tag{25}$$

Since A is of the form (22), one can write $F_0(\sigma)$ in the form

$$F_0(\sigma) = V + \sum_{j=1}^{n} \frac{C_j}{\sigma - \alpha_j} , \tag{26}$$

where C_j are constant matrices. The expression (26) is obtained by expanding in simple fractions every entry of $F_0(\sigma)$. Using this relation and (4) (proved above) one obtains from (66, § 5) the relation

$$l_j b_j^* = \bar{F}_0'(-\alpha_j) C_j . \tag{27}$$

But the matrix $\bar{F}_0'(-\alpha_j)$ is nonsingular since otherwise (see (4) and (56, § 5)) one would have $\pi(-\alpha_j, \alpha_j) = 0$, contradicting (23). Therefore (27) yields

$$C_j = w_j b_j^* , \quad \text{where} \quad w_j = (\bar{F}_0'(-\alpha_j))^{-1} l_j . \tag{28}$$

Writing

$$W = (w_1 \, w_2 \ldots w_n), \; B^* = (b_1 \, b_2 \ldots b_n)$$

and using the expression (22) of matrix A one sees that $F_0(\sigma)$, (26), can be written in the form (5).

We shall now eliminate the restrictive hypotheses used above. Assume first that $\pi(-\sigma, \sigma) \not\equiv 0$ (dropping the assumptions that $M = 0$ and that A has the form (22)). Then one can always find n real, distinct, strictly positive numbers α_j, $j = 1, 2, \ldots, n$ satisfying Conditions (23). Since our system is completely controllable, there exists another system $\tilde{\mathcal{S}}$: $(\tilde{J}, \tilde{K}, \tilde{L}, \tilde{M}: \tilde{A}, \tilde{B})$, belonging to the same class (as shown in Section 5, § 5) and whose matrix $\tilde{A}$ is of the form (22) (where the numbers α_j have been defined above) while the matrix $\tilde{M}$ is zero. Since the characteristic polynomial $\pi(-\sigma, \sigma)$ is invariant (see Section 4, § 5), Relation (3) remains valid in the same form for system $\tilde{\mathcal{S}}$. Now all the assumptions admitted before are satisfied and therefore the characteristic matrix $\tilde{H}$ of the system $\tilde{\mathcal{S}}$ satisfies the relations (of the type (4)—(6))

$$\tilde{H}(-\sigma, \sigma) = \bar{\tilde{F}}_0'(-\sigma)\, \tilde{F}_0(\sigma), \tag{29}$$

where

$$\tilde{F}_0(\sigma) = \tilde{V} + W^*(\sigma I - \tilde{A})^{-1} \tilde{B}, \tag{30}$$

$$\det \tilde{F}_0(\sigma) = \sqrt{\tilde{\nu}}\, \frac{\psi(\sigma)}{\det(\sigma I - \tilde{A})}. \tag{31}$$

Since the characteristic function H of the initial system is related to $\tilde{H}$ by (44, § 5), we obtain from (29) Eq. (4), where (see also (30))

$$F_0(\sigma) = \tilde{F}_0(\sigma)\, PC(\sigma) = (\tilde{V} + \tilde{W}^*(\sigma I - \tilde{A})^{-1} \tilde{B})\, PC(\sigma). \tag{32}$$

Taking into account the identity (see the footnote 3, § 5)

$$(\sigma I - \tilde{A})^{-1} \tilde{B}\, PC(\sigma) = R(\sigma I - A)^{-1} B \tag{33}$$

and using the expression (40, § 5) of $C(\sigma)$, one obtains from (32)

$$F_0(\sigma) = \tilde{V}\, P(I_m - Q^*(\sigma I - A)^{-1} B) + \tilde{W}^*\, R(\sigma I - A)^{-1} B,$$

a relation which is identical to (5) if one defines

$$V = \tilde{V} P \quad \text{and} \quad W^* = -\tilde{V} P Q^* + \tilde{W}^* R.$$

From the formula $P = I_m$ (72, § 5) it follows that det $P = 1$ and therefore Relations (31), (32) and (48, § 5) imply

$$\det F_0(\sigma) = \sqrt{\tilde{\nu}}\,\frac{\psi(\sigma)}{\det(\sigma I - A)}, \tag{34}$$

which takes the form (6). The equality $\tilde{\nu} = \nu$ results (under our assumption that $\pi(-\sigma, \sigma) \not\equiv 0$) from Relations (38), (4), (3) and (56, § 5). Thus we have also obtained Relations (3)—(6) in the case $\pi(-\sigma, \sigma) \not\equiv 0$.

Assume now that $\pi(-\sigma, \sigma) \equiv 0$. Then $\psi(\sigma) \equiv 0$ (see (3)). Replace $H(-\sigma, \sigma)$ by $H_\varepsilon(-\sigma, \sigma) = H(-\sigma, \sigma) + \varepsilon I$ where ε is an arbitrary positive number. Then the corresponding polynomial

$$\pi_\varepsilon(-\sigma, \sigma) = \frac{1}{\nu} \det(-\sigma I - A^*) \det(\sigma I - A) \det H_\varepsilon(-\sigma, \sigma)$$

(see (56, § 5)) is not identically zero. Consequently, as we have seen, one can find matrices V_ε and W_ε such that

$$H(-\sigma, \sigma) + \varepsilon I = \bar{F}'_\varepsilon(-\sigma) F_\varepsilon(\sigma), \tag{35}$$

where

$$F_\varepsilon(\sigma) = V_\varepsilon + W^*_\varepsilon (\sigma I - A)^{-1} B. \tag{36}$$

From (35) we deduce, as in the case of Relation (24), the equality

$$(H(-i\omega, i\omega) + \varepsilon I)_{jj} = \sum_{k=1}^{n} |(F_\varepsilon(i\omega))_{kj}|^2 \quad (\omega \text{ real}); \tag{37}$$

whence it follows that if ε is bounded (e.g. satisfying the inequalities $0 < \varepsilon < 1$), the absolute values of all the entries of the matrix $F_\varepsilon(i\omega)$ are bounded by a constant which does not depend on ε. Therefore, one deduces from (36) that the absolute values of all the entries of the matrix $V_\varepsilon = \lim\limits_{\sigma\to\infty} F_\varepsilon(\sigma)$ are also bounded by a constant which does not depend on ε. The same conclusion applies to the expression $W^*_\varepsilon (i\omega_j I - A)^{-1} B = F_\varepsilon(i\omega_j) - V_\varepsilon$, for every ω_j. Taking n real and distinct numbers ω_j, the expression

$$W^*_\varepsilon((i\omega_1 I - A)^{-1}B \quad (i\omega_2 I - A)^{-1} B \ldots (i\omega_n I - A)^{-1}B) \tag{38}$$

will have the same property. By Property 13° of Theorem 1, § 34 (appendix A), the rank of the matrix which multiplies W_ε^* in (38) equals n. Hence there exists a nonsingular matrix S_0 composed of n adequately chosen columns of the matrix which multiplies W_ε^* in (38), such that the product $W_\varepsilon^* S_0 = K(\varepsilon)$ represents a matrix whose entries are bounded (in absolute value) by a constant that does not depend on ε. From this it follows that $W_\varepsilon^* = S_0^{-1} K(\varepsilon)$. Therefore the entries of the matrix W_ε (as well as the entries of matrix V_ε) are bounded (in absolute value) by a constant which does not depend on ε. Consider a sequence of positive numbers ε smaller than one and approaching zero. Under the condition specified above one can find a subsequence ε_k with the property that the matrices V_{ε_k} and W_{ε_k} are convergent. From Relations (35) and (36) it follows that $V = \lim_{k\to\infty} V_{\varepsilon_k}$ and $W = \lim_{k\to\infty} W_{\varepsilon_k}$ satisfy Relations (4) and (5). From relations (4) and (56, § 5) and condition $\pi(-\sigma, \sigma) \equiv 0$ it follows that Condition (6) is satisfied for $\psi(\sigma) \equiv 0$ (i.e. for the only polynomial ψ which satisfies (3)). Thus Relations (3)—(6) have been obtained in the general case.

Consider now together with the system (1) the system $\check{\mathcal{S}}$ consisting from equation $dx/dt = Ax + Bu$ (the same as in (1)) and the integral (8). It can be easily seen that the matrix $\check{D}$ in the expression (8) is of the same form as the right-hand member of Relation (7) (see Section 1, § 5) and namely we have

$$\check{K} = V^* V, \quad \check{L} = WV, \quad \check{M} = WW^*. \tag{39}$$

It follows from (36, § 5) and (31, § 5) that the characteristic matrix $\check{H}$ of the system $\check{\mathcal{S}}$ satisfies the relation $\check{H}(-\sigma, \sigma) = \overline{F}_0'(-\sigma) F_0(\sigma)$, where $F_0(\sigma)$ is given by (5). Therefore (see (4)) we have the equality $\check{H}(-\sigma, \sigma) = H(-\sigma, \sigma)$. Using Proposition 2, § 5, we deduce from (61, § 5) and (39) the relations

$$V^*V = K, \quad WV = L + NB, \quad WW^* = M + NA + A^*N.$$

which prove (7). Now (7) shows that the matrix $\check{D}$ of the system $\check{\mathcal{S}}$ is obtained by adding to D a matrix of the form $T^*F + F^*T$, where F and T are of the form (6, § 5), (7, § 5). From the equality $z^*Dz = z^*(D + T^*F + F^*T)z$ (see Section 1, § 5) it follows that the integral $\eta(0, t_1)$ can be indeed written in the form (8).

The fact that if system (1) and ψ have real coefficients then V, W and N can be chosen real, is proved as at the end of Section 4, § 8.

3. Generalization of the Yakubovich-Kalman lemma

From the equivalence of Properties 2° and 4° of Theorem 1 one obtains a generalization of the Yakubovich-Kalman lemma (see Section 5, § 8).

Lemma 1. *(Generalized Yakubovich-Kalman lemma).* *If the pair (A, B) is completely controllable, then the condition that the matrix $H(-i\omega, i\omega)$ be positive semidefinite for every real number ω which satisfies the condition $\det(i\omega I - A) \neq 0$ is equivalent to the following property: "There exist three matrices V, W and N such that Lur'e generalized equations*

$$K = V^* V, \tag{40}$$

$$L + NB = WV, \tag{41}$$

$$M + NA + A^* N = W W^*, \; (N^* = N) \tag{42}$$

are satisfied".

Lemma 1 is proved by simply observing that Eqs. (40)—(42) represent only another way of writing (7).

4. Special forms for multi-input positive systems

As in the case of single-input positive systems we can bring any multi-input positive systems to a form in which the expression of the integral η is particularly simple. We consider first the case in which the system is completely controllable and $\det K \neq 0$ or (see (40)) $\det V \neq 0$. Then V^{-1} exists, a fact which will be used in the course of the proof.

To be able to use with slight modifications, some of the formulas previously established it is convenient to apply the distinctive sign $\sim$ over the symbols of system (1) and retain the simple symbols (without the sign $\sim$) for the special form into which we shall bring the system. Since the system is positive, for every scalar factorization of the form (3) there exists a matrix factorization of the form (29)—(31). As shown before, the matrix $H(\lambda, \sigma)$ of any other system belonging to the same class [1]) admits a factorization of the form (4), where (see (32) and the subsequent formulas):

$$F_0(\sigma) = \tilde{V} P + (-\tilde{V} P Q^* + \tilde{W}^* R)(\sigma I - A)^{-1} B. \tag{43}$$

[1]) The relations are valid for every system of the form (1) (having the matrix $H(\lambda, \sigma)$) from which the system marked by "$\sim$" is derived by transformations of the form (15, § 5)—(29, § 5).

We shall now choose the matrices P and Q so that expression (43) becomes as simple as possible. Since, by hypothesis, $\tilde{V}^{-1}$ exists, we may take $P = \tilde{V}^{-1}$ and $Q^* = \tilde{W}^* R$; then (43) becomes equal to the identity matrix I_m and hence, by (4), we also have the equality $H(-\sigma, \sigma) = I_m$. Assume that system (1) has been obtained from the initial system (marked with the sign $\sim$) in the manner described. Beside system (1) consider the system $\check{\mathcal{S}}$

$$\frac{dx}{dt} = Ax + Bu, \quad \eta(0, t_1) = [x^* \tilde{U} x]_0^{t_1} + \int_0^{t_1} \|u\|^2 \, dt, \quad (44)$$

where the hermitian matrix $\tilde{U}$ is as yet undetermined. Obviously, the characteristic matrix $\check{H}$ of system (44) satisfies the relation $\check{H}(-\sigma, \sigma) = I_m$ and hence we have the equality $\check{H}(-\sigma, \sigma) = H(-\sigma, \sigma)$. Applying Proposition 2, § 5 we conclude that there exists a hermitian matrix $\tilde{U}$ such that system (1) (and at the same time the initial system) can be brought to the form (44) (cf. (51, § 8)). From the relations $F_0(\sigma) = I_m$ and (34) it follows that the characteristic equation of matrix A of the system (44) satisfies

$$\det(\sigma I - A) = \sqrt{\tilde{\nu}} \; \psi(\sigma) \quad (45)$$

and hence has the same roots as the polynomial $\psi(\sigma)$ from the relation of factorization (3), from which we have started.

We now consider the general case and assume that the system (1) is positive and the pair (A, B) is completely controllable. Then, by Property 3° of Theorem 1 the integral η can be written in the form (8). Consider the equation

$$\frac{dx}{dt} = Ax + Bu \quad (46)$$

and define the "output"

$$v = Vu + W^* x. \quad (47)$$

Eqs. (46) and (47) define a "block" to which one can apply Theorem 1, § 36 (Appendix A). Thus, by an invertible transformation of the type (8, § 5), (9, § 5) Eq. (46) can be written in the form (69, § 36)—(71, § 36), while the integral (8) takes the form (see the last condition in (6, § 36))

$$\eta(0, t_1) = [\Phi(y_{kj}, z, w)]_0^{t_1} + \int_0^{t_1} \left(\|u_0\|^2 + \sum_{i=1}^{s} \|y_{1i}\|^2 + \|C_1^* z\|^2 \right) dt, \quad (48)$$

where Φ is a quadratic form in the variables y_{kj}, z and w obtained from $x^*(J - N)x$ (see (8)) by applying the transformation $\widetilde{x} = Rx$ (see (5, § 36)) and then expressing the vector x in terms of its components y_{kj}, z and w. The other symbols used have the same meaning as in Theorem 1, § 36. Obviously, some of these symbols may be absent (see footnote 5, § 36). Assume now that $\pi(-\sigma, \sigma) \not\equiv 0$. Then from (3) it follows that $\psi(\sigma) \not\equiv 0$. Consequently, by (6), det $F_0(\sigma) \not\equiv 0$. Notice that the matrix $F_0(\sigma)$ given by (5) is exactly the transfer matrix of the block (46)--(47) and that all the conditions of part B of Theorem 1, § 36 are satisfied. We apply the theorem quoted above to system (1) under the following assumptions: the system is positive, the pair (A, B) is completely controllable and $\pi(-\sigma, \sigma) \not\equiv 0$. In this way we find that, using a transformation of the type studied in § 5, we can bring system (1) into the form (cf. (52, § 8)—(54, § 8))

$$\left.\begin{aligned} \frac{dy_{kj}}{dt} &= \sum_{i=1}^{j-1} T_{kji}\, y_{1i} + y_{(k+1)j} + B_{kj}u_0, \text{ for } 0 < k < p_j \ \underset{(\text{if } p_j > 1)}{} \\ \frac{dy_{p_j j}}{dt} &= \sum_{i=1}^{j-1} T_{p_j ji}\, y_{1i} + B_{p_j j}\, u_0 + u_j \end{aligned}\right\} \begin{array}{l} \text{for} \\ 0 < j \leqslant s \\ (\text{if } s > 0) \end{array} \tag{49}$$

$$\frac{dw}{dt} = \sum_{i=1}^{s} F_{2i}\, y_{1i} + A_{22} W + B_{20}\, u_0, \ (\text{if } n_2 > 0) \tag{50}$$

$$\eta\,(0, t_1) = [\Phi(y_{kj}, w)]_0^{t_1} + \int_0^{t_1} \left(\|u_0\|^2 + \sum_{i=1}^{s} \|y_{1i}\|^2 \right) dt. \tag{51}$$

Furthermore, from Relations (73, § 36) and (6), since we have $G(\sigma) = F_0(\sigma)$, it follows that

$$\det\,(\sigma I_2 - A_{22}) = \varkappa \psi\,(\sigma), \tag{52}$$

where $\varkappa$ is a constant and I_2 is the identity matrix having the same dimensions as matrix A_{22}. The symbols used in the above expressions have the same meaning as in Theorem 1, § 36, except Φ which designates a quadratic form in the variables y_{kj} and w. The above special forms of our systems will be used in several proofs in the next chapter.

§ 10. Discrete positive systems

1. The theorem of positiveness for discrete systems

Let us consider a multi-input discrete system of the form

$$x_{k+1} = Ax_k + Bu_k, \quad \eta\,(0, k_1) = \sum_{k=0}^{k_1} z_k^* D z_k, \tag{1}$$

where we use the same notations as in § 6. As in § 9 we can state :

Definition 1. *System (1) is said to be positive if the expression $\eta\,(0, k_1)$ can be written in the form*

$$\eta\,(0, k_1) = \alpha\,(x_{k+1}) - \alpha\,(x_0) + \sum_{k=0}^{k_1} \beta\,(x_k, u_k), \text{where } \beta\,(x_k, u_k) \geqslant 0. \tag{2}$$

Theorem 1. *If the pair (A, B) is completely controllable, then the following properties are equivalent:*

1° *System (1) is positive (Definition (1)).*

2° *The matrix $H\left(\frac{1}{\sigma}, \sigma\right)$ (31, §6) is positive semidefinite for every complex number σ which satisfies the conditions $|\sigma| = 1$ and $\det\,(\sigma I - A) \neq 0$.*

3° *There exists at least one polynomial ψ which satisfies the conditions $\psi(0) \neq 0$ and (see (32, §6)):*

$$\pi\left(\frac{1}{\sigma}, \sigma\right) = \overline{\psi}\left(\frac{1}{\sigma}\right)\psi(\sigma), \quad (\psi(0) \neq 0). \tag{3}$$

Moreover, for every polynomial ψ with these properties, there exists an $m \times m$-matrix V, and an $n \times m$-matrix W, such that

$$H\left(\frac{1}{\sigma}, \sigma\right) = \overline{F}_0'\left(\frac{1}{\sigma}\right) F_0\,(\sigma), \tag{4}$$

where

$$F_0\,(\sigma) = V + W^*\,(\sigma I - A)^{-1}\,B, \tag{5}$$

$$\det F_0\,(\sigma) = \sqrt{\nu}\,\frac{\psi\,(\sigma)}{\det\,(\sigma I - A)} \tag{6}$$

Furthermore, there exists an $n \times n$-hermitian matrix N, such that the equality

$$\begin{pmatrix} J & 0 & 0 \\ 0 & K & L^* \\ 0 & L & M-J \end{pmatrix} + \begin{pmatrix} -N & 0 & 0 \\ 0 & B^*NB & B^*NA \\ 0 & A^*NB & A^*NA \end{pmatrix} =$$

$$\begin{pmatrix} J-N & 0 & 0 \\ 0 & V^*V & V^*W^* \\ 0 & WV & WW^*-J+N \end{pmatrix}, \tag{7}$$

is satisfied and η can be written as

$$\eta\,(0, k_1) = [x^*\,(J - N)\,x]_0^{k_1+1} + \sum_{k=0}^{k_1} \|\, Vu_k + W^*x_k \,\|^2. \tag{8}$$

4° *There exists a hermitian matrix N and two matrices V and W such that (7) is satisfied.*

5° *There exists a hermitian matrix N such that the matrix*

$$\begin{pmatrix} K & (L + A^*NB)^* \\ L + A^*NB & M + A^*NA - N \end{pmatrix} \tag{9}$$

is positive semidefinite.

2. Proof of the theorem

The proof of Theorem (1) is entirely similar to the proof of Theorem 1 of § 9 and consequently stress will be laid only on some specific features. We again use the scheme 3° → 4° → → 5° → 1° → 2° → 3° the first implications of which result almost immediately. Property 4° is explicitly stated in Property 3°. From (7) it follows that the matrix (9) can be written in the form $(V\ W^*)^*(V\ W^*)$ and hence Property 5° results from 4°. Adding to the matrix D of (1) the second matrix of the left hand member of equality (7) (this leaves the value of the expresion z^*Dz unaltered), we obtain the matrix D of an expression $\eta(0, k_1)$ of the form (2, § 6) where the matrix of the quadratic form under the sum symbol is identical to (9). Consequently, Property 1° follows from 5°. To prove the implication 1° → 2°, we consider the polarized system (18, § 6) and the particular control functions (23, § 6) and (26, § 6), for $\lambda = \frac{1}{\sigma}$, where σ is a number with the following property : "there exists a positive integer k_1 such that $\sigma^{k_1} = 1$". Obviously, any number σ of the form $\sigma = e^{2\pi\xi i}$, where ξ is a rational number, has this property. Proceeding as in the proof of implication $1° \to 2°$ of § 9, one sees that the matrix $H\left(\frac{1}{\sigma}, \sigma\right)$ is positive semidefinite for all numbers σ of the form $e^{2\pi\xi i}$ (ξ rational), whence, by continuity, Property 2° follows.

It remains to prove the implication $2° \to 3°$. By Property 2° one obtains (for $|\sigma| = 1$) $\det H\left(\frac{1}{\sigma}, \sigma\right) \geqslant 0$ and therefore (see (32, § 6)) : $\pi\left(\frac{1}{\sigma}, \sigma\right) \geqslant 0$. By Proposition 1, § 37 (Appendix B), $\pi\left(\frac{1}{\sigma}, \sigma\right)$ can be factorized in the form (3). The matrix

$$X(\sigma) = H\left(\frac{1}{\sigma}, \sigma\right) \det\left(\frac{1}{\sigma} I - A^*\right) \det(\sigma I - A) \tag{10}$$

satisfies all the conditions of Theorem 1, § 38 (Appendix B). From (3) and (32, § 6), we see that $\det X(\sigma)$ can be factorized in the form

$$\det X(\sigma) = \bar{\rho}\left(\frac{1}{\sigma}\right) \rho(\sigma), \tag{11}$$

where

$$\rho(\sigma) = \sqrt{\nu}\, \psi(\sigma) (\det(\sigma I - A))^{m-1}. \tag{12}$$

By Theorem 1, § 38, for every polynomial ψ that satisfies (3), there exists a matrix polynomial D such that

$$X(\sigma) = \bar{D}'\left(\frac{1}{\sigma}\right) D(\sigma), \quad \det D(\sigma) = \rho(\sigma). \tag{13}$$

Using (10) and (12) we obtain (4) and (6) where

$$F_0(\sigma) = \frac{D(\sigma)}{\det(\sigma I - A)}. \tag{14}$$

The fact that $F_0(\sigma)$ can be written in the form (5) will be shown introducing temporarily the hypotheses $\pi\left(\frac{1}{\sigma}, \sigma)\right) \not\equiv 0$ and $M = 0$ and assuming that A has the form

$$A = \operatorname{diag}(\alpha_1, \alpha_2, \ldots, \alpha_n), \tag{15}$$

where the numbers α_j are real, distinct, greater than 1, and satisfy the conditions $\pi\left(\frac{1}{\alpha_k}, \alpha_k\right) \neq 0, k = 1, 2, \ldots, n$. Then one can proceed as in Section 5, § 5, with the specific changes for discrete systems mentioned in Section 3, § 6. From $\psi(0) \neq 0$ it follows that the polynomial ρ of (12) satisfies the condition $\rho(0) \neq 0$ and hence, by Theorem 1, § 38, the degree of the matrix polynomial D does not exceed n.

Expanding each entry of $F_0(\sigma)$, (14), into simple fractions, one obtains the expression

$$F_0(\sigma) = V + \sum_{k=1}^{n} \frac{D_k}{\sigma - \alpha_k}, \tag{16}$$

where the D_k are constant matrices. As in § 9 one finds that the matrices D_k are of the form $D_k = w_k b_k^*$ and hence, by introducing the matrix $W^* = (w_1 \, w_2 \ldots w_k)$, one can write (16) in the form (5). The case in which the conditions $M = 0$ and (15) are not satisfied, or $\pi\left(\frac{1}{\sigma}, \sigma\right) \equiv 0$ are treated by remarking that in every class which contains a comlpetely controllable system, there exists a system which satisfies these conditions. The other statements of Property 3° are proved as in § 9, using Proposition 2, § 6.

3. Generalization of the Kalman-Szegö lemma

The equivalence of the properties 2° and 4° of Theorem 1 is a generalization of a result established by Kalman and Szegö [1] in the case of single-input discrete systems ($m=1$).

Lemma 1. *(Generalized Kalman-Szegö lemma). If the pair (A, B) is completely controllable, the condition that the matrix $H\left(\frac{1}{\sigma}, \sigma\right)$ be positive semidefinite for every number σ satisfying the conditions $\det(\sigma I - A) \neq 0$ and $|\sigma| = 1$ is equivalent to* *the condition: "There exist three matrices V, W and N such that the following equations are satisfied:*

$$K + B^* NB = V^* V, \tag{17}$$

$$L + A^* NB = WV, \tag{18}$$

$$M + A^* NA - N = W W^*, \ (N^* = N)." \tag{19}$$

Indeed, from Eqs. (17)—(19) one obtains Eq. (7) and conversely.

§ 11. Positive systems with time-dependent coefficients

Some of the results obtained in the preceding paragraphs of this chapter can be extended to systems with time-dependent coefficients (see § 7)

$$\frac{\mathrm{d}x}{\mathrm{d}t} = A\,(t)\,x + B\,(t)\,u, \quad \eta\,(t_0, t_1) = \int_{t_0}^{t_1} z^*\,(t)\,D\,(t)\,z\,(t)\,\mathrm{d}t. \tag{1}$$

A direct generalization of Definition 1, § 9 gives:

Definition 1. *System (1) is said to be positive if the integral η can be written as*

$$\eta\,(t_0, t_1) = [\alpha\,(x\,(t), t)]_{t_0}^{t_1} + \int_{t_0}^{t_1} \beta\,(x\,(t),\ u\,(t),\ t)\,\mathrm{d}t, \tag{2}$$

where $\beta(x(t), u(t), t) \geqslant 0$.

Theorem 1, § 9 and Lemma 1, § 9 cannot be fully extended to systems of the form (1). Instead of the necessary and sufficient conditions of positiveness given by Theorem 1, § 9, we obtain the following sufficient conditon:

Proposition 1. *If there exist a hermitian and differentiable matrix-function N, $n \times n$, and two matrix functions V, $m \times m$, and W, $n \times m$, such that:*

$$K(t) = V^*(t)\,V(t), \tag{3}$$

$$L(t) + N(t)\,B(t) = W(t)\,V(t), \tag{4}$$

$$M(t) + N(t)\,A(t) + A^*(t)\,N(t) + \frac{\mathrm{d}N(t)}{\mathrm{d}t} = W\,(t)\,W^*\,(t) \tag{5}$$

then system (1) is positive.

Indeed, Relations (3)—(5) can be also written in matrix form

$$\begin{Vmatrix} 0 & 0 & J(t) \\ 0 & K(t) & L^*(t) \\ J(t) & L(t) & M(t) + \dfrac{\mathrm{d}J(t)}{\mathrm{d}t} \end{Vmatrix} +$$

$$+ \begin{Vmatrix} 0 & 0 & -N(t) \\ 0 & 0 & B^*(t)N(t) \\ -N(t) & N(t)B(t) & N(t)A(t) + A^*(t)N(t) \end{Vmatrix} = \qquad (6)$$

$$= \begin{Vmatrix} 0 & 0 & J(t) - N(t) \\ 0 & V^*(t)V(t) & V^*(t)W^*(t) \\ J(t) - N(t) & W(t)V(t) & W(t)W^*(t) + \dfrac{\mathrm{d}(J(t) - N(t))}{\mathrm{d}t} \end{Vmatrix}.$$

Therefore, if to the matrix $D(t)$ we add a matrix of the form $T^*(f)F(t) + F^*(t)T(t)$, where $F(t)$ and $T(t)$ are of the form (6, § 7) and (7, § 7) (an operation which does not alter the value of the expression $z^*(t)D(t)z(t)$ since $F(t)z(t) = 0$), then the resulting matrix $\widetilde{D}(t)$ corresponds to the integral

$$\eta(t_0, t_1) = [x^*(t)(J(t) - N(t))x(t)]_{t_0}^{t_1} + \int_{t_0}^{t_1} \| V(t)u(t) + W^*(t)\, x(t) \|^2 \,\mathrm{d}t. \qquad (7)$$

This expression is obviously of the form (2).

System (3)—(5) can be written in a more compact form if one assumes further that $K(t)$ is nonsingular for every $t \geqslant t_0$. Then from (3) it follows that $V(t)$ is also nonsingular for every $t \geqslant t_0$. Multiplying Eq. (4) on the right by $V^{-1}(t)$ one obtains the expression of matrix $W(t)$. Replacing this expression in (5) and using the equality $V^{-1}(t)(V^{-1}(t))^* = K^{-1}(t)$ (which follows from (3)), one obtains the differential equation

$$\frac{\mathrm{d}N(t)}{\mathrm{d}t} = N(t)\,A_0(t)\,N(t) + N(t)\,B_0(t) + B_0^*(t)\,N(t) + C_0(t), \qquad (8)$$

where

$$A_0(t) = B(t)\,K^{-1}(t)\,B^*(t), \qquad (9)$$

$$B_0(t) = B(t)\,K^{-1}(t)\,L^*(t) - A(t), \qquad (10)$$

$$C_0(t) = L(t)\,K^{-1}(t)\,L^*(t) - M(t). \qquad (11)$$

If all the quantities from equation (8) are scalars, we obtain an equation of the Riccati type. In its general form (8), this equation occurs in an investigation by R. E. Kalman [1] where several properties of the solutions of equation (8) are established, under certain assumptions concerning the coefficients. Other results referring to Eq. (8) considered from the point of view adopted in this work, may be found in the author's paper [17].

§ 12. Nonlinear positive systems

We now consider the system of the general form

$$\frac{\mathrm{d}x}{\mathrm{d}t} = f(x, u, t), \tag{1}$$

$$\eta(t_0, t_1) = [\alpha(x(t), t)]_{t_0}^{t_1} + \int_{t_0}^{t_1} \beta(x(t), u(t), t)\, \mathrm{d}t, \tag{2}$$

where, as before, u is an m-dimensional vector-valued function whose components, representing the "control functions", are piecewise continuous functions of time; x is an n-dimensional vector-valued function such that $x(t)$ represents the state of the system under investigation at time t. We need, of course, to introduce some assumptions on the function f, from the right hand member of the differential equation (1). We can assume, for instance, that the function $(y, v, t) \mapsto f(y, v, t)$ is piecewise continuous in all the variables. We also assume that the function f satisfies conditions ensuring the existence of the solutions of the differential equation (1) for every initial condition and in every interval of the form $t_0 \leqslant t \leqslant t_1$ [1]). Moreover, we assume that the functions $(y, t) \mapsto \alpha(y, t)$ and $(y, v, t) \mapsto \beta(y, v, t)$ which occur in the expression of the integral (2) are piecewise continuous with respect to all the variables.

As in the previous paragraph, the system (1) is said to be positive if the inequality $\beta(x(t), u(t), t) \geqslant 0$ is satisfied, or if the integral can be written in a form similar to (2) (but with different functions α and β so that the above condition is satisfied.

For systems of the general form (1), (2), we cannot give any general theorem similar to Theorem 1, § 9, nor even a

---

[1]) A concise discussion of such conditions can be found, for instance, in A. Halanay, [1].

result of the form of Proposition 1, § 11. Therefore, the conditions under which a system of the form (1) (2) is positive have to be established in each separate case.

As an example, consider the system

$$\frac{dx}{dt} = u, \qquad \eta(t_0, t_1) = \int_{t_0}^{t_1} \varphi(x)\,(\alpha_0 x + \beta_0 u)\,dt. \tag{3}$$

All the quantities involved are assumed to be real scalars. It is further assumed that φ is a continuous function, defined for every x. Introducing the function ψ as

$$\psi(x) = \int_0^x \varphi(\tilde{x})\,d\tilde{x},$$

and using the first equation of (3), one obtains

$$\int_{t_0}^{t_1} \varphi(x)\,u\,dt = [\psi(x)]_{t_0}^{t_1}$$

for every function x satisfying the first equation in (4). Using this relation, one can write system (3) in the form

$$\frac{dx}{dt} = u, \qquad \eta(t_0, t_1) = [\beta_0 \psi(x)]_{t_0}^{t_1} + \int_{t_0}^{t_1} \alpha_0 \varphi(x) x\,dt. \tag{4}$$

In order that system (4) (and hence also (3)) be positive, it is sufficient that the inequality

$$\alpha_0 \varphi(x)\,x \geqslant 0 \tag{5}$$

be satisfied for every x.

Other examples of nonlinear positive systems which play an important part in applications will be found in the next chapters. However, the examined systems will possess other properties besides positiveness and will belong to the family of hyperstable systems.

CHAPTER 4

HYPERSTABLE SYSTEMS AND BLOCKS

In contradistinction to the previous chapters, where only systems of some general forms were considered, a part of the present chapter deals also with "blocks" having one or more inputs and outputs. We first study the general properties of the hyperstable blocks and systems (§ 13). The case of single-input systems is studied in §§ 14 and 15 where we give several necessary and sufficient criteria of hyperstability, emphasizing especially the frequency-type criteria. The same problems are then reexamined for multi-input (§§ 16 and 17) and discrete (§ 18) systems. A more general concept of hyperstability, for abstract systems, is given in § 19. This concept is then studied in § 20 where the hyperstability of integral blocks is treated. Asymptotic stability forms the object of §§ 21—23. In § 24 we reexamine the blocks from § 15 showing that the property of hyperstability can be characterized by the simultaneous stability of a collection of systems with negative feedback.

In this chapter we use a series of well-known concepts which, however, acquire a new significance when associated with the concept of hyperstability. Thus, in § 13 the property of hyperstability is characterized using the concept of positive real function — introduced in the theory of electrical networks for purposes quite different from ours. Also, the hyperstability of the integral blocks (§ 20) can be expressed with the help of the "positive functions" which play such an important part in the theory of probability and for which hyperstability is a completely new field of application. In this chapter we use extensively the material treated in the foregoing sections and hence the results obtained by the authors quoted. We point out once more that the concepts of hyperstability and absolute stability (see Chapter 1) are closely related. In the present work, the main tool for the investigation of asymptotic stability (§ 24) is the Lemma of I. Barbălat [1]. The study of the integral blocks (§ 20) is completed from different points of view by the works of A. Halanay [2—5] and C. Corduneanu [1]. Other related results have been obtained by Viorel Barbu [1], N. Luca [1] and T. Morozan [1], [2]. We also mention the papers of

H. Smets [1—3], I. Kudrevich [1], [2], G. Zames [1] and I. W. Sandberg [1], [2] among many other works which have points in common with our approach.

The definition of the property of hyperstability adopted in the present chapter is a generalization of the definitions given by the author in his previous papers, where various particular forms of systems have been considered.

§ 13. General properties of the hyperstable systems

1. Linear systems of class $\mathcal{H}$

We shall introduce the concept of hyperstability starting from the following problem: what are the most general conditions which secure the boundedness of the solution of an equation of the form [1])

$$\frac{\mathrm{d}x}{\mathrm{d}t} = Ax + Bu \tag{1}$$

whenever these solutions satisfy an integral restriction of the type $\eta(0, t_1) \leqslant 0$ (for every $t_1 > 0$), where

$$\eta(0, t_1) = [x^* J x]_0^{t_1} + \int_0^{t_1} (u^* K u + u^* L^* x + x^* L u + x^* M x)\, \mathrm{d}t \tag{2}$$

(the symbols used in (1) and (2) have the same meaning as in § 5). As shown in the introductory chapter (see also Chapter 5) there are many stability problems occurring in applications, which conform to the above scheme.

The problem may be stated more exactly as follows: to determine a class $\mathcal{H}$ of systems of the form (1)—(2) with the property specified in

Definition 1. *The system (1)—(2) is said to belong to class $\mathcal{H}$ if there exists a positive constant δ such that for every pair of functions u and x which satisfy Eq. (1) and the inequality*

$$\eta(0, t_1) \leqslant 0, \text{ for every } t_1 \geqslant 0, \tag{3}$$

one has the inequality

$$\| x(t) \| \leqslant \delta \| x(0) \|, \text{ for every } t \geqslant 0. \tag{4}$$

---

[1]) In this paragraph the case of multi-input systems is treated from the outset since no simplification results by first considering the case of single-input systems. The reader who, on a first reading, is interested only in single-input systems, may replace the vector $u(t)$ by a scalar $\mu(t)$ and the matrix B by a vector b wherever they occur.

In the above definition we have introduced the simplest kind of integral restriction (of the form (3)). However, we shall see that the systems which belong to class $\mathcal{H}$ retain Property (4) even in the case of integral restrictions of a more general form. This remark is important for applications and will be taken into account later in defining the general property of hyperstability.

2. Hypotheses concerning the systems of class $\mathcal{H}$

In order to obtain general results it is obviously necessary to carry out the study of systems of the form (1), (2) without introducing restrictive hypotheses about them. At the same time it is convenient to leave out cases which are not important for applications and which would considerably complicate the proofs or the statement of results. Taking into account these requirements we shall assume in this chapter, — whenever we are dealing with system (1), (2) — , that certain conditions are satisfied; most of these conditions are needed in order that the problem treated be of interest.

Thus, in order that Definition 1 be meaningful it is obviously necessary to assume that for every initial condition $x(0) = x_0$, there exists at least one pair of functions u, x (with $x(0) = x_0$) such that Eq. 1 and restriction (3) are satisfied. (Otherwise, Condition (2) would be so restrictive that it would not be satisfied by any pair of functions u, x which satisfy Eq. (1) and the initial condition $x(0) = x_0$).

By slightly strengthening this condition we shall require the system (1)—(2) to possess a property of "minimal stability" [1]) in the sense of the following definition:

Definition 2. *The system (1)—(2) is said to possess the property of minimal stability (or to be minimally stable) if for any initial condition $x(0) = x_0$ there exists a pair of continuous functions x and u, defined for $t \geqslant 0$, and satisfying (α) Eq. (1) (for $x(0) = x_0$), (β) restriction (3) and (γ) the condition*

$$\lim_{t \to \infty} x(t) = 0. \tag{5}$$

We remark that if system (1), (2) belongs to class $\mathcal{H}$ the function x which appears in Definition 2 must in any case satisfy inequality (4) (this results from Definition 1). Definition 2

---

[1]) The concept of minimal stability may be considered as an extension of the concept of stability-in-the limit introduced by M. A. Aizerman and F. R. Gantmacher [1].

requires in addition that Condition (5) be satisfied as well. The condition is acceptable since in applications we want to satisfy conditions of the form (5) (i.e. we are particularly interested in asymptotic stability). If the system (1)—(2) is not minimally stable, then asymptotic stability in the presence of integral restrictions of the form (3) is impossible.

The second hypothesis required is the condition

$$B \neq 0. \tag{6}$$

If this condition is not satisfied the solutions x of Eq. (1) do not depend on the control function u and, therefore, the problem treated is very particular and easy to solve directly.

Finally, the last condition concerning the system (1)—(2) is that its characteristic polynomial π should not be identically zero. In the case of single-input completely controllable systems, the meaning of this condition follows easily from Proposition 1, § 4: If $\pi(\lambda, \sigma) \equiv 0$, there exists a system belonging to the same class as the system considered (of the form (1, § 2)—(2, § 2)), for which all the coefficients J, $\varkappa$, l, M in the expression of $\eta(0, t_1)$ are zero. Consequently the integral $\eta(0, t_1)$ is always identically zero and the introduction of restriction of the form (3) is meaningless. In the general case of multiple systems the meaning of the condition $\pi(\lambda, \sigma) \not\equiv 0$ is more complex. In § 16 we show that if the condition is not satisfied, the system can be written in a form in which some of the components of the vector u do not affect the values of the integral $\eta(0, t_1)$ and, therefore, inequality (3) does not introduce any restriction concerning these components. If these components do not affect the solutions of Eq. (1), they can be eliminated and the problem takes a simpler form. If, on the other hand, these components do affect the solution of Eq. (1) they can be always chosen, such that Condition (4) is not fulfilled although inequality (3) is satisfied and therefore the corresponding system does not belong to class $\mathcal{H}$.

The three conditions specified above are not the only ones that will be admitted in the general study of the hyperstability of systems of the form (1)—(2). In this chapter we shall sometimes introduce some additional hypotheses, but only to obtain results which do not remain valid in the same form in the general case.

3. Other properties of the systems belonging to class $\mathscr{H}$

This section is specially intended as an introduction to the general concept of hyperstability ; to this end we shall mention some properties whose general proof will be given only in § 16 (or, in the case of single-input systems, in § 14). Consequently, this section can be omitted without affecting the logical flow of the exposition. However, it should be noted that the Definitions 3 and 4 given below are frequently used in this chapter.

As will be shown in § 16 [1]), if the hypotheses stated in the previous section are satisfied, then the property "the system (1)—(2) belongs to class $\mathscr{H}$" is equivalent to any of the properties h_s and h_p stated in the following definitions :

Definition 3. *System (1)—(2) is said to have property h_s if there exists a constant $\delta \geqslant 0$ such that for every pair of functions u, x which satisfy Eq. (1), the following statement is true: For every triplet of constants $\beta_1 \geqslant 0$, $\beta_2 \geqslant 0$ and $T > 0$ for which one has the inequality*

$$\eta(0, t_1) \leqslant \beta_1^2 + \beta_2 \sup_{0 \leqslant t \leqslant t_1} \| x(t) \|, \text{ for every } t \in [0, T] \tag{7}$$

one also has the inequality

$$\| x(t) \| \leqslant \delta(\beta_1 + \beta_2 + \| x(0) \|), \text{ for every } t_1 \in [0, T]. \tag{8}$$

Definition 4. *System (1)—(2) is said to have property h_p if there exist two constants $\beta_3 \geqslant 0$ and $\beta_4 \geqslant 0$ such that for every pair of functions u and x which satisfy Eq. (1), one has the inequality*

$$\eta(0, t_1) \geqslant -\beta_3 \| x(0) \|^2 - \beta_4 \| x(0) \| \sup_{0 \leqslant t \leqslant t_1} \| x(t) \|, \text{ for all } t_1 \geqslant 0. \tag{9}$$

Let us denote by (h_{so}) the property obtained from h_s (Definition 3) for $\beta_1 = \beta_2 = 0$. From Definition 1 one sees at once that the statement "system (1)—(2) has property (h_{so})" is only another way of saying that the system belongs to class $\mathscr{H}$. Hence, the property of Definition 1 is obviously included in Property h_s. The fact (proved in § 16) that these properties are equivalent shows that, without restricting the class $\mathscr{H}$, one can replace the condition of Definition 1 by the more general

---

[1]) This results from Theorem 1, § 16, by noting that the property "the system (1)—(2) belong to class $\mathscr{H}$" is identical to the Property (h_{so}) stated in the theorem quoted.

property h_s; clearly, in this way the range of application is extended. Moreover, taking into account also property h_p, we shall show that the systems of class $\mathcal{H}$ can be combined one with another according to certain rules, thus obtaining new systems belonging to the same class $\mathcal{H}$. This property — of special interest for applications — will be proved in Section 8 of this paragraph.

In order to obtain all these properties we shall define in the next section the general property of hyperstability by a generalization of the properties h_s and h_p defined above.

4. Definition of the property of hyperstability

Let us consider the system

$$\frac{\mathrm{d}x}{\mathrm{d}t} = f(x, u, t) \tag{10}$$

$$\eta(t_0, t_1) = [\varphi(x(t), t)]_{t_0}^{t_1} + \int_{t_0}^{t_1} \psi(x(t), u(t), t)\,\mathrm{d}t \tag{11}$$

under assumptions similar to those in § 12.

In order to extend Conditions (7)—(9) to systems of the form (10)—(11), we shall use (instead of expressions of the form $\delta\|x\|$) certain functions with the properties specified in the following definitions:

Definition 5. *The function $y \mapsto \alpha(y)$ is said to belong to class M_i if it has the following properties:*

a) *The function $y \mapsto \alpha(y)$ is defined and has real and nonnegative values for all the values of the complex p-vector y.*

b) *There exists a function $\varkappa \mapsto \rho(\varkappa)$ vanishing for $\varkappa = 0$, defined, continuous and with real and nonnegative values for all $\varkappa \geqslant 0$ such that if the inequality $\alpha(y) \leqslant \varkappa_0$ is satisfied for $\varkappa_0 \geqslant 0$, then the inequality $\|y\| \leqslant \rho(\varkappa_0)$ is also satisfied.*

Definition 6. *The function $y \mapsto \beta(y)$ is said to belong to class M_s if it has the following properties:*

a) *The function $y \mapsto \beta(y)$ is defined and has real and nonnegative values for all the values of the complex p-vector y.*

b) *There exists a function $\rho \mapsto \varkappa(\rho)$ vanishing for $\rho = 0$, defined, continuous and taking real and nonnegative values for all $\rho \geqslant 0$, such that if the inequality $\|y\| \leqslant \rho_0$ is satisfied (for $\rho_0 \geqslant 0$), then the inequality $\beta(y) \leqslant \varkappa(\rho_0)$ is also satisfied.*
In the above definitions p can be an arbitrary positive integer, although in this paragraph we take $p = n$.

A function defined by $\alpha(y) = \alpha_0\|y\|$ where α_0 is a strictly positive constant (> 0) belongs at the same time to class M_i and to class M_s. (In the case $\alpha_0 = 0$, it belongs only to class M_s).

If α is of the form $\alpha(y) = \tilde{\varkappa}(\|y\|)$, where $\tilde{\varkappa}(\rho)$ vanishes for $\rho = 0$ and is defined, continuous, strictly increasing for all $\rho \geqslant 0$ and has the property $\lim\limits_{\rho\to\infty} \tilde{\varkappa}(\rho) = \infty$, then α belongs at the same time to class M_i and to class M_s. Furthermore, the function in Definition 5 is equal to the inverse function $\varkappa^{-1}$ and the function $\varkappa$ in Definition 6 is equal to the function $\rho \mapsto \tilde{\varkappa}(\rho)$.

We also remark that if the function $y \mapsto \beta(y)$ is continuous, then Condition b of Definition 6 is automatically satisfied. It follows also that any function belonging to class M_i also belongs to class M_s.

For simplicity and in order to state the definition of the property of hyperstability in a form valid for more general cases (see § 19), we use the word "solution" of a system of the form (10)—(11) in the following sense:

Definition 7. *By "solution" of a system of the form (10)—(11) in the interval $[t_0, T_0]$, ($T_0 > t_0$) we mean any set consisting of: a) a function u defined and piecewise continuous in the interval $[t_0, T_0]$; b) a function x defined and continuous in the same interval and satisfying, together with u, equation (1); c) a function $t_1 \mapsto \eta(t_0, t_1)$ defined in the interval $t_0 \leqslant t_1 \leqslant T_0$ and connected with functions u and x by Relation (11).*

We can now formulate the general definition of the property of hyperstability.

Definition 8. *System (10)—(11) is said to be hyperstable if there exists a function α of class M_i and the functions β, γ and δ of class M_s, such that the following two properties are satisfied:*

Property H_s. For every interval $[t_0, T_0]$ ($T_0 > t_0$) and for every solution of system (10)—(11) in the interval $[t_0, T_0]$ (see Definition 7) the following statement is true:

For every constant $\beta_0 \geqslant 0$ for which one has the inequality

$$\eta(t_0, t) \leqslant \beta_0^2, \text{ for every } t \in [t_0, T_0] \tag{12}$$

one also has the inequality

$$\alpha(x(t)) \leqslant \beta_0 + \beta(x(t_0)), \tag{13}$$

for every $t \in [t_0, T_0]$.

Property H_p. For every interval $[t_0, T_0]$, $(T_0 > t_0)$ and every solution of the system (10)—(11) in this interval the inequality

$$\eta(t_0, t) \geqslant -[\gamma^2(x(t_0))]^2 - \delta(x(t_0)) \sup_{t_0 \leqslant \tau \leqslant t} \alpha(x(\tau)), \tag{14}$$

is satisfied for every t in the interval $[t_0, T_0]$.

Properties H_s and H_p are closely related to Properties h_s and h_p and we have the following simple proposition:

Proposition 1. *If system (10)—(11) is of the form (1)—(2) and if Properties h_s and h_p are satisfied (see Definitions 3 and 4), then the system is hyperstable in the sense of Definition 8.*

Since system (1)—(2) is autonomous it is sufficient to check properties H_s and H_p for $t_0 = 0$ (then these properties hold for any other t_0). Property H_s results from h_s if one takes $\beta_2 = 0$, $\beta_1 = \beta_0$, $\alpha(x) = \|x\| / \delta$ and $\beta(x) = \|x\|$. Property H_p results from h_p if we take $\gamma(x) = \sqrt{\beta_3}\, \|x\|$ and $\delta(x) = \delta\beta_4 \|x\|$.

5. A consequence of property H_s

Inequality (12) of Property H_s is not as general as the corresponding inequality (7) of property h_s since the term which contains the expression $\sup \|x(t)\|$ is absent. However, one has

Proposition 2. *If system (10)—(11) has Property H_s, then for every solution of the system the following statement is true: For every pair of constants, $\widetilde{\beta}_1 \geqslant 0$ and $\widetilde{\beta}_2 \geqslant 0$, for which one has*

$$\eta(t_0, t) \leqslant \widetilde{\beta}_1^2 + \widetilde{\beta}_2 \sup_{t_0 \leqslant \tau \leqslant t} \alpha(x(\tau)), \textit{ for every } t \in [t_0, T] \tag{15}$$

one also has

$$\alpha(x(t)) \leqslant 2(\widetilde{\beta}_1 + \widetilde{\beta}_2 + \beta(x(t_0))), \textit{ for every } t \in [t_0, T]. \tag{16}$$

Proof. We shall prove first a lemma (which will be used throughout this chapter).

Lemma 1. *In the function $t \to \xi(t)$ is defined, real-valued and continuous in the interval $[t_0, T]$ and has the property that there exist two constants $\alpha_1 \geqslant 0$ and $\alpha_2 \geqslant 0$ such that one has*

$$[\xi(t)]^2 \leqslant \alpha_1^2 + \alpha_2 \sup_{t_0 \leqslant \tau \leqslant t} |\xi(\tau)| \tag{17}$$

then one also has

$$|\xi(t)| \leqslant \alpha_1 + \alpha_2. \tag{18}$$

Indeed, there always exists a number t_1 in the interval $[t_0, T]$ such that the relations

$$|\xi(t_1)| = \sup_{t_0 \leqslant \tau \leqslant t_1} |\xi(\tau)| = \sup_{t_0 \leqslant \tau \leqslant T} |\xi(\tau)| = \nu \tag{19}$$

are satisfied (the last equality defines the number ν). Writing inequality (17) for $t = t_1$ and using (19) one obtains the inequality $\nu^2 \leqslant \alpha_1^2 + \alpha_2 \nu$ whence

$$\nu \leqslant \alpha_2/2 + \sqrt{\alpha_2^2 + 4\alpha_1^2}\,/\,2 \leqslant \alpha_1 + \alpha_2 \tag{20}$$

(we have also used the inequality $\sqrt{\alpha_2^2 + 4\alpha_1^2} \leqslant \alpha_2 + 2\alpha_1$, which is obviously true for $\alpha_1 \geqslant 0$ and $\alpha_2 \geqslant 0$). From (20) and (19) one obtains (18).

We shall now use Lemma 1 to prove Proposition 2. From (15) it follows that for every number t_2 from the interval $[t_0, T]$ one has :

$$\eta(t_0, t) \leqslant \beta_0^2 \text{ for every } t \in [t_0, t_2] \tag{21}$$

where β_0 is given by

$$\beta_0^2 = \tilde{\beta}_1^2 + \tilde{\beta}_2 \sup_{0 \leqslant \tau \leqslant t_2} \alpha(x(\tau)) \tag{22}$$

Inequality (21) is of the form (12). Applying Property H_s (for $T_0 = t_2$) one obtains the relation (13), and in particular (for $t = T_0 = t_2$) :

$$[\alpha(x(t_2))]^2 \leqslant [\beta_0 + \beta(x(t_0))]^2.$$

Using the obvious inequality

$$(\beta_0 + \beta(x(t_0)))^2 \leqslant 2\beta_0^2 + 2[\beta(x(t_0))]^2$$

we obtain, taking into account (22) and replacing the symbol t_2 by the symbol t

$$[\alpha(x(t))]^2 \leqslant 2\tilde{\beta}_1^2 + 2[\beta(x(t_0))]^2 + 2\tilde{\beta}_2 \sup_{t_0 \leqslant \tau \leqslant t} \alpha(x(\tau))$$

for every t from the interval $[t_0, T]$. The obtained inequality is of the form from Lemma 1, for $\xi(t) = \alpha(x(t))$, $\alpha_1^2 = 2\tilde{\beta}_1^2 + 2[\beta(x(t_0))]^2$, $\alpha_2 = 2\tilde{\beta}_2$. Applying Lemma 1 and remarking that α_1 satisfies the inequality $\alpha_1 \leqslant 2\tilde{\beta}_1 + 2\beta(x(t_0))$, we obtain (16).

In the particular case of systems of the form (1)—(2) one can restate Proposition 2 as follows:

Proposition 3. *Assume that there exists a constant $\delta_0 \geqslant 0$ such that for every pair of functions u, x which satisfy equation (1), the following statement is true:*

For every constant $\beta_0 \geqslant 0$ for which one has

$$\eta(0, t) \leqslant \beta_0^2 \text{ for every } t \in [0, T] \tag{23}$$

one also has

$$\| x(t) \| \leqslant \delta_0(\beta_0 + \| x(0) \|) \text{ for every } t \in [0, T]. \tag{24}$$

Then system (1)—(2) has Property h_s (see Definition 3).

Indeed, proceeding as at the end of Section 4, one sees that under the condition of Proposition 3 system (1), (2) has property H_s for $\alpha(y) = \|(y)\| / \delta_0$ and $\beta(y) = \|y\|$. Consider now two constants $\beta_1 \geqslant 0$ and $\beta_2 \geqslant 0$ for which inequality (7) is satisfied. Introducing the function α, specified above, one obtains an inequality of the form (15) for $\widetilde{\beta}_1 = \beta_1$ and $\widetilde{\beta}_2 = \delta_0\beta_2$. By Proposition 2, Relation (16) is satisfied. Replacing $\alpha(x)$, $\beta(x)$, $\widetilde{\beta}_1$ and $\widetilde{\beta}_2$ by the values indicated above, Relation (16) can be written as

$$\| x(t) \| \leqslant 2\delta_0(\beta_1 + \delta_0\beta_2 + \| x(0) \|)$$

whence (8) follows immediately, with $\delta = \max(2\delta_0, 2\delta_0^2)$.

6. A sufficient condition of hyperstability

In Section 4 we have defined the property of hyperstability with the help of two distinct properties: H_s and H_p. We now show that — if one is willing to accept some loss of generality — one can replace Properties H_s and H_p by a single condition, which in many cases can be easily checked.

Proposition 4. *Assume that there exist a function $\widetilde{\alpha}$ of class M_i and the functions $\widetilde{\beta}$ and $\widetilde{\gamma}$ of class M_s such that for every solution of the system (10)—(11) the inequality*

$$[\widetilde{\alpha}(x(t))]^2 \leqslant \eta(t_0, t) + [\widetilde{\beta}(x(t_0))]^2 + \widetilde{\gamma}(x(t_0)) \sup_{t_0 \leqslant \tau \leqslant t} \widetilde{\alpha}(x(\tau)) \tag{25}$$

is satisfied for all $t \geqslant t_0$. Then system (10)—(11) is hyperstable (Definition 8). Furthermore if system (10)—(11) has the form

(1)–(2) and $\tilde{\alpha}(y)$, $\tilde{\beta}(y)$ and $\tilde{\gamma}(y)$ have the form $\varkappa_j \|y\|$, where $\varkappa_j$ are constants [1]*), then Properties h_s and h_p stated in Definitions 3 and 4 are also satisfied.*

First we prove that Property H_s is satisfied. From (25) and (12) one obtains an inequality of the form (17) (see Lemma 1) for $\xi(t) = \tilde{\alpha}(x(t))$, $\alpha_1^2 = \beta_0^2 + [\tilde{\beta}(x(t_0))]^2$, $\alpha_2 = \tilde{\gamma}(x(t_0))$. Applying Lemma 1 and using the inequality $\alpha_1^2 \leqslant 4(\beta_0 + \tilde{\beta}(x(t_0))^2$, one obtains from (18) the inequality

$$\tilde{\alpha}(x(t)) \leqslant 2(\beta_0 + \tilde{\beta}(x(t_0)) + \tilde{\gamma}(x(t_0))/2).$$

In order to obtain Condition (13) of Property H_s it is sufficient to define $\alpha(y)$ and $\beta(y)$ as $\alpha(y) = \tilde{\alpha}(y)/2$ and $\beta(y) = \tilde{\beta}(y) + \tilde{\gamma}(y)/2$. (One can easily see that the function α belongs to class M_i and β to class M_s).

Consider now Property H_p and note that by using the obvious inequality $[\tilde{\alpha}(x)]^2 \geqslant 0$ one obtains from (25) the inequality

$$\eta(t_0, t) \geqslant [-\tilde{\beta}^2(x(t_0))]^2 - 2\tilde{\gamma}(x(t_0)) \sup_{t_0 \leqslant \tau \leqslant t} \alpha(x(\tau))$$

(we have used the previously established relation $\alpha(y) = \tilde{\alpha}(y)/2$). The above inequality can be written in the form (14) if one defines the functions γ and δ (belonging to class M_s) by the relations $\gamma(y) = \tilde{\beta}(y)$ and $\delta(y) = 2\,\tilde{\gamma}(y)$.

Examining the last statement of Proposition 4 we find that if $\tilde{\alpha}(y)$, $\tilde{\beta}(y)$ and $\tilde{\gamma}(y)$ are of the form $\varkappa_j \|y\|$, then $\alpha(y)$, $\beta(y)$, $\gamma(y)$ and $\delta(y)$ defined in the foregoing proof, are of the same form. Therefore, Property H_p of Definition 4 and Property h_s of Definition 3 — for $\beta_2 = 0$ — are consequences of properties H_p and H_s respectively. Then, by using Proposition 3 one obtains the general form of Property h_s (for every $\beta_2 \geqslant 0$). This concludes the proof of Proposition 4.

Thus inequality (25) gives a sufficient criterion of hyperstability. It is natural to ask whether inequality (25) gives a necessary and sufficient condition of hyperstability. As will be shown in § 14, this is not true even for the simplest systems. However, in §§ 14 and 16, we shall show that inequality (25) gives a necessary and sufficient condition of hyperstability, if the system investigated satisfies certain additional non-degeneracy conditions.

---

[1]) Obviously, if $\tilde{\alpha}(y) = \varkappa \|y\|$, $\tilde{\beta}(y) = \varkappa_2 \|y\|$ and $\tilde{\gamma}(y) = \varkappa_3 \|y\|$, then the conditions : $\tilde{\alpha}$ belongs to class M_i and $\tilde{\beta}$ and $\tilde{\gamma}$ belong to class M_s, are satisfied if and only if $\varkappa_1 > 0$, $\varkappa_2 \geqslant 0$ and $\varkappa_3 \geqslant 0$.

7. Hyperstability of systems which contain "memoryless elements"

So far we have used to a very small extent the assumption that the system under investigation is of the form (10)—(11). Therefore it is to be expected that the results obtained can be extended to much more general systems. In § 19 the problem will be treated in more details. We examine here only the case in which besides Relations (10) and (11), one has one or more relations of the form

$$h(x(t), u(t), t) = 0 \tag{26}$$

where $(y, v, t) \mapsto h(y, v, t)$ is a vector-valued function defined and piecewise continuous for all the values of the arguments. Then all the definitions and propositions given before remain valid if we replace system (10)—(11) by "system (26), (10), (11)" and if we add to the conditions of Definition 7 the condition that Relation (26) should also be satisfied. It is clear that the set of the solutions of system (26), (10), (11) is a subset of the set of the solutions of system (10)—(11). Furthermore, it is necessary to assume that the restrictions introduced by Relation (26) are not excessive and therefore that the set of the solutions of the system (26), (10), (11) has sufficiently many elements. In Section 9 of this paragraph one considers various examples of systems containing restrictions of the form (26).

8. The "sum" of two hyperstable systems

Consider two systems of the form (10), (11)

$$\frac{\mathrm{d}x_j}{\mathrm{d}t} = f_j(x_j, u_j, t), \quad j = 1, 2, \tag{27}$$

$$\eta_j(t_0, t_1) = [\varphi_j(x_j, u_j, t)]_{t_0}^{t_1} + \int_{t_0}^{t_1} \psi_j(x_j, u_j, t)\, \mathrm{d}t, \qquad j = 1, 2. \tag{28}$$

By definition the "sum" of these two systems is another system of the form (10)—(11) obtained as follows: Eq. (10) of the sum-system is given by Eqs. (27) for $j = 1, 2$ and the integral (11) of the sum-system is given by the sum

$$\eta(t_0, t_1) = \eta_1(t_0, t_1) + \eta_2(t_0, t_1) \tag{29}$$

where $\eta_j(t_0, t_1)$, $j = 1, 2$ are given by (28). To the above relations one may also add some restrictions of the form

$$h(x_1, x_2, u_1, u_2, t) = 0 \tag{30}$$

with the same properties as (26). As pointed out in the previous section, the introduction of the restrictions (30) means that one considers a subset of the set of solutions of the sum-system defined above (one selects the subset of these solutions which, in addition to the equations of the sum-system, also satisfy (30)). If to the initial systems (27), (28) one associates from the beginning some conditions of the form (26), these may be included in (30).

The introduction of sum-systems is useful because of the following property :

Proposition 5. *The sum of two hyperstable systems is a hyperstable system.*

Proof. Let S_j, $j = 1, 2$ be two hyperstable systems. Then there exist some functions α_j, β_j, γ_j and δ_j, $j = 1, 2$ with the properties specified in Definition 8. In particular, from (14) one obtains

$$\eta_j(t_0, t) \geqslant [-\gamma_j(x_j(t_0))]^2 - \delta_j(x_j(t_0)) \sup_{t_0 \leqslant \tau \leqslant t} \alpha_j(x_j(\tau)), \quad j = 1, 2. \tag{31}$$

We show that Property H_s is satisfied by the sum-system. Assume that inequality (12) is satisfied for η given by (29). Using (29) and (31) for $j = 2$, one obtains

$$\eta_1(t_0, t) \leqslant \beta_0^2 + [\gamma_2(x_2(t_0))]^2 + \delta_2(x_2(t_0)) \sup_{t_0 \leqslant \tau \leqslant t} \alpha_2(x_2(\tau)) \tag{32}$$

for every t in the interval $[t_0, T]$. Consider an arbitrary number T_1 in the interval $[t_0, T]$. By (32) one has

$$\eta_1(t_0, t) \leqslant \beta_0^2 + [\gamma_2(x_2(t_0))]^2 + \delta_2(x_2(t_0)) \sup_{t_0 \leqslant \tau \leqslant T_1} \alpha_2(x_2(\tau)), \quad \text{for every } t \in [t_0, T_1] \tag{33}$$

where the right hand member does not depend on t. Denoting the right hand member by $\tilde{\beta}_0^2$ one obtains an inequality of the form (12) (for $T_0 = T_1$). Since system S_1 is hyperstable it follows that inequality (13) must be also satisfied. In particular, for $t = T_1$ one obtains $\alpha_1(x_1(T_1)) \leqslant \tilde{\beta}_0 + \beta_1(x_1(t_0))$ which further implies $\alpha_1^2(x_1(T_1)) \leqslant 2\tilde{\beta}_0^2 + 2\beta_1^2(x_1(t_0))$. Replacing here $\tilde{\beta}_0^2$ by the right hand member of (33) and replacing further in the inequality thus obtained T_1 by t, one obtains

$$[\alpha_1(x_1(t))]^2 \leqslant 2\beta_0^2 + 2[\beta_1(x_1(t_0))]^2 + 2[\gamma_2(x_2(t_0))]^2 + \\ + 2\delta_2(x_2(t_0)) \sup_{t_0 \leqslant \tau \leqslant t} \alpha_2(x_2(\tau)), \quad \text{for every } t \in [t_0, T]. \tag{34}$$

Since S_1 and S_2 play symmetric rôles we can interchange the indices 1 and 2 thus obtaining

$$[\alpha_2(x_2(t))]^2 \leqslant 2\beta_0^2 + 2[\beta_2(x_2(t_0))]^2 + 2[\gamma_1(x_1(t_0))]^2 + \\ + 2\delta_1(x_1(t_0)) \sup_{t_0 \leqslant \tau \leqslant t} \alpha_1(x_1(\tau)), \text{ for every } t \in [t_0, T]. \tag{35}$$

From the inequalities $\alpha_j(x_j(\tau)) \geqslant 0$ one obtains

$$\sup_{t_0 \leqslant \tau \leqslant t} \alpha_j(x_j(\tau)) \leqslant \sup_{t_0 \leqslant \tau \leqslant t} (\alpha_1(x_1(\tau)) + \alpha_2(x_2(\tau))), \; j = 1, 2$$

whence one also has

$$\delta_1(x_1(t_0)) \sup_{t_0 \leqslant \tau \leqslant t} \alpha_1(x_1(\tau)) + \delta_2(x_2(t_0)) \sup_{t_0 \leqslant \tau \leqslant t} \alpha_2(x_2(\tau)) \leqslant \\ \leqslant \delta_1(x_1(t_0)) + \delta_2(x_2(t_0))) \sup_{t_0 \leqslant \tau \leqslant t} (\alpha_1(x_1(\tau)) + \alpha_2(x_2(\tau))). \tag{36}$$

Adding (34) and (35) and using (36) and the obvious inequality $[\alpha_1(x_1(t))]^2 + [\alpha_2(x_2(t))]^2 \geqslant (\alpha_1(x_1(t)) + \alpha_2(x_2(t))^2/2$ one obtains the inequality

$$(\alpha_1(x_1(t))) + \alpha_2(x_2(t)))^2/2 \leqslant 4\beta_0^2 + 2[\beta_1(x_1(t_0))]^2 + \\ + 2[\beta_2(x_2(t_0))]^2 + 2[\gamma_1(x_1(t_0))]^2 + 2[\gamma_2(x_2(t_0))]^2 + \\ + 2(\delta_1(x_1(t_0)) + \delta_2(x_2(t_0))) \sup_{t_0 \leqslant \tau \leqslant t} (\alpha_1(x_1(\tau)) + \alpha_2(x_2(\tau))) \\ \text{for every } t \in [0, t].$$

The above inequality is of the form (17) of Lemma 1 for

$$\xi(t) = (\alpha_1(x_1(t)) + \alpha_2(x_2(t)))/2 \tag{37}$$

$$\alpha_1^2 = 2\beta_0^2 + [\beta_1(x_1(t_0))]^2 + [\beta_2(x_2(t_0))]^2 + [\gamma_1(x_1(t_0))]^2 + [\gamma_2(x_2(t_0))]^2 \tag{38}$$

$$\alpha_2 = 2(\delta_1(x_1(t_0)) + \delta_2(x_2(t_0))) \tag{39}$$

From (38) it also follows that

$$\alpha_1^2 \leqslant 4(\beta_0 + \beta_1(x_1(t_0)) + \beta_2(x_2(t_0)) + \gamma_1(x_1(t_0)) + \gamma_2(x_2(t_0)))^2.$$

Using instead of α_1^2 the right hand member of this inequality and applying Conclusion (18) of Lemma 1 one can write the inequality thus obtained in the form (13), where (if we introduce the n_1-vector y_1, the n_2-vector y_2 and the $n_1 + n_2$-vector $y = \begin{pmatrix} y_1 \\ y_2 \end{pmatrix}$)

$$\left.\begin{aligned} \alpha(y) &= (\alpha_1(y_1) + \alpha_2(y_2))/4 \\ \beta(y) &= \beta_1(y_1) + \beta_2(y_2) + \gamma_1(y_1) + \gamma_2(y_2) + \delta_1(y_1) + \delta_2(y_2) \end{aligned}\right\}. \quad (40)$$

One sees easily that the function α belongs to class M_i and the function β to class M_s. Thus we have established that the sum-system has Property H_s for the functions α and β given by (40). We shall now show that it also possesses Property H_p. Adding the inequalities (31) for $j = 1, 2$ and using (36) we conclude that η(see (29)) satisfies the inequality

$$\eta(t_0, t) \geqslant -[\gamma_1(x_1(t_0))]^2 - [\gamma_2(x_2(t_0))]^2 -$$

$$-(\delta_1(x_1(t_0)) + \delta_2(x_2(t_0)) \sup_{t_0 \leqslant \tau \leqslant t} (\alpha_1(x_1(\tau)) + \alpha_2(x_2(\tau))).$$

The above inequality is of the form (14), where α is given by (40) and γ and δ are defined by the relations

$$[\gamma(y)]^2 = [\gamma_1(y_1)]^2 + [\gamma_2(y_2)]^2$$

$$\delta(y) = 4(\delta_1(y_1) + \delta_2(y_2)).$$

The proof of Proposition 5 ends with the remark that γ and δ belong to the class M_s.

9. Hyperstable blocks and their principal properties

If we associate to the equation

$$\frac{dx}{dt} = f(x, u, t) \quad (41)$$

the "output"

$$v = g(x, u, t) \tag{42}$$

we obtain a "block" with the input u and the output v. Such blocks are of interest not only because they are largely used in technical problems but also because of their remarkable properties. We assume that the functions f and g satisfy conditions similar to those required in § 12. We shall further assume that v has the same dimension as u (i.e. $v(t)$ is an m-vector).

Besides Eqs. (41) and (42) we sometimes also have a relation of the form

$$h(x, u, v, t) = 0, \tag{43}$$

where the function h is defined and piecewise continuous for all the values of the arguments.

For any block of the form (41), (42) we define an "associated system" described by the relations

$$\frac{dx}{dt} = f(x, u, t), \tag{44}$$

$$\eta(t_0, t_1) = \text{Re}\left(\int_{t_0}^{t_1} u^*(t)\, v(t)\, dt\right), \tag{45}$$

which takes the form (10)—(11) after replacing v by relation (42) and after using the relation $2\text{Re}\, u^*v = u^*v + v^*u$. When there also exist relations of the form (43), these are connected with Eqs. (44), (45) as shown in Section 8.

After introducing the associated system (44)—(45), all the definitions and results obtained in the foregoing sections are immediately extended to blocks of the form (41)—(42). In particular, block (41)—(42) is said to be hyperstable if its associated system (44)—(45) is hyperstable (Definition 8).

By a natural extension of Definition 7 we shall call "solution" of the block (41), (42) (and possibly (43)) a set of functions u, x, v (u : piecewise continuous and x : continuous), which satisfy the equations of the block : (41), (42) (and (43) respectively). We remark that a block is defined whenever the set of all its solutions is known (this allows a considerable generalization of the concept of block).

Blocks may be combined together in various ways thus obtaining from two or more distinct blocks new and more complex ones. Some of those combinations are of particular interest for us since they have the property that if the component blocks are hyperstable, then the resulting block is also hyperstable.

For example, consider two blocks B_j, $j = 1, 2$, described by the relations (of the form (41), (42))

$$B_j : \begin{cases} \dfrac{dx_j}{dt} = f_j(x_j, u_j, t) \\ v_j = g_j(x_j, u_j, t) \end{cases} \tag{46}$$

Assume that all the vectors u_1, u_2, v_1 and v_2 have the same number of components (equal to m). By introducing two new vector-valued functions u and v with m components one can define in several ways a new block with the input u and the output v. For instance, one can define the block B_p by means of

$$B_p : \begin{cases} \dfrac{dx_j}{dt} = f_j(x_j, u_j, t), \; j = 1, 2 \\ v = g_1(x_1, u, t) + g_2(x_2, u, t), \; (u = u_1 = u_2) \end{cases} \tag{47}$$

namely by describing its solutions as follows: the functions u_j, x_j, v_j, $(j = 1, 2)$, u and v are said to be a solution of the block B_p if they satisfy Eqs. (46), for $j = 1, 2$ as well as the relations

$$u_1 = u_2 = u\,; \quad v = v_1 + v_2. \tag{48}$$

Using the usual way of representing block diagrams one sees that Relations (48) correspond to the connections of Fig. 4.1.b. Hence we may say that the defined above block B_p

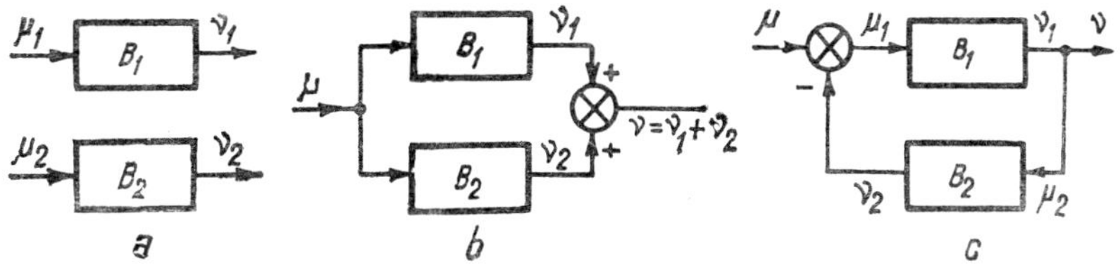

Fig. 4.1

is obtained by the parallel connection of blocks B_1 and B_2.

Relation (48) implies that

$$\mathrm{Re}(v^* u) = \mathrm{Re}(v_1^* u_1) + \mathrm{Re}(v_2^* u_2). \tag{49}$$

Besides blocks B_1, B_2 and B_p we introduce now their associated systems (of the form (44)—(45)) which we shall denote by S_1, S_2 and S_p respectively.

By (49) we have

$$\eta_p(t_0, t_1) = \eta_1(t_0, t_1) + \eta_2(t_0, t_1), \tag{50}$$

where $\eta_j(t_0, t_1)$ $(j = 1, 2)$ and $\eta_p(t_0, t_1)$ are the integrals of the form (45) for the systems $S_j (j = 1, 2)$ and S_p respectively. It can be easily seen that any solution of system S_p is also a solution of the system obtained by adding the systems S_1 and S_2 as has been shown in the previous sections. (Relations (48) play the part of Relations (30)). Using Proposition 5 one obtains

Proposition 6. *Any block resulting from the parallel connection of two hyperstable blocks is hyperstable.*

A similar property also holds when the blocks B_1 and B_2 are combined in a different way. Assume that the equations

$$v_1 = g_1(x_1, u - v_2, t), \tag{51}$$

$$v_2 = g_2(x_2, v_1, t) \tag{52}$$

have a solution of the form

$$v_2 = g_0(x_1, x_2, u, t) \tag{53}$$

where the function g_0 is defined and piecewise continuous for all the values of the variables involved. (Obviously, once one knows v_2 one can find v_1 by means of Relation (51)). We also assume that the function g_0 is such that the existence of the solutions of Eqs. (54) below is ensured, for every initial condition and in every interval $[t_0, t_1]$.

We define the block B_f by the relations

$$B_f: \left\{ \begin{aligned} &\frac{dx_1}{dt} = f_1(x_1, u - g_0(x_1, x_2, u, t), t) \\ &\frac{dx_2}{dt} = f_2(x_2, g_1(x_1, u - g_0(x_1, x_2, u, t), t), t) \\ &v = g_1(x_1, u - g_0(x_1, x_2, u, t), t) \end{aligned} \right\}. \tag{54}$$

As in the case of block B_p, we actually define the block B_f by giving a precise description of its solutions: the functions x_j, u_j, v_j $(j = 1, 2)$, u and v are said to be a solution of the block B_f if they satisfy Eqs. (46) for $j = 1, 2$, and the relations

$$u_1 = u - v_2\,;\ u_2 = v_1\,;\ v = v_1. \tag{55}$$

Notice that Relations (55) correspond to the connection c of Fig. 4.1 and hence one can say that block B_f is a block with "negative feedback" consisting of the blocks B_1 and B_2. From Relations (55) one obtains once again Eq. (49). By introducing the associated systems S_f, S_1 and S_2 as we did before we obtain instead of (50) the similar relation:

$$\eta_f(t_0, t_1) = \eta_1(t_0, t_1) + \eta_2(t_0, t_1). \tag{56}$$

It follows that the solutions of system S_f constitute once again a subset of the set of the solutions of the sum of the systems S_1 and S_2. (This subset is determined by the feedback Relations (55) — which can be also written as in (30)). Consequently from Proposition 5 one obtains the following result:

Proposition 7. *Any block with negative feedback consisting of two hyperstable blocks is hyperstable.*

The "parallel" and "negative feedback" connections are not the only ones with the property that the resulting block is hyperstable whenever the component blocks are hyperstable. Indeed, by examining the foregoing arguments it is clear that this property holds if one replaces the relations of the form (48) or (55) (which characterize the type of connection) by any other relation which implies (49).

Some consequences resulting from the propositions established above will be frequently used in Chapter 5. Thus, from the form of the integral η in the case of the associated system (44)—(45), and from Property H_8 of Definition 8 one obtains immediately the following result:

Proposition 8. *Assume that block (41)—(42) is hyperstable. Then the equation*

$$\frac{\mathrm{d}x}{\mathrm{d}t} = f(x, 0, t) \tag{57}$$

(obtained from (41) for $u = 0$) admits the trivial solution $x(t) \equiv 0$, and this solution is stable in the sense of Liapunov. Furthermore, all the solutions of equation (57) are bounded.

Indeed, recall we always assume f to be such that the existence if the solutions of equation (57) is secured. (This happens e.g. if f is continuous in all the variables). Relation

(45), for $u(t) \equiv 0$, becomes $\eta(t_0, t_1) \equiv 0$. Hence Condition (12) of Property H_s is satisfied for $\beta_0 = 0$. Using inequality (13) (which holds by virtue of Property H_s) one finds that all the solutions of Eq. (57) are bounded. This fact can be used to relax the conditions securing the existence of the solutions in every interval $[t_0, t_1]$, for every $t_1 > t_0$. From the condition $\beta_0 = 0$ and inequality (13), written for $x(0) = 0$, it follows that the solution of Eq. (57) for the initial condition $x(0) = 0$ is precisely the trivial solution $x = 0$. The fact that this solution is stable in the sense of Liapunov follows also from inequality (13) and the condition $\beta_0 = 0$, using the properties of the functions α and β.

Combining Proposition 8 with Proposition 7, we see for instance, that the block with negative feedback corresponding to the block diagram *c*) of Fig. 4.1 has a stable trivial solution for the input $u = 0$, if blocks B_1 and B_2 are hyperstable and if the existence of the solutions of the system for every initial condition is ensured.

A similar but more general result is given by

Proposition 9. *Consider a differential equation of the form*

$$\frac{dx}{dt} = f(x, t) \tag{58}$$

and assume that it has a solution for every initial condition $x(0)$. Assume further that one can find one or more hyperstable blocks of the form (46) for $j = 1, 2, \ldots, k$ together with a relation of the form

$$h(x_1, x_2, \ldots, x_k, u_1, u_2, \ldots, u_k, v_1, v_2, \ldots, v_k, t) = 0 \tag{59}$$

with the following properties: (i): (59) implies the identity

$$\mathrm{Re}\left(\sum_{j=1}^{k} u_j^* v_j\right) = 0 \tag{60}$$

and (ii): for every solution x of equation (58) the equations (46), $j = 1, 2, \ldots, k$ and (59) have a solution x_j such that $x(t) = \sum_{j=1}^{k} T_j x_j(t)$, for some suitably chosen matrices T_j.

Then the following conclusions are true: (a) all the solutions of Eq. (58) are bounded; (b) Eq. (58) has the trivial solution $x(t) = 0$; (c) the trivial solution is stable in the sense of Liapunov.

The proof is based on the remark that all the systems associated with the blocks (46), for $j = 1, 2, \ldots, k$, are hyperstable and hence the system obtained by adding all these systems is also hyperstable (see Proposition 5). Owing to Condition (60) the integral $\eta(0, t_1)$ of the sum-system is identically zero. Hence, by using Property H_s of Definition 8, one obtains all the conclusions of Proposition 9. The proof shows that Proposition 9 remains valid even if Condition (60) is replaced by

$$\mathrm{Re}\left(\sum_{j=1}^{k} u_j^* v_j\right) \leqslant 0.$$

Proposition 9 supplies a method for obtaining stability criteria for equations of the form (58). In Chapter 5 the method will be repeatedly applied to deal with particular problems. For the moment we give here two immediate extensions of Proposition 9.

In applications one frequently meets blocks which are described *only* by equations of the form

$$v = g(u, t) \tag{61}$$

and where the differential equation is absent. Proposition 9 can be directly extended to include this case if one introduces the following definition.

Definition 9. *The block (61) is said to be a memoryless hyperstable block if the following inequality is satisfied:*

$$Re\,(u^*v) = Re\,(u^*g(u, t)) \leqslant 0, \text{ for every } m\text{-vector } u \text{ and every } t \geqslant t_0. \tag{62}$$

Under the same conditions a system consisting of Eq. (61) and the integral $\eta(t_0, t_1) = Re\left(\int_{t_0}^{t_1} u^*v\,dt\right)$ *is said to be a memoryless hyperstable system.*

We remark that Proposition 5 remains valid for memoryless hyperstable systems, whence we immediately obtain the following generalization of Proposition 9:

Proposition 10. *Proposition 9 remains valid when besides the hyperstable blocks (46)* $j = 1, 2, \ldots, k$, *one also has memoryless hyperstable blocks of the form (62), the other conditions being the same.*

The second extension refers to the possibility of "inverting" a hyperstable block by interchanging the input and the output. (This operation corresponds to the inversion of the directions of the arrows, e.g. in Fig. 4.1.*a*). If we proceed as we did before and define a block by the set of its solutions, then the problem of defining the inverse block is immediately solved, since any set of functions which satisfy the equations of an arbitrary block, evidently satisfy also the equations of the inverse block. However, the part played by the input and output functions in the equations of the block (of the form (41), (42)) is completely different; this accounts both for the fact that in general inverse blocks cannot be expressed in the same form as the direct blocks and for the interest generated by the introduction of inverse blocks for extending the range of applications.

It can be easily seen that Proposition 9 also remains valid when we admit inverse hyperstable blocks. Indeed, any set of hyperstable blocks of the form (47) to which one adds inverse hyperstable blocks of the same form can be also considered as an association consisting only of hyperstable blocks of the form (47). To this end it is sufficient to replace all the inverse blocks by the corresponding direct blocks, — by interchanging the "input" and "output" — , and to make the same permutation in the Relations (59). After affecting this permutation, Condition (61) is still satisfied and hence the problem reduces to the case dealt with in Proposition 9. The result remains valid if we also admit blocks of the form (61):

Proposition 11. *Proposition 9 remains valid if besides the hyperstable blocks (46) and the memoryless hyperstable blocks (62) there occur also the inverses of some hypeɹeable blocks of the form (46), the other conditions being the same.*

The above arguments show that the use of inverse blocks can be avoided since they can be replaced by the corresponding direct blocks; however, inverse blocks are often useful because they allow a more unitary and intuitive treatment of the problems. In § 15 an example will be given of the use of inverse hyperstable blocks. For the moment we remark that the block of Fig. 4.1.*c* can be considered as the inverse of a block obtained by the parallel connection of the block B_2 with the inverse of block B_1 (see block diagram Fig. 4.1.*b*). Thus, the block diagrams 4.1.*b* and 4.1.*c* which seem to be completely different, can be obtained from each other if one introduces inverse blocks.

§ 14. Single-input hyperstable systems

In this paragraph we shall examine systems of the form

$$\frac{\mathrm{d}x}{\mathrm{d}t} = Ax + b\mu \tag{1}$$

$$\eta(0, t_1) = \Big[x^* J x\Big]_0^{t_1} + \int_0^{t_1} (\varkappa\mu^*\mu + \mu^* l^* x + x^* l\mu + x^* M x)\,\mathrm{d}t, \tag{2}$$

where the notations have the same meaning as in § 2. We assume that the conditions specified in Section 2 of § 13 are satisfied, namely : (i) the system is minimally stable (see Definition 2, § 13) (ii) the vector b is different from zero and (iii) the characteristic polynomial π (see (18, § 3)) is not identically zero.

The principal result which will be established in this paragraph is that the hyperstability of the system under investigation is equivalent to a certain condition on the frequency expressed by means of the function

$$\chi(-i\omega,\ i\omega) = k + 2\,\mathrm{Re}\,(l^*\,(i\omega\,I - A)^{-1}\,b) + \\ + b^*\,(-\,i\omega\,I - A^*)^{-1}\,M\,(i\omega\,I - A)^{-1}\,b \tag{3}$$

which is obtained from the expression (13, § 3) for the characteristic function χ with $\sigma = -\lambda = i\omega$. Furthermore we shall also establish other auxiliary properties which give a more comprehensive characterization of the property of hyperstability for the systems under investigation.

All these results are contained in the following theorem :

Theorem 1. *Assume that the following conditions are satisfied : (i) the system (1), (2) is minimally stable; (ii) $b \neq 0$ and (iii) the characteristic polynomial $\pi(\lambda,\ \sigma)$ is not identically zero.*

Then the following properties are equivalent

(H) : "The system (1)—(2) is hyperstable (see Definition 8, §13)"

(F) : "The inequality $\chi(-\,i\omega,\ i\omega) \geqslant 0$ is satisfied for every real number ω which satisfies the condition det $(i\omega I - A) \neq 0$".

(h) : "Properties h_s and h_p of Definitions 3, § 13 and 4, § 13 are satisfied".

(h_{so}) : "Property h_s of Definition 3, § 13 is satisfied in the particular case $\beta_1 = 0$, $\beta_2 = 0$, $T \to \infty$".

(h_{po}) : "Property h_p of Definition 4, § 13 is satisfied in the particular case $x(0) = 0$".

(f) : "The inequality $\chi(\sigma,\ \sigma) \geqslant 0$ is satisfied for every number σ which satisfies the conditions Re $\sigma > 0$ and det $(\sigma I - A) \neq 0$".

If besides Conditions (i), (ii) and (iii) the system satisfies also the additional Condition (iv): "the pair $(A,\ b)$ is completely controllable" as well as one of the following conditions (v′) "$\pi(-\,\sigma,\ \sigma) \not\equiv 0$" or (v″) "there exists no constant σ_0 such that $\pi(\lambda,\ \sigma_0) \equiv 0$", then each of the properties stated above is equivalent to the following property :

(h_+) : "The integral $\eta(0,\ t_1)$ can be written in the form similar to (12, § 8)

$$\eta\,(0,\ t_1) = [x^*\ U\,x]_0^{t_1} + \int_0^{t_1} |\,\gamma\mu + w^*\,x\,|^2\ dt, \tag{4}$$

where U is a positive definite matrix".

Before giving the proof of Theorem 1 we examine some of its consequences. The equivalence of the property of hyperstability and each of the properties (h_{so}) and (h_{po}) shows that in the case under investigation, Properties H_s and H_p of the definition of hyperstability (Definition 8, § 13) are not independent but one follows from the other. Furthermore, it follows that each of these properties can be also replaced by Property (h_{so}) or (h_{po}) (which are more particular than H_s and H_p respectively). Property (h_+) — which is also equivalent to the property of hyperstability if the more restrictive conditions mentioned in Theorem 1 are satisfied — is useful because in certain cases it allows us to establish Liapunov-functions of the type "quadratic form plus integral".

It should be noted that the hypotheses as well as the various properties stated in Theorem 1 remain unchanged when one applies a transformation of the type studied in §§ 2—4. For Property H this results directly from Definition 8, § 13. Property (f) can be also expressed in the form $\pi(-i\omega, i\omega) \geqslant 0$ (see Relation (18, § 3)). Since the characteristic polynomial is invariant under the transformations considered (see Proposition 1, § 3) the invariance of Property (f) follows at once. Similarly, one can easily check the invariance of the hypotheses and of all the properties stated in Theorem 1. This remark plays an important part in the proof which follows since — without loss of generality — it allows us to assume that the system under investigation has been brought to a special form by a transformation of the type studied in §§ 2—4.

We shall prove Theorem 1 according to the following scheme of implications

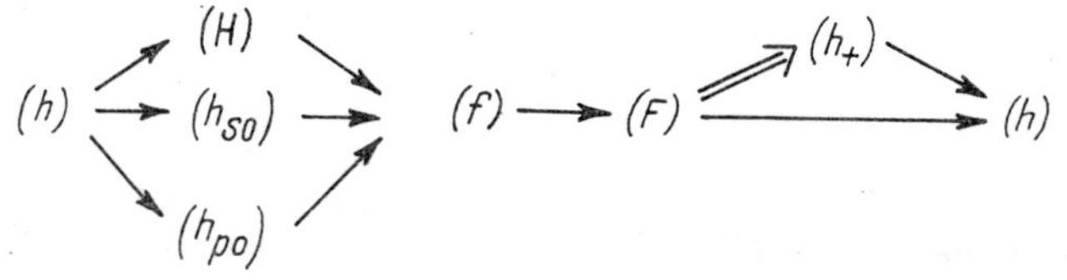

$$(h) \begin{array}{c} \nearrow (H) \searrow \\ \rightarrow (h_{so}) \rightarrow \\ \searrow (h_{po}) \nearrow \end{array} (f) \rightarrow (F) \begin{array}{c} \Rightarrow (h_+) \searrow \\ \rule{3cm}{0.4pt} \end{array} (h) \qquad (5)$$

where (H), (F), (h), (h_{so}), (h_{po}) and (h_+) denote the respective properties stated in Theorem 1 and the arrow signifies "implies". The double arrow $\Rightarrow$ used only in the implication $(F) \Rightarrow (h_+)$ means that the implication is true only under the additional conditions *(iv)* and *(v′)* (or *v″*), required in the final part of Theorem 1.

We now prove the implications (5). The implications $(h) \rightarrow (H)$, $(h) \rightarrow (h_{so})$ and $(h) \rightarrow (h_{po})$ follow at once from the definitions of the respective properties. The implications $(H) \rightarrow (f)$, $(h_{so}) \rightarrow (f)$ and $(h_{po}) \rightarrow (f)$ are proved

by simply showing that if property *(f)* does not take place then none of the properties *(H)*, (h_{so}) and (h_{po}) is satisfied. To show this we shall first establish some consequence resulting from the assumption that property *(f)* is not satisfied, i.e. "There exists a number σ_0 for which the conditions Re $\sigma_0 > 0$, det $(\sigma_0 I - A) \neq 0$ and $\chi(\overline{\sigma}_0, \sigma_0) < 0$ are satisfied". (6)

Consider the control function

$$\check{\mu}(t) = e^{\sigma t_0} \tag{7}$$

for which Eq. (1) has the particular solution

$$\check{x}(t) = (\sigma_0 I - A)^{-1} b\, e^{\sigma t_0}. \tag{8}$$

From the conditions Re $\sigma_0 > 0$ and det $(\sigma_0 I - A) \neq 0$ and the hypothesis (ii) : $b \neq 0$, it follows that

$$\lim_{t\to\infty} \| \check{x}(t) \| = \infty. \tag{9}$$

Substituting (7) and (8) in (2) one obtains the equality

$$\check{\eta}(0, t_1) = \chi(\overline{\sigma}_0, \sigma_0) \int_0^{t_1} e^{(\overline{\sigma}_0 + \sigma_0)t}\, dt$$

(see Relation (11, § 3)), whence the inequality

$$\check{\eta}(0, t_1) < 0, \quad \text{for every } t_1 > 0. \tag{10}$$

It also follows that, for the considered solution, we have, for every t_0

$$\lim_{t_1\to\infty} \check{\eta}(t_0, t_1) = -\infty \tag{11}$$

where the expression $\check{\eta}(t_0, t_1)$ is obtained by substituting the solution considered above in Relation (2) and replacing $\int_0^{t_1}$ by $\int_{t_0}^{t_1}$.

We now show that these conditions contradict each of the Properties *(H)*, (h_{so}) and (h_{po}). Indeed, under the conditions stated above system (1)—(2) is satisfied by functions (7) and (8) which also satisfy Conditions (9) and (10), thus contradicting both Property (H_s) of Definition 8, § 13 and Property (h_{so}). Concerning Property (h_{po}) we remark that by Proposition 2, § 31 there exists a control function μ_0 defined and continuous in the interval (0, 1), such that the corresponding solution x_0 of Eq. (1) for $x(0) = 0$ has at the point $t = 1$ the same value as solution (8), namely

$$x_0(1) = \check{x}(1) = (\sigma_0 I - A)^{-1} b\, e^{\sigma_0}.$$

Furthermore, the corresponding integral $\eta_0(0, 1)$ is obviously finite. If we define the control function μ as

$$\mu(t) = \begin{cases} \mu_0(t) & \text{for } 0 \leqslant t \leqslant 1 \\ \check{\mu}(t) & \text{for } t > 1, \end{cases} \tag{12}$$

then the corresponding solution of Eq. (1) for $x(0) = 0$ is given by

$$x(t) = \begin{cases} x_0(t) & \text{for } 0 \leqslant t \leqslant 1 \\ \check{x}(t) & \text{for } t > 1\,. \end{cases} \tag{13}$$

The corresponding integral for $t_1 > 1$ is equal to

$$\eta(0, t_1) = \eta_0(0, 1) + \check{\eta}(1, t_1), \quad (t_1 > 1). \tag{14}$$

By Condition (11) one has the property $\lim\limits_{t_1 \to \infty} \eta(0, t_1) = -\infty$ which obviously contradicts Property (h_{po}).

To prove the other implications of diagram (5) we remark that implication *(f)* $\to$ *(F)* results immediately from the fact that the function $\sigma \to \chi(\bar{\sigma}, \sigma)$ is continuous in the neighbourhood of every point $\sigma = i\omega$ at which the condition $\det(i\omega I - A) \neq 0$ is satisfied.

Thus of scheme (5) it remains to prove only the implications *(F)* $\Rightarrow$ (h_+) $\to$ h and *(F)* $\to$ *(h)*.

The implication *(F)* $\Rightarrow$ (h_+) has to be proved assuming that the additional Conditions *(iv)* and *(v')* (or (v'')) are satisfied. By Conditions (*F*) and *(iv)* system (1), (2) is completely controllable and positive (see Property 2° of Theorem 1, § 8). By virtue of Property 3° of Theorem 1, § 8 the integral η can be written in the form (12, § 8), i.e. (using the notation $U = J - N$) in the form (4). From the hypothesis of minimal stability it follows that the matrix U must be positive semi-definite. Indeed, according to the property of minimal stability (see Definition 2, § 13), for every initial condition $x(0) = x_0$, there exists a pair of continuous functions x and μ which satisfy Eq. (1), (for $x(0) = x_0$), the condition $\lim\limits_{t \to \infty} x(t) = 0$

and the inequality $\eta(0, t_1) \leqslant 0$ for all $t_1 \geqslant 0$. Since the integral $\eta(0, t_1)$ has the form (4), from the last inequality one obtains the inequality $x_0^* U x_0 \geqslant x^*(t_1)\, U x(t_1)$. Since this inequality holds for all $t_1 \geqslant 0$ and we have $\lim_{t \to \infty} x(t) = 0$ while x_0 is arbitrary one obtains the inequality $x^* U x \geqslant 0$ for every x, which shows that the matrix U is indeed positive semidefinite [1]).

We now prove that if, in addition, one of the Conditions (v') or (v'') is satisfied, then one can choose U even strictly positive definite. We assume first that $\pi(-\sigma, \sigma) \not\equiv 0$ (see Condition (v')). Then, as shown in Section 6, § 8, the system under investigation can be written in one of the forms (51, § 8) or (52, § 8)—(54, § 8). Furthermore, the polynomial ψ which occurs in Condition (8, § 8) (and whose choice determined both the integral (12, § 8) and the respective canonical systems), may be chosen so as to have no root with the real part strictly negative. Then, by Relations (49, § 8) and (55, § 8), the matrices A_c and A_{11} (which occur in Relations (51, § 8) and (52, § 8)) have no characteristic value with the real part strictly negative. Under these conditions the matrix U is strictly positive definite. Indeed, otherwise — since U is positive semidefinite (see footnote 1) — there exists a vector $x_0 \neq 0$ such that $x_0^* U x_0 = 0$. We consider the control function μ and the corresponding solution x, which occur in the property of minimal stability for the initial condition $x(0) = x_0$. We have then $\lim_{t \to \infty} x(t) = 0$ and $\eta(0, t_1) \leqslant 0$ for every $t_1 \geqslant 0$. Hence, in the case of the system (51, § 8) the function μ_c must be identically zero [2]) while in the case of the system (52, § 8) — (54, § 8) the function $\xi_1(t)$ must be identically zero. Hence the

---

[1]) Taking into account the hypotheses which have been used in this proof we find that we have proved in fact the following statement: "for every minimally stable system of the form (1), (4) the matrix U is positive semidefinite".

[2]) This simple conclusion is obtained as follows: Using condition $\overset{*}{x_0} U x_0 = 0$ we see that the inequality $\eta(0, t_1) \leqslant 0$ reduces to

$$x_c^*(t_1)\, U x_c(t_1) + \int_0^{t_1} |\mu_c(t)|^2\, dt \leqslant 0, \quad \text{for all } t_1 \geqslant 0.$$

Since U is positive semidefinite (see footnote 1) both terms of the left side of the inequality are nonnegative. Hence, the inequality can be satisfied only if both terms are identically zero. Thus we obtain the condition

$$\int_0^{t_1} |\mu_c(t)|^2\, dt = 0 \text{ for all } t_1 \geqslant 0.$$

Since the function μ_c is continuous it follows that $\mu_c(t) \equiv 0$. A similar reasoning applies to the system (52, § 8)—(54, § 8).

function x (which has the properties $\lim_{t\to\infty} x(t) = 0$ and $x(0) \neq 0$) must satisfy in the case of system (51, § 8) the equation

$$\frac{\mathrm{d}x}{\mathrm{d}t} = A_c\, x\,. \tag{15}$$

Similarly, in the case of system (52, § 8)—(54, § 8) we see that the function x, partitioned according to Eqs. (52, § 8) and (53, § 8) in the form $x' = (y'_0\ \xi_1\ \xi_2 \ldots \xi_{n-q})$, must satisfy the equations

$$\frac{\mathrm{d}y_c}{\mathrm{d}t} = \mathrm{A}_{11}\, y_c\,, \qquad \xi_j = 0,\ j = 1, 2, \ldots, n - q\,. \tag{16}$$

As we have seen, however, the matrices A_0 and A_{11} have no characteristic value with a strictly negative real part and hence both Eqs. (15) and (16) do not have any solution with the properties $\lim_{t\to\infty} x(t) = 0$ and $x(0) \neq 0$. Thus we obtain a contradiction which proves that under the hypotheses stated above the matrix U is positive definite. This concludes the proof of implication *(F)* $\Rightarrow$ *(h₊)* under hypothesis (v').

The implication $(h_+) \rightarrow$ *(h)* is almost obvious: since the matrix U is positive definite it follows that there exist two strictly positive constants $\varkappa_1$ and $\varkappa_2$ such that

$$\varkappa_1^2\, \|y\|^2 \leqslant y^*\, Uy \leqslant \varkappa_2^2\, \|y\|^2\,.$$

for every n-vector y. Consequently, from (4) one obtains an inequality of the form (25, § 13), where $\widetilde{\alpha}(y) = \varkappa_1\|y\|$, $\widetilde{\beta}(y) = \varkappa_2\,\|y\|$ and $\widetilde{\gamma}(y) = 0$. Using the final part of Proposition 4, § 13 one obtains Property (h).

We proved the equivalence of the properties stated in Theorem 1 under the additional Assumptions (iv) and (v'). If the reader feels happy with these assumptions, he may add them to the general Assumptions (i)—(iii) of Theorem 1 — and in this form the theorem is proved.

We prove now the implication $(F) \Rightarrow (h_+)$ under the Assumption (v''). Obviously, we may assume that Condition (v') is not satisfied (since otherwise the implication is true, as it has just been proved). Therefore we assume that $\pi(-\sigma, \sigma) \equiv 0$, i.e., that $\pi(\lambda, \sigma)$ vanishes for all the values of λ and σ which satisfy the condition $\lambda + \sigma = 0$. Then $\pi(\lambda, \sigma)$ is divisible by $(\lambda + \sigma)$. This together with Conditions (iii) and (v'') show that $\pi(\lambda, \sigma)$ satisfies the conditions of Proposition 2, § 8. As before, from Conditions (iv) and (F) it follows that the system under investigation is positive. Hence one can apply Proposition 2, § 8 and deduce that the system (1)—(2) can be written in the form (72, § 8), (73, § 8) having the properties specified in the proposition quoted. Furthermore, from condition (*i*) it follows that U is positive semidefinite (see footnote 1). We shall show that U is even positive definite. Assume there exists a vector $x_0 \neq 0$ such that we have $x_0^* U x_0 = 0$, and consider the continuous functions μ and x which occur in the property of minimal stability for the initial condition $x(0) = x_0$. For these functions one has $\lim_{t \to \infty} x(t) = 0$, $x(0) = x_0 \neq 0$ and $\eta(0, t_1) \leqslant 0$ for every $t_1 \geqslant 0$. From the last condition and from (73, § 8) we deduce (since $x_0^* U x_0 = 0$) the inequality $x^*(t)\, U x(t) \leqslant 0$ for all $t \geqslant 0$. Since U is positive semidefinite this equality is possible only if $U x(t) = 0$ [1]). Hence, by Proposition 2, § 8 one obtains $x(t) \equiv 0$ which contradicts the condition $x_0 \neq 0$. Thus the implication $(F) \Rightarrow (h_+)$ is also proved under the hypothesis (v''). Since the implication $(h_+) \rightarrow (h)$ has been proved above under general conditions, we have established here the equivalence of all the properties stated in Theorem 1 under the assumption that Conditions (*iv*) and (v'') are satisfied.

Thus, it remains to prove the implication $(F) \rightarrow (h)$ of scheme (5) (without assuming that Conditions (*iv*) and (v') or (v'')) are satisfied). The proof will be given in two steps, assuming first that Hypothesis (*iv*) is satisfied and then renouncing to this assumption.

Assuming that Condition (*iv*) is satisfied, we have to examine only the case in which neither (v') nor (v'') are satisfied (since otherwise the implication is true — as the foregoing reasonings show). We assume therefore that Conditions (*i*), (*ii*), (*iii*) and (*iv*) as well as the Property F of Theorem 1 are satisfied and that both Conditions (v') and (v'') are not satisfied. Then we easily see that $\pi(\lambda, \sigma)$ can be written in the form (63, § 8) and that the conditions of Proposition 1, § 8 are satisfied; hence system (1)—(2) can be brought into the form (64, § 8)—

[1]) In general for any positive semidefinite hermitian matrix U the inequality $x^* U x \leqslant 0$ entails the inequality $x^* U x = 0$. One sees easily that this leads to $Ux = 0$. Indeed, we have $(x^* + \varepsilon^* y^*) U (x + \varepsilon y) \geqslant 0$ for any scalar ε and any vector y. Since $x^* U x = 0$, the inequality can be also written as $\varepsilon^* y^* U x + \varepsilon x^* U y + \varepsilon^* \varepsilon y^* U y > 0$. We may take $\varepsilon = -\varepsilon_0 y^* U x$, where ε_0 is an arbitrary positive number. Then the last inequality becomes $-\varepsilon_0 |y^* U x|^2 (2 - \varepsilon_0 y^* U y) \geqslant 0$. One can choose ε_0 such that $2 - \varepsilon_0 y^* U y > 0$ and then the inequality implies $y^* U x = 0$. Since this is true for any vector y we obtain $Ux = 0$, as claimed.

(66, § 8), that is after rearranging the equations):

$$\frac{\mathrm{d}y}{\mathrm{d}t} = A_{11}y + b_1\mu(t),\ (A_{11},\ b_1)\text{: a completely controllable pair} \tag{17}$$

$$\eta\,(0, t_1) = [y^*\,U\,y]_0^{t_1} \tag{18}$$

$$\frac{\mathrm{d}z}{\mathrm{d}t} = \mathrm{d}c^*\,y + A_{22}\,z. \tag{19}$$

It can be easily seen that Relations (17) and (18) define a system for which Condition (v'') is satisfied. Therefore, making use of the previous proofs, we find that the matrix U of (18) is positive definite. We shall now show that from the property of minimal stability (see (i)) it follows that in (19) A_{22} is a Hurwitz matrix. Consider the initial condition $y(0) = 0$ and $z(0) = z_0$. From the property of minimal stability it follows that there exist three continuous functions μ, y and z satisfying the conditions $y(0) = 0$, $z(0) = z_0$, $\lim\limits_{t\to\infty} y(t) = 0$, $\lim\limits_{t\to\infty} z(t) = 0$ and $\eta(0, t_1) \leqslant 0$ for all $t_1 \geqslant 0$. From the last inequality, and from $y(0)=0$ one obtains, using (18), the inequality $y^*(t)\,Uy(t) \leqslant 0$ or – since the matrix U is positive definite – $y(t) \equiv 0$. Thus, from Eq. (19) it follows that z satisfies the equation $\mathrm{d}z/\mathrm{d}t = A_{22}z$. Since $\lim\limits_{t\to\infty} z(t) = 0$ and since the initial condition $z(0)$ is arbitrary, we conclude that A_{22} is a Hurwitz matrix.

We now show that system (17)—(19) has Properties h_s and h_p. Property h_p results immediately from Relation (18) and by the fact that U is positive definite. To prove Property h_s, it will be sufficient to prove the property in the particular case $\beta_2 = 0$. (see Proposition 3, § 13). Therefore we consider a solution of the system which satisfies the inequality (23, § 13). Since U is positive definite, we see immediately that there exists a constant $\delta_1 \geqslant 0$ such that $\|y(t)\|^2 \leqslant \delta_1^2(\|y(0)\|^2 + \beta_0^2)$ whence we obtain

$$\|\,y(t)\,\| \leqslant \delta_1\,(\beta_0 + \|\,y(0)\|) \text{ for } 0 \leqslant t \leqslant T. \tag{20}$$

Since this inequality is of the form (24, § 13) we have only to establish a similar relation for $z(t)$. This is obtained by using Relation (19) which gives

$$z(t) = \mathrm{e}^{A_{22}t}z\,(0) + \int_0^t \mathrm{e}^{A_{22}(t-\tau)}\,\mathrm{d}c^*\,y\,(\tau)\,\mathrm{d}\tau. \tag{21}$$

Since A_{22} is a Hurwitz matrix, there exist two constants $\alpha_1 > 0$ and $\alpha_2 > 0$ such that

$$\| e^{A_{22}t} \| \leqslant \alpha_1 e^{-\alpha_2 t}.$$

Therefore, we obtain from (21) the inequality

$$\| z(t) \| \leqslant \alpha_1 \| z(0) \| + \alpha_1 \| \mathrm{d} \| \cdot \| c \| \int_0^t e^{-\alpha_2 (t-\tau)} \| y(\tau) \| \, d\tau$$

and hence

$$\| z(t) \| \leqslant \alpha_1 \| z(0) \| + \alpha_3 \sup_{0 \leqslant \tau \leqslant t} \| y(\tau) \|,$$

where $\alpha_3 = \alpha_1 \| \mathrm{d} \| \, \| \mathrm{c} \| \int_0^\infty e^{-\alpha_2 \tau} d\tau$. Using also inequality (20) one obtains

$$\| z(t) \| \leqslant \alpha_1 \| z(0) \| + \alpha_3 \, \delta_1 (\beta_0 + \| y(0) \|) \text{ for } 0 \leqslant t \leqslant T. \tag{22}$$

The inequalities (20) and (22) can be easily expressed, in the form (24, § 13). Define the norm $\| x \|$ of the vector $x = \begin{pmatrix} y \\ z \end{pmatrix}$ as $\| x \| = \sqrt{\| y \|^2 + \| z \|^2}$. From the inequality $\| x \| \leqslant \| y \| + \| z \|$ from (20), (22), $\| y \| \leqslant \| x \|$ and $\| z \| \leqslant \| x \|$, one obtains

$$\| x(t) \| \leqslant \delta_1 (1 + \alpha_3) \, \beta_0 + (\delta_1 + \alpha_1 + \alpha_3 \, \delta_1) \| x(0) \|.$$

This inequality implies the Inequality (24, § 13) for $\delta_0 = \delta_1 + \alpha_1 + \alpha_3 \delta_1$. We proved the implication $(F) \rightarrow (h)$ (and hence Theorem 1) also in the case of systems satisfying the Conditions (*i*), (*ii*), (*iii*) and (*iv*).

We now drop Assumption (iv). We assume therefore that the pair (A, b) is not completely controllable. Since Condition (*ii*) : $b \neq 0$ is satisfied, we can apply Proposition 1, § 31 and write Eq. (1) in the form

$$\frac{d\tilde{y}}{dt} = \tilde{A}_{11} \tilde{y} + A_{12} w + b_1 \tilde{\mu}(t), \; (\tilde{A}_{11}, b_1): \text{a completely controllable pair} \tag{23}$$

$$\frac{dw}{dt} = A_{22} \, w. \tag{24}$$

Using a transformation of the type studied in §§ 2—4, one can write Eqs. (23)—(24) in a simpler form. Since the pair $(\tilde{A}_{11}, b_1)$ is completely controllable one can find a vector q such that the matrix $\tilde{A}_{11} + b_1 q^*$ has arbitrarily chosen characteristic values (see Property 9° of Theorem 1, § 31). Hence these values may be chosen different from the characteristic values of the matrix A_{22}. By effecting the transformation $\mu = \tilde{\mu} - q^* y$, $\tilde{y} = y - Kw$ (where the matrix K will be determined below) and putting $A_{11} = A_{11} + b_1 q^*$, Eq. (24) remains unaltered and instead of Eq. (23) one obtains

$$\frac{dy}{dt} = A_{11} y + (A_{11} K - K A_{22} + A_{12}) w + b_1 \mu .$$

Since the matrices A_{11} and A_{22} have no common characteristic value, the matrix K can be chosen such that the coefficient of the term in w of the above equation vanishes (see F. R. Gantmacher [1] Chap. VIII, § 3). We also remark that the pair (A_{11}, b_1) is completely controllable (see Theorem 2, § 31).

From the foregoing it follows that, without loss of generality, we can assume that Eq. (1) can be partitioned into the following two equations :

$$\frac{dy}{dt} = A_{11} y + b_1 \mu, \ (A_{11}, b_1) : \text{a completely controllable pair} \quad (25)$$

$$\frac{dw}{dt} = A_{22} w. \quad (26)$$

If one partitions similarly Relation (2) according to the formulas

$$x = \begin{pmatrix} y \\ w \end{pmatrix}, \ l = \begin{pmatrix} l_1 \\ l_2 \end{pmatrix}, \ M = \begin{pmatrix} M_{11} & M_{12} \\ M_{21} & M_{22} \end{pmatrix}, \ J = \begin{pmatrix} J_{11} & J_{12} \\ J_{21} & J_{22} \end{pmatrix},$$

one can write the integral $\eta(0, t_1)$ as

$$\eta(0, t_1) = \eta_c(0, t_1) + \eta_r(0, t_1) \quad (27)$$

where

$$\eta_c(0, t_1) = [y^* J_{11} y]_0^{t_1} + \int_0^{t_1} (\varkappa |\mu|^2 + \mu^* l_1^* y + y^* l_1 \mu + y^* M_{11} y) \, dt \quad (28)$$

$$\eta_r(0, t_1) = [y^* J_{12} w + w^* J_{21} y + w^* J_{22} w]_0^{t_1} + \int_0^{t_1} (\mu^* l_2^* w + w^* l_2 \mu + $$

$$+ y^* M_{12} w + w^* M_{21} y + w^* M_{22} w) \, dt. \quad (29)$$

We easily see that the characteristic function χ of the system (25)—(29) (see (13, § 3)) is identical to the characteristic function of the "completely controllable part" of the system, i.e. to the characteristic function of the system consisting of Eqs. (25) and (28). Hence, if Condition (F) is satisfied for the whole system, then this condition is satisfied also for the completely controllable part of the system. A similar statement is clearly valid for Conditions (ii) and (iii). We prove that the same is true for Condition (i), i.e. if system (25)—(29) is minimally stable then so is also its completely controllable part. Consider an initial condition $y(0) = y_0$ (where y_0 is arbitrary) and $w(0) = 0$. Since the system is minimally stable, there exist three continuous functions μ, y and w which satisfy : a) equations (25) and (26) for the initial conditions $y(0) = y_0$, $w(0) = 0$; b) the conditions $\lim_{t\to\infty} y(t) = 0$ and $\lim_{t\to\infty} w(t) = 0$ and c) the inequality $\eta(0, t_1) \leqslant 0$ for all $t_1 \geqslant 0$. In order to prove that the completely controllable part is also minimally stable it is sufficient to show that for the functions considered above the inequality $\eta_c(0, t_1) \leqslant 0$ is also satisfied for all $t_1 \geqslant 0$. But this inequality results immediately since by (26) and the initial condition $w(0) = 0$ we have $w(t) \equiv 0$, whence, using (29), we deduce $\eta_r(0, t_1) \equiv 0$ and hence (see (27))

$$\eta(0,)t_1 \equiv \eta_c(0, t_1).$$

Thus we have obtained the following result : if the initial system satisfies Conditions (i)—(iii) and (F), then its completely controllable part satisfies the same conditions. Since for completely controllable systems Theorem 1 has been proved above we conclude that Properties h_s and h_p are satisfied for the system consisting of Eqs. (25) and (28).

In order to extend the Properties h_s and h_p from the completely controllable part of the system to the whole system, we shall establish first an inequality which is satisfied by the integral (29). This integral can be written in a simpler form as follows : multiplying Eq. (25) on the left by the vector $b_1^*/b_1^*b_1$ (this is possible since the vector b_1 is different from zero) one obtains a relation of the form

$$\mu = q_1^* \frac{dy}{dt} + q_2^* y,$$

where the exact values of the vectors q_1 and q_2 are not needed for the evaluations which follow. We obtain further the equalities

$$\int_0^{t_1} w^* l_2 \mu \, dt = \int_0^{t_1} \left(w^* l_2 q_1^* \frac{dy}{dt} + w^* l_2 q_2^* y\right) dt =$$

$$= [w^* l_2 q_1^* y]_0^{t_1} + \int_0^{t_1} (-w^* A_{22}^* l_2 q_1^* + w^* l_2 q_2^*) y \, dt.$$

(we have integrated by parts the term containing dy/dt, using also Eq. (26)). Obviously a similar expression is also obtained for the integral $\int \mu^* l_2^* w \, dt$. Introducing the expressions obtained in (29) one can write the integral in the form

$$\eta_r(0, t_1) = [y^* \widetilde{J}_{12} w + w^* \widetilde{J}_{21} y + w^* J_{22} w]_0^{t_1} + \int_0^{t_1} (y^* \widetilde{M}_{12} w +$$

$$+ w^* \widetilde{M}_{21} y + w^* M_{22} w) \, dt, \tag{30}$$

where the exact values of the matrices $\widetilde{J}_{12}$, $\widetilde{J}_{21}$, $\widetilde{M}_{12}$ and $\widetilde{M}_{21}$ are not needed in the sequel. We remark that from the property of minimal stability it follows that for any initial condition $w(0)$, there exists a solution of Eq. (26) with the property $\lim_{t \to \infty} w(t) = 0$. In other words A_{22} is a Hurwitz matrix. Hence there exist two constants $\alpha_1 > 0$ and $\alpha_2 > 0$ such that $\| w(t) \| \leqslant \alpha_1 e^{-\alpha_2 t} \| w(0) \|$. Therefore we easily deduce from (30) that there exist two constants $\gamma_1 \geqslant 0$ and $\gamma_2 \geqslant 0$ such that

$$-\eta_r(0, t_1) \leqslant \gamma_1 \| w(0) \|^2 + \gamma_2 \| w(0) \| \sup_{0 \leqslant t \leqslant t_1} \| y(t) \|. \tag{31}$$

Using also the inequalities $\| w \| \leqslant \| x \|$ and $\| y \| \leqslant \| x \|$ (where the norm of the vector $x = \begin{pmatrix} y \\ w \end{pmatrix}$ is defined as $\| x \| = \sqrt{\| y \|^2 + \| w \|^2}$) one obtains

$$-\eta_r(0, t_1) \leqslant \gamma_1 \| x(0) \|^2 + \gamma_2 \| x(0) \| \sup_{0 \leqslant t \leqslant t_1} \| x(t) \|. \tag{32}$$

From the above inequality one obtains Property h_p as follows: Since Property h_p is satisfied for the completely controllable part of the system, there exist two constants $\gamma_3 \geqslant 0$ and $\gamma_4 \geqslant 0$ such that

$$\eta_c(0, t_1) \geqslant -\gamma_3 \| y(0) \|^2 - \gamma_4 \| y(0) \| \sup_{0 \leqslant t \leqslant t_1} \| y(t) \| \geqslant$$

$$\geqslant -\gamma_3 \| x(0) \|^2 - \gamma_4 \| x(0) \| \sup_{0 \leqslant t \leqslant t_1} \| x(t) \|.$$

Taking into account (27) and (32) we immediately find that the integral $\eta(0, t_1)$ satisfies an inequality of the form (32) (where η_r is replaced by η, γ_1 is replaced by $\gamma_1 + \gamma_3$ and γ_2 by $\gamma_2 + \gamma_4$). Hence system (25)–(29) has Property h_p.

We prove now that the system has also Property h_s; We use Proposition 2, § 13 and assume that $\eta(0, t_1)$ satisfies (23, § 13). Adding the inequalities (23, § 13) and (31) and using (27) one obtains

$$\eta_c(0, t_1) \leqslant \beta_1^2 + \beta_2 \sup_{0 \leqslant t \leqslant t_1} \| y(t) \| \tag{33}$$

where

$$\beta_1^2 = \beta_0^2 + \gamma_1 \| w(0) \|^2 \leqslant (\beta_0 + \sqrt{\gamma_1} \| w(0) \|)^2$$

$$\beta_2 = \gamma_2 \| w(0) \|.$$

Since, as we have seen, system (25), (28) has Property h_s we deduce from (33) and Definition 3, § 13 (where x is replaced by y) the inequality

$$\| y(t) \| \leqslant \delta(\beta_1 + \beta_2 + \| y(0) \|) \leqslant \delta(\beta_0 + (\sqrt{\gamma_1} + \gamma_2) \| w(0) \| + \| y(0) \| \leqslant \tag{34}$$

$$\leqslant \delta(1 + \sqrt{\gamma_1} + \gamma_2)(\beta_0 + \| x(0) \|).$$

Thus we have obtained for $y(t)$ an inequality of the form (4, § 13) (required in Proposition 3, § 13). Concerning $w(t)$, from (26) and the fact that A_{22} is a Hurwitz matrix, one obtains

$$\| w(t) \| \leqslant \alpha_1 \| w(0) \|. \quad (\alpha_1 > 0), \tag{35}$$

From (34) and (35) and the inequality $\| x \| \leqslant \| w \| + \| y \|$ we deduce for $\| x(t) \|$ an inequality of the form (24, § 13) where $\delta_0 = \delta(1 + \sqrt{\gamma_1} + \gamma_2) + \alpha_1$. Therefore system (25)–(29) has also Property h_s. This completes the proof of the implication $(F) \to (h)$ and concludes the proof of Theorem 1.

We remark that from Theorem 1 it also follows that the property of hyperstability is equivalent to the property:

"system (1)—(2) belongs to class $\mathscr{H}$ (see Definition 1, § 13)" (this has been already stated in Section 3, § 13). Indeed, the property from Definition 1, § 13 has the same content as the Property (h_{so}) of Theorem 1.

We conclude this paragraph with two additional remarks.

Remark I: There exist hyperstable systems for which no inequality of the form (25, § 13) is satisfied (This fact was already mentioned at the end of Section 6, § 13).

To show this, consider the system

$$\frac{d\xi}{dt} = \mu \tag{36}$$

$$\frac{d\rho}{dt} = -\alpha\rho + \xi + \mu \quad (\alpha > 0) \tag{37}$$

$$\eta(0, t_1) = [\xi^2]_0^{t_1} = 2\int_0^{t_1} \xi\mu\, dt. \tag{38}$$

This system is minimally stable. Indeed consider an arbitrary initial condition $\xi(0) = \xi_0$, $\rho(0) = \rho_0$ and the control function $\mu(t) = e^{-t}\xi_0$. Then Eqs. (36) and (37) are satisfied by the functions

$$\xi(t) = e^{-t}\,\xi_0$$

$$\rho(t) = e^{-\alpha t}\,\rho_0$$

and the integral (38) can be written as

$$\eta(0, t_1) = \xi^2(t_1) - \xi^2(0) = -\xi_0^2(1 - e^{-2t_1}),$$

whence we have the properties $\lim\limits_{t\to\infty} \xi(t) = 0$, $\lim\limits_{t\to\infty} \rho(t) = 0$ and $\eta(0, t_1) \leqslant 0$ for all $t_1 \geqslant 0$, which are of the form required in the definition of the property of minimal stability (see Definition 2, § 13). Hence, Condition (*i*) of Theorem 1 is satisfied. Clearly Condition (*ii*) is also satisfied. Using Relation (13, § 3) we find that the characteristic function of the system (36)—(38) is

$$\chi(\lambda, \sigma) = \frac{(\lambda + \sigma)}{\lambda\sigma}$$

and hence Condition (*iii*) is also satisfied, Condition (*F*) is clearly satisfied since $\chi(-\sigma, \sigma) \equiv 0$. By Theorem 1 (36)—(38) is hyperstable. Assume now that an inequality of the form (25, § 13) is satisfied for this system. Then (taking $t_0 = 0$) we see that for all the solutions for which $x(0) = 0$ and $\eta(0, t_1) = 0$ (for a given $t_1 > 0$) we must also have $x(t_1) = 0$. In particular, this must happen for the following solution of the system (36)—(38) : we consider the control function

$$\mu(t) = \begin{cases} e^{-} & \text{for } 0 \leqslant t \leqslant 1 \\ -e^{-(t-1)} & \text{for } 1 < t \leqslant 2 \end{cases} \tag{39}$$

to which corresponds the following solution of Eqs. (36) and (37) (for vanishing initial conditions)

$$\xi(t) = \begin{cases} 1 - e^{-t} & \text{for } 0 \leqslant t \leqslant 1 \\ e^{-(t-1)} - e^{-1} & \text{for } 1 < t \leqslant 2 \end{cases} \tag{40}$$

$$\rho(t) = \begin{cases} 1 - e^{-\alpha t} & \text{for } 0 \leqslant t \leqslant 1 \\ (1 + e^{-1} - e^{-\alpha})\, e^{-\alpha(t-1)} - e^{-1} & \text{for } 1 < t \leqslant 2. \end{cases} \tag{41}$$

From (40) we have $\xi(0) = 0$ and $\xi(2) = 0$, whence $\eta(0, 2) = 0$ (see (38)). By the foregoing remarks we must also have $\rho(2) = 0$ (for any $\alpha > 0$). But from (41) one sees that there exist positive numbers α for which $\rho(2) \neq 0$. The contradiction shows that (25, § 13) is a sufficient condition of hyperstability but not a necessary one (see Proposition 4, § 13). If, however, the Conditions (*iv*) and (*v*′) (or (*v*″) are also satisfied, then, by Theorem 1 we have the Property (h_+), from which one can easily obtain an inequality of the form (25, § 13).

Remark II. The assumption of minimal stability can be eliminated, but then the necessary and sufficient conditions of hyperstability become more complicated. To show this, we shall examine the system

$$\frac{dx}{dt} = Ax + b\mu, \qquad \eta(t_0, t_1) = \int_{t_0}^{t_1} z^* D z \, dt, \quad t_1 \geqslant t_0, \tag{42}$$

where the lower integration limit, t_0, is left arbitrary for simplifying subsequent reasonings. Now the condition of minimal stability is removed, but we make the following assumptions :

"the pair (A, b) is completely controllable" (43)

"there exist n distinct numbers $\sigma_j, j=1, 2,\ldots, n$, such that

$$\mathrm{Re}\ \sigma_j > 0,\ \det(\sigma_j I - A) \neq 0,\ \pi(-\sigma_j, \sigma_j) = 0\text{''}, \tag{44}$$

where π is the characteristic polynomial of system (1) (see § 3).

For simplicity we shall confine ourselves to the sufficient condition of hyperstability, stated in Proposition 4, § 13.

Theorem 2. *If Conditions (43) and (44) are satisfied, the following properties are equivalent:*

1° *System (42) satisfies the sufficient condition of hyperstability stated in Proposition 4, § 13.*

2° *The integral η can be written as*

$$\eta(t_0, t_1) = [x^*(J-N)x]_{t_0}^{t_1} + \int_{t_0}^{t_1} |\gamma\mu + w^* x|^2 \mathrm{d}t, \tag{45}$$

where the matrix $J - N$ is positive definite (see (12, § 8)).

3° *There exists a constant $\delta \geqslant 0$ such that the inequality*

$$\| x(t_1) \| \leqslant \delta \left(\sqrt{\eta(t_0, t_1) + |\eta(t_0,t_1)|} + \|x(t_0)\|\right), \tag{46}$$

is satisfied for every pair of real numbers t_0 and $t_1 \geqslant t_0$ and every pair of functions μ and x which satisfy Relations (42).

4° *The characteristic function of the system (1) (see § 3) satisfies the condition*

"*the matrix having the (j, k)-th entry* $\alpha_{jk} = \dfrac{\chi(\bar{\sigma}_j, \sigma_k)}{\bar{\sigma}_j + \sigma_k}$ (47)

$(j, k = 1, 2,\ldots, n)$ is positive definite,"
for every set of n distinct numbers $\sigma_1, \sigma_2, \ldots, \sigma_n$ satisfying the conditions $Re\ \sigma_j > 0$ and $\det(\sigma_j I - A) \neq 0$[1]).

5° *The characteristic function of system (42) satisfies the condition*

"*$\chi(-iw, iw) \geqslant 0$ for every real ω with the property*

$$\det(i\omega I - A) \neq 0\text{''}. \tag{48}$$

as well as Condition (47) where the numbers σ_j are the same as in (44).

We shall prove Theorem 2 by proving the implications

$$2° \to 3° \to 4° \to 5° \to 2°, \tag{49}$$

$$2° \to 1° \to 4°. \tag{50}$$

---

[1]) From (14, § 3) it follows that the matrix $\{\alpha_{jk}\}$ is hermitian ($\alpha_{jk} = \bar{\alpha}_{kj}$).

2° → *3°*. From 2° it follows that there exist $\beta_1 > 0$ and $\beta_2 > 0$ such that

$$\beta_1 \|x(t_1)\|^2 \leqslant \eta(t_0, t_1) + \beta_2 \|x(t_0)\|^2.$$

We also consider the obvious inequality $0 \leqslant |\eta(t_0, t_1)|$. Adding the two foregoing inequalities, then extracting the square root of both sides and conveniently majorizing the right-hand member one can write the inequality thus obtained in the form (46), where the number δ can be taken equal to $\max(1, \sqrt{\beta_2/\beta_1})$.

3° → *4°*. If Property 4° is not true, then there exists a set of n distinct numbers σ_j, $j = 1, 2, \ldots, n$ satisfying the conditions

$$\operatorname{Re} \sigma_j > 0, \quad \det(\sigma_j I - A) \neq 0 \tag{51}$$

and for which Condition (47) is not satisfied. In other words, there exists a set of numbers δ_j, $j = 1, 2, \ldots, n$, not all equal to zero, such that the number $\rho(\sigma_j)$ given by the relation

$$\rho(\sigma_j) = \sum_{j=1}^{n} \sum_{k=1}^{n} \bar{\delta}_j \delta_k \frac{\chi(\bar{\sigma}_j, \sigma_k)}{\bar{\sigma}_j + \sigma_k}, \quad (\max |\delta_j| \neq 0), \tag{52}$$

is negative. If we take

$$\mu(t) = \sum_{j=1}^{n} \delta_j e^{\sigma t}, \tag{53}$$

then the differential Eq. of (42) has the particular solution

$$x(t) = \sum_{j=1}^{n} \delta_j (\sigma_j I - A)^{-1} b\, e^{\sigma_j t}. \tag{54}$$

Using Relation (13, § 3), the expression (4, § 2) of z and the form (6, § 2) of the matrix D, one obtains the relation

$$z^*(t) Dz(t) = \sum_{j=1}^{n} \sum_{k=1}^{n} \bar{\delta}_j \delta_k \chi(\bar{\sigma}_j, \sigma_k) e^{(\bar{\sigma}_j + \sigma_k)t},$$

whence (see (51) and (52))

$$\lim_{t_0 \to -\infty} \eta(t_0, 0) = \rho(\sigma_j) \tag{55}$$

where $\rho(\sigma_j)$ has the expression (52) and hence is negative as we have seen. Consequently, Relation (46) of Property 3° takes the form $\|x(0)\| \leqslant \delta \lim\limits_{t_0 \to -\infty} \|x(t_0)\|$. But from (54) and (51) we have $\lim\limits_{t_0 \to -\infty} \|x(t_0)\| = 0$ hence

$$x(0) = 0. \tag{56}$$

But since the pair (A, b) is completely controllable, from Property 13° of Theorem 1, § 31, one obtains the conclusion:

"the vectors $(\sigma_j I - A)^{-1} b$, $j = 1, 2, \ldots, n$ are linearly independent". (57)

Since the numbers δ_j are not all zero, from (57) and (54) it follows that $x(0) \neq 0$. This contradicts (56).

4° → 5°. From Property 4° it follows in particular that $\chi(\bar{\sigma}, \sigma) > 0$ for any number σ with $\operatorname{Re} \sigma > 0$. This entails, by continuity, Condition (48). The rest of Property 5° is an obvious particular case of Property 4°.

5° → 2°. Comparing Condition (48) with Property 2° of Theorem 1, § 8 we deduce that system (1) is positive and hence that Property 3° of the same theorem is also satisfied. From Condition (44) and the equality $\pi(-\sigma_j, \sigma_i) = \pi(\bar{\sigma}_j, -\bar{\sigma}_i)$ (see 19, § 3) it follows that the polynomial $\sigma \mapsto \pi(-\sigma, \sigma)$ has $2n$ distinct roots : σ_j and $-\bar{\sigma}_j$, $j = 1, 2, \ldots, n$. Since the degree of this polynomial is $\leqslant 2n$ (see 20, § 3), we deduce that the factorization (8, § 8) may be chosen such that the roots of the polynomial ψ are σ_j, thus obtaining the relations

$$\psi(\sigma_j) = 0, \quad j = 1, 2, \ldots, n \tag{58}$$

for all the numbers σ_j of (44). By Property 3° of Theorem 1, § 8 the integral η can be written in the form (12, § 8) and at the same time the Relations (8, § 8)—(11, § 8) are satisfied. Obviously, in the integral (12, § 8) the lower integration limit may be replaced by an arbitrary number t_0, thus obtaining formula (45). It remains to show that the matrix $J - N$ is positive definite. Using Relations (11, § 8) and (58) we deduce that if the numbers σ_j satisfy Conditions (44), then the functions (53) and (54) [1]) satisfy the equality $\gamma\mu(t) + w^* x(t) = 0$ and hence the integral (45), written for $t_1 = 0$, becomes

$$\eta(t_0, 0) = x^*(0)\,(J - N)\,x(0) - x^*(t_0)\,(J - N)\,x(t_0). \tag{59}$$

---

[1]) In the proof which follows the numbers δ_j are arbitrary and to have to satisfy only the condition $\max |\delta_j| \neq 0$.

But, by virtue of Property 5° the expression $\rho(\sigma_j)$, (52), is positive if at least one of the numbers δ_j is different from zero. Therefore one obtains from (55)

$$\lim_{t_0 \to -\infty} \eta\ (t_0,\ 0) > 0,\ \text{if}\ \max\ (|\ \delta_j\,|) \neq 0. \tag{60}$$

Relations (59) and (60) together with the condition $\lim\limits_{t_0 \to -\infty} x(t_0) = 0$ which results from (54), lead to the inequality

$$x^*(0)\ (J - N)\ x(0) > 0, \tag{61}$$

where

$$x(0) = \sum_{j=1}^{n} \delta_j\ (\sigma_j\ I - A)^{-1}b,\ \max\ (|\ \delta_j\,|) \neq 0. \tag{62}$$

Since the vectors $(\sigma_j I - A)^{-1}$ are linearly independent (see (57)), and the numbers δ_j are arbitrary, Relations (61), (62) show that the matrix $J - N$ is positive definite.

We shall now prove the implications (50). The first of them is obvious. To prove the implication 1° → 4° we remark that if Property 4° is not satisfied, then one can introduce the formulas (51)—(55) with the same meaning as above. One obtains as before the conditions $\lim\limits_{t_0 \to -\infty} \eta(t_0,\ 0) \leqslant 0$ and $\lim\limits_{t_0 \to -\infty} x(t_0) = 0$.

On the other hand, owing to the conditions $\operatorname{Re} \sigma_j > 0$, the vector $x(t)$ is bounded for all $t \leqslant 0$. It follows that the right hand member of inequality (25, § 13), for $t = 0$, tends to zero when t_0 tends to $-\infty$; hence (for $t = 0$) the left hand member must vanish, thus entailing Condition (56) which is contradictory (as shown above immediately after Formula (56)). This concludes the proof of Theorem 2.

§ 15. Simple hyperstable blocks

In this paragraph we shall examine the blocks described by the equations

$$\frac{\mathrm{d}x}{\mathrm{d}t} = Ax + b\mu \tag{1}$$

$$\nu = a^*\ x + \alpha\mu. \tag{2}$$

The symbols from Eq. (1) have the same meaning as in § 2. In Relation (2), a is an n-vector and α, a scalar constant. Besides block (1)—(2) we shall also consider the "associated system" (see Section 9, § 13)

$$\frac{dx}{dt} = Ax + b\mu . \tag{3}$$

$$\eta(0, t_1) = \mathrm{Re}\left(\int_0^{t_1} \mu^* \nu \, dt\right) = \int_0^{t_1} \left(\mathrm{Re}\,\alpha\, |\mu|^2 + \frac{1}{2}\mu^* a^* x + \frac{1}{2} x^* a \mu\right) dt,$$

which has the form given in § 2 with the particular values $\varkappa = \mathrm{Re}\,\alpha$, $l = \frac{1}{2}\, a$, $J = M = 0$. According to the definitions given in Section 9, § 13 the block (1)—(2) is hyperstable if and only if the associated system (3) is hyperstable (Definition 8, § 13). Similarly, block (1)—(2) is said to be minimally stable if and only if the associated system (3) is minimally stable (Definition 2, § 13). Thus the study of the properties of hyperstable blocks reduces practically to the simple application of the results obtained in § 14. We have only to observe that the characteristic function of the associated system (3) can be written [1] as

$$\chi(\lambda, \sigma) = \frac{1}{2}(\bar{\gamma}(\lambda) + \gamma(\sigma)), \tag{4}$$

where

$$\gamma(\sigma) = \alpha + a^* (\sigma I - A)^{-1} b. \tag{5}$$

is the transfer function of block (1)—(2).

Furthermore, in the case of blocks of the form (1)—(2) there exist also other equivalent ways of expressing the property of hyperstability as the following theorem shows.

Theorem 1. *Assume that: (i) the block (1)—(2) is minimally stable and that the conditions (ii) $b \neq 0$ and (iii): $2\,\mathrm{Re}\,\alpha + a^*(I - A)^{-1} b \not\equiv 0$ are satisfied. Then the following properties are equivalent*

(H): "The block (1)—(2) is hyperstable"

(F): "The transfer function of the block (1)—(2) satisfies the inequality

$$\mathrm{Re}\,\gamma(i\omega) \geqslant 0, \tag{6}$$

---

[1] This results easily by using formula (13, § 3).

for every real number ω, *which satisfies the condition* $\det(i\omega I - A) \neq 0$".

(H'): "The associated system (3) satisfies one of the properties (h), (h_{so}), (h_{po}) *and (f), stated in Theorem 1, § 14".*

(h): "All the feedback systems defined by the relations

$$\frac{dx}{dt} = Ax + b\mu, \quad \nu = a^*x + \alpha\mu, \quad \frac{d\nu}{dt} = -\lambda\mu \tag{7}$$

and

$$\frac{dx}{dt} = Ax + b\mu, \quad \nu = a^* x + \alpha\mu, \quad \frac{d\mu}{dt} = -\lambda\nu \tag{8}$$

have stable trivial solutions for every real and strictly positive number λ, *except in the case of system (7), for* $\alpha = 0$, *the value* $\lambda_0 = -a^*b$[1]".

(f') "The function γ *is positive real (see Appendix C)". If besides the assumptions specified at the beginning of the theorem, one has also one of the following conditions* (α): *"the transfer function is irreducible" or* (β) *"the pair (A, b) is completely controllable and the expression* $\bar{\gamma}(-\sigma) + \gamma(\sigma)$ *is not identically zero", then each of the above properties is also equivalent to the property:*

(h'): "The integral $\eta(0, t_1)$ *can be written in the form (4, § 14) where the matrix U is positive definite".*

Before proving Theorem 1 we compare it with Theorem 1, § 14. We first remark that some of the properties stated in Theorem 1 are expressed with the help of the transfer function of the block and are simpler than the corresponding statements of Theorem 1, § 14. The equivalence of the property of hyperstability and Property (*f'*) gives a new, particularly concise characterization of hyperstability [2]). Property (*h*) is interesting because it connects the property of hyperstability with the stability of the trivial solution of a collection of linear systems with constant coefficients. We remark that the systems (7) and (8) have a structure which corresponds to the block diagram of

---

[1]) This exception is introduced because when $\alpha = \lambda + a^*b = 0$ the system (7) cannot be always written in the form in which ordinary differential equations are generally studied. As shown later, if Property (h_λ) is satisfied, then the condition $\alpha = 0$ entails the inequality $a^*b \geqslant 0$. Therefore the exceptional value $\lambda_0 = -a^*b$ is not positive and hence this case is automatically eliminated.

[2]) This equivalence also points to certain analogies between hyperstable blocks and electrical networks. (Thus, hyperstable blocks can be characterized by the fact that their transfer function has the same properties as the driving point impedance of a passive one-port).

Fig. 4.1.*c* where the input is zero, the block B_1 is described by the equations [1])

$$B_1 : \mathrm{d}x/\mathrm{d}t = Ax + b\mu_1, \quad \nu_1 = a^* x + \alpha\mu_1$$

and the block B_2 is, in the case of system (7), described by the relation

$$B_2' : \nu_2 = \frac{1}{\lambda} \frac{\mathrm{d}\mu_2}{\mathrm{d}t}, \quad \lambda > 0$$

(and performs a pure differentiation) whereas in the case of system (8), B_2 is described by the relation

$$B_2'' : \frac{\mathrm{d}\nu_2}{\mathrm{d}t} = \lambda\mu_2, \quad \lambda > 0$$

and performs a pure integration).

In these relations we have used the notations μ_1, μ_2, ν_1 and ν_2 instead of the symbols u_1, u_2, v_1 and v_2 of Fig. 4.1.*c*. Note also the relations $\mu_1 = -\nu_2$, $\mu_2 = \nu_1$.

We introduce now a class of blocks, $\mathcal{B}$, consisting of hyperstable blocks or inverses of hyperstable blocks satisfying also the condition that any feedback system of the form shown in Fig. 4.1.*c* where the block B_2 belongs to class $\mathcal{B}$, can be written in the usual form of a system of ordinary differential equations. Then, by Proposition 11, § 13, if the block B_1 is hyperstable, all the systems defined above have a stable trivial solution for every block B_2 belonging to class $\mathcal{B}$. As we shall show in the course of the proof block B_2'' is hyperstable. One can directly see that the block B_2' is the inverse of the hyperstable block B_2'' [2]). Hence, all the blocks B_2' and B_2'' belong to class $\mathcal{B}$ for $\lambda > 0$. They constitute a subclass of class $\mathcal{B}$ which will be denoted by $\mathcal{B}_\lambda$. By Theorem 1, in order that the block B_1 be hyperstable — i.e. in order that all the systems of the type shown in Fig. 4.1.*c* have a stable trivial solution for every block B_2 of class $\mathcal{B}$ — it is necessary and sufficient that this stability property should occur for all the blocks of the subclass $\mathcal{B}_\lambda$. Thus, the stability of a general family of systems of the type shown in Fig. 4.1.*c* is equivalent to the stability of a linear

---

[1]) To avoid confusion, the symbols μ and ν from (1), (2), (7) and (8) have been replaced by μ_1 and ν_1.

[2]) It follows that in order to prove the implication $(H) \rightarrow (h_\lambda)$ it is sufficient to show that block B''_2 is hyperstable and that Eqs. (7) and (8) can be written in the usual form of a system of ordinary differential equations.

subfamily of the family considered [1]). Concluding these preliminary considerations we further remark that Property (h_λ) can also be verified experimentally.

To prove Theorem 1, we shall first compare the assumptions required in Theorem 1 and in Theorem 1, § 14. One can easily see that if the conditions required at the beginning of Theorem 1 are satisfied, then Conditions (*i*), (*ii*) and (*iii*) of Theorem 1, § 14 are also satisfied. Concerning Conditions (α) and (β) we remark that $\gamma(\sigma)$ can be written in the form

$$\gamma(\sigma) = \frac{\rho(\sigma)}{\det\,(\sigma I - A)}, \tag{9}$$

where ρ is a polynomial. Hence, using (4) and (18, § 3), one obtains

$$\pi(\lambda, \sigma) = \frac{1}{2\nu}(\bar{\rho}\,(\lambda) \det\,(\sigma I - A) + \rho(\sigma) \det\,(\lambda I - A^*)). \tag{10}$$

If Condition (α) is satisfied then the pair (A, b) is completely controllable (see Theorem 1, § 33). Hence, Condition (*iv*) of Theorem 1, § 14 is satisfied. We shall now show that condition (v'') of the theorem quoted is also satisfied. Indeed, otherwise there exists a number σ_0 such that $\pi(\lambda, \sigma_0) \equiv 0$. Using Relation (10) we obtain

$$\bar{\rho}(\lambda) \det\,(\sigma_0 I - A) + \rho\,(\sigma_0) \det\,(\lambda I - A^*) \equiv 0. \tag{11}$$

If $\det\,(\sigma_0 I - A) \neq 0$, we obtain from (11) the identity $\bar{\rho}(\lambda) \equiv \varkappa_0 \det\,(\lambda I - A^*)$ (where the constant $\varkappa_0$ may be zero), whence it follows that (9) is reducible, which contradicts Condition (α). If $\det\,(\sigma_0 I - A) = 0$, then from (11) one obtains $\rho(\sigma_0) = 0$, which again means that (9) is reducible. We have thus proved that if block (1)—(2) satisfies Condition (α), then the associated

[1]) Thus we obtain a result of the type required by Aizerman's problem (see also Section 2, chap. 1). However, whereas in the initial problem of Aizerman the general answer is negative, the modified problem solved in this paragraph admits a general affirmative solution. Notice however the essential differences between the problem treated here and Aizerman's problem: the concept of absolute stability is replaced by the concept of hyperstability and the subfamily of linear systems for which the stability of the system is required, is larger. (In the corresponding problem of Aizerman the linear subfamily contains only blocks B_2 described by relations of the form $\mu = -\lambda\nu$, $\lambda > 0$, whereas in the case treated in this paragraph one also admits the blocks belonging to class $\mathcal{B}_\lambda$. The restriction to the "semi-infinite sector" ($\lambda > 0$), adopted in this paragraph, can be eliminated by effecting a "control transformation" as in Section 3, § 2.

system (3) satisfies Conditions (iv) and (v'') of Theorem 1, § 14. Finally, one can easily see, that if Condition (β) of Theorem 1 is satisfied, then Conditions (iv) and (v') of Theorem 1, § 14 are also satisfied (for Condition (v') we use again Relation (10)).

The foregoing remarks show that under the assumptions of Theorem 1 one can apply Theorem 1, § 14 to the associated system (3). By definition, the hyperstability of block (1) – (2) simply means the hyperstability of system (3) and hence, by Theorem 1, § 14, this property is equivalent to each of the properties of the theorem quoted, except Property (h_+) for which additional hypotheses are required. Using Relation (4) the equivalence of Properties (H), (F) and (H') of Theorem 1 follows immediately. Furthermore these properties are equivalent to Property (f') since (see Relation 4) Property (f) of Theorem 1, § 14 expresses the fact that the transfer function is a positive real function (see Appendix C). Taking into account the relations between Conditions (α) and (β) and Conditions (iv), (v') and (v'') established above, one also obtains the equivalence of Properties (H), (F), (H'), (f') and Property (h_+) (of course, if one of the conditions (α) or (β) is satisfied). Hence, it only remains to prove, (without using Hypotheses (α) or (β)) the equivalence of Property (h_λ) and the (equivalent) Properties (H), (F), (H') and (f'). This will be done by proving the implications

$$(H) \rightarrow (h_\lambda) \rightarrow (f'). \tag{12}$$

The proof of the implication $(H) \rightarrow (h_\lambda)$ is simple and (as mentioned in the remarks following the statement of Theorem 1) amounts to showing that the block B_2'', defined above, is hyperstable and that Eqs. (7) and (8) can be written in the usual form. We shall first show that Eqs. (7) and (8) can be written in the usual form. Replacing in the last equation of (7) the variable ν by its value given by (2), and using Eq. (1) one can write system (7) in the form

$$\frac{dx}{dt} = Ax + b\mu, \quad \alpha \frac{d\mu}{dt} + (\lambda + a^* b)\,\mu + a^* A x = 0. \tag{13}$$

Assuming $\alpha \neq 0$, system (13) becomes

$$\frac{dx}{dt} = Ax + b\mu, \quad \frac{d\mu}{dt} = -\frac{1}{\alpha} a^* A x - \frac{\lambda + a^* b}{\alpha}\,\mu. \tag{14}$$

If the conditions $\alpha = 0$ and $\lambda + a^* b \neq 0$ are satisfied, the last equation of (13) gives $\mu = -a^* A x/(\lambda + a^* b)$. Introducing this relation in the first equation, the system can be written as

$$\frac{dx}{dt} = \left(A - \frac{1}{\lambda + a^* b}\, b\, a^*\, A\right) x. \tag{15}$$

Thus, system (7) is expressed in one of the usual forms (14) and (15). (The exceptional case $\alpha = 0$ and $\lambda + a^* b = 0$ has been excluded by the conditions required in the statement of property (h_λ)). Concerning system (8), if in the last equation of (8) one replaces the variable ν by its value given by (2), one obtains the system

$$\frac{dx}{dt} = Ax + b\mu, \quad \frac{d\mu}{dt} = -\lambda\,(a^* x + \alpha\mu). \tag{16}$$

By the remarks made immediately after the statement of Theorem 1 (see also footnote 4) all we have to prove is that the block

$$B_2'' : \frac{d\nu_2}{dt} = \lambda\,\mu_2, \quad \lambda > 0$$

is hyperstable. From the equation of the block one obtains immediately the relation

$$\eta_2(0, t_1) = \mathrm{Re}\left(\int_0^{t_1} \mu_2^*\, \nu_2\, dt\right) = \frac{1}{\lambda}\,\mathrm{Re}\left(\int_0^{t_1} \frac{d\nu_2^*}{dt}\, \nu_2\, dt\right) =$$

$$= \frac{1}{\lambda}\,|\nu_2(t_1)|^2 - \frac{1}{\lambda}\,|\nu_2(0)|^2.$$

We have thus obtained a relation of the form (25, § 13) whence, using Proposition 4, § 13, we deduce that B_2'' is hyperstable. This proves the implication $(H) \to (h_\lambda)$.

We now prove $(h_\lambda) \to (f')$. If Property (f') is not satisfied, then the equivalent property (B) of Theorem 1, Appendix C is not true. Hence, there exists a number σ_0 satisfying the conditions

$$\frac{\gamma(\sigma_0)}{\sigma_0} < 0, \quad \mathrm{Re}\,\sigma_0 > 0, \quad \det(\sigma_0 I - A) \neq 0 \tag{17}$$

or the conditions

$$\gamma(\sigma_0)\,\sigma_0 < 0, \quad \mathrm{Re}\,\sigma_0 > 0, \quad \det(\sigma_0 I - A) \neq 0. \tag{18}$$

(The above relations imply the fact that in the case (17) the number $\gamma(\sigma_0)/\sigma_0$ is real and finite, while in the case (18) the number $\gamma(\sigma_0)\sigma_0$ is real and finite). If the Conditions (17) are satisfied, it is easy to see that the functions defined by

$$\mu(t) = e^{\sigma_0 t}, \quad x(t) = (\sigma_0 I - A)^{-1} b\, e^{\sigma_0 t}, \quad \nu(t) = \gamma(\sigma_0)\, e^{\sigma_0 t} \tag{19}$$

satisfy Eqs. (8) for $1/\lambda = -\gamma(\sigma_0)/\sigma_0 > 0$. From Condition $\mathrm{Re}\,\sigma_0 > 0$ it follows that $\lim_{t \to \infty} \|x(t)\| = \infty$. Hence the trivial solution of Eq. (8), for $1/\lambda = -\gamma(\sigma_0)/\sigma_0 > 0$, is not stable, contrary to Property (h_λ). In the case of Conditions (18) the reasonings are similar, but with a slight complication arising from the need to eliminate the exceptional case $\alpha = 0$, $\lambda + a^*b = 0$. Assume that the number σ_0 which appears in (18) has also the property

$$\gamma(\sigma_0)\,\sigma_0 \neq a^* b. \tag{20}$$

Then the functions (19) satisfy Eqs. (7) (or one of Eqs. (14) and (15)) for $\lambda = \gamma - (\sigma_0)\sigma_0 > 0$; furthermore, the condition $\lambda \neq -a^*b$ is also satisfied. Hence one obtains as before a contradiction of property (h_λ).

Let us now assume that the inequality (20) is not satisfied and that $\gamma(\sigma)\sigma$ reduces to a constant

$$\gamma(\sigma)\,\sigma = \varkappa_0. \tag{21}$$

In particular we have $\varkappa_0 = \gamma(\sigma_0)\sigma_0$ and hence, by Condition (18) the constant $\varkappa_0$ must be real and strictly negative. We thus obtain from (21) the equality

$$\gamma(\sigma)/\sigma = \varkappa_0/\sigma^2, \qquad (\varkappa_0 < 0).$$

If follows at once that Conditions (17) are satisfied for every positive number σ_0 (except the eigenvalues of the matrix A). But we have shown before that from (17) it follows that Property (h_λ) is not satisfied. Hence, in this case too, Property (h_λ) is contradicted. Reexamining the proof one finds that one obtains the same contradiction when inequality (21) is satisfied only for the points σ of a certain open and connected set D which includes the point σ_0 and an interval of the real positive semiaxis of the plane (σ) and which contains no point σ satisfying the condition $\det(\sigma I - A) = 0$ or the inequality $\mathrm{Re}\,\sigma \leqslant 0$.

Finally, let us assume that none of the Conditions (20) and (21) is satisfied. We can always find an open and connected set D in the plane of the complex variable σ, with the properties stated above. Then the function $\gamma(\sigma)\sigma$ is analytic in D. As we have seen, if $\gamma(\sigma)\sigma$ is constant in D, Property (h_λ) is contradicted. Hence, it is sufficient to assume that $\gamma(\sigma)\sigma$ is not constant in D. We introduce a new complex variable ψ and consider the mapping $\sigma \mapsto \psi = \gamma(\sigma)\sigma$. The image of the domain D — in the plane of the complex variable ψ — under the function $\sigma \mapsto \gamma(\sigma)\sigma$ (which is analytic and non-constant) is also an open and connected set (see S. Stoilow [1], vol. 1, Chapter 2, Section 24). Moreover Δ contains the point $\psi_0 = \gamma(\sigma_0)\sigma_0$ located on the real negative semi-axis (see (18)). It follows immediately that there exists in D at least one point σ_0' which satisfies Conditions (18) and (20). Hence, as shown above, Property (h_λ) is not satisfied. This proves the implication $(h_\lambda) \to (f')$ and completes the proof of Theorem 1.

Before concluding this paragraph we prove the following statement (mentioned in footnote 1): "If Property (h_λ) is satisfied and if $\alpha = 0$, then $a^*b \geqslant 0$". We use Corollary 1 of Appendix C. From Relation (5) and the condition $\alpha = 0$ it follows that in our case the numbers β and α, defined in the statement of the corollary quoted, are zero and hence, by the quoted corollary, one can only have one of the cases (*iii*) $\lim_{\sigma\to\infty} \gamma(\sigma)\sigma > 0$ and (*iv*) $\gamma(\sigma) = \alpha = 0$. Moreover, from the (easily established) identity

$$a^* (\sigma I - A)^{-1} b = \frac{a^* b}{\sigma} + \frac{1}{\sigma} a^* A (\sigma I - A)^{-1} b$$

one obtains $\lim_{\sigma\to\infty} \gamma(\sigma)\sigma = a^*b$. Therefore, in the case (*iii*) one has the inequality $a^*b > 0$ and in the case (*iv*), the equality $a^*b = 0$. Thus one always has $a^*b \geqslant 0$.

§ 16. Multi-input hyperstable systems

In this paragraph, which generalizes the results of § 14, we study the hyperstability of systems of the form

$$\frac{dx}{dt} = Ax + Bu \tag{1}$$

$$\eta(0, t_1) = [x^* Jx]_0^{t_1} + \int_0^{t_1} (u^* Ku + u^* L^* x + x^* Lu + x^* M x)\, dt, \tag{2}$$

where the symbols have the same meaning as in § 5. We assume that the general hypotheses mentioned in Section 2, § 13 are satisfied.

Beside the system (1)—(2) we also consider a family of single-input systems

$$\frac{dx}{dt} = Ax + b\mu \tag{3}$$

$$\eta(0, t_1) = [x^* Jx]_0^{t_1} + \int_0^{t_1} (\varkappa \mu^* \mu + \mu^* l^* x + x^* l \mu + x^* M x)\, dt \tag{4}$$

where

$$b = Bc, \quad \varkappa = c^* Kc, \quad l = Lc, \quad (\|c\| = 1) \tag{5}$$

and where c is an arbitrary n-vector with $\|c\| = 1$. System (3)—(5) is obtained from system (1)—(2) in the particular case of control functions of the form

$$u(t) = c\, \mu(t). \tag{6}$$

Using formulas (13, § 13) and (37, § 5) one finds that the characteristic function χ of system (3)—(5) is related to the characteristic function H of system (1)—(2) by

$$\varkappa(\lambda, \sigma) = c^* H(\lambda, \sigma) c. \tag{7}$$

From the definition of the property of hyperstability (Def. 8, § 13) it immediately follows that if system (1)—(2) is hyperstable, then so is also every corresponding single-input systems of the form (3)—(5). The fact that the converse is also true, is one of the results stated in the following theorem.

Theorem 1. *Assume that (i) the system (1)—(2) is minimally stable (Definition (2, § 13)), (ii) $B \neq 0$ and (iii): $\pi(\lambda, \sigma) \not\equiv 0$. Then the following properties are equivalent*

(H): "System (1)—(2) is hyperstable".

(F): "The matrix $H(-i\omega, i\omega)$ is positive semidefinite for every real number ω which satisfies the condition $\det(i\omega I - A) \neq 0$".

(h): "Properties (h_s) and (h_p) of Definitions 3, § 13 and 4, § 14 are satisfied".

(h_{so}): "Property h_s of Definition 3, § 13 is satisfied in the particular case $\beta_1 = 0$ and $\beta_2 = 0$".

(h_{po}): "Property h_p of Definition 4, § 13 is satisfied in the particular case $x(0) = 0$."

(h_c): "All single-input systems of the form (3) – (5) are hyperstable".

(f): "The matrix $H(\bar{\sigma}, \sigma)$ is positive semidefinite for every number σ which satisfies the conditions $Re\ \sigma > 0$ and $\det(\sigma I - A) \neq 0$" [1].

If in addition to (i), (ii) and (iii) one also assumes that (iv) "the pair (A, B) is completely controllable" and (v) $\pi(-\sigma, \sigma) \not\equiv 0$ then the above properties are also equivalent to the property

(h_+) "The integral $(0, t_1)$ can be written in the form (similar to (8, § 9))

$$\eta(0, t_1) = [x^* Ux]_0^{t_1} + \int_0^{t_1} \|Vu + W^* x\|^2 \, dt, \tag{8}$$

where the matrix U is positive definite."

The greatest part of the above theorem is similar to Theorem 1, § 14 and the general remarks of § 14 which follow immediately after the statement of the theorem quoted, remain valid. In Theorem 1 we have in addition the specific Property (h_c).

We prove Theorem 1 in accordance with the following scheme of implications

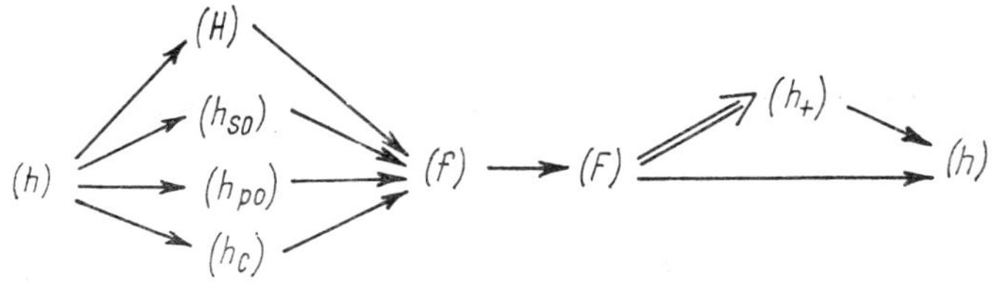

(9)

which is similar to the scheme (5, § 14).

Here, the simple arrow $\rightarrow$ means that the implication is true if Conditions (*i*), (*ii*) and (*iii*) are satisfied; the double arrow $\Rightarrow$ means that the implication is true if Conditions (*iv*) and (*v*) are also satisfied.

The implications $(h) \rightarrow (H)$, $(h) \rightarrow (H_{so})$, $(h) \rightarrow (h_{po})$ and $(h) \rightarrow (h_c)$ follow directly from the definitions of the respective properties. The implications $(H) \rightarrow (f)$, $(h_{so}) \rightarrow (f)$, $(h_{po}) \rightarrow (f)$, and $(h_c) \rightarrow (f)$ are proved by showing that if property (f) is not satisfied, then none of the properties (H), (h_{so}), (h_{po}) and (h_c) is satisfied. To prove this we first obtain some consequences

[1]) From Relation (38, § 5) it follows that $H(\bar{\sigma}, \sigma)$ is a hermitian matrix.

of the property (non-f) (the negation of Property (f): "There exists a vector c_0 and a scalar σ_0 such that one has

$$\|c_0\| = 1, \quad \text{Re}\,\sigma_0 > 0, \quad \det(\sigma_0 I - A) \neq 0 \tag{10}$$

$$c_0^* H(\bar{\sigma}_0, \sigma_0)\, c_0 < 0. \tag{11}$$

From (11) one obtains (by continuity)

$$c^* H(\bar{\sigma}_0, \sigma_0)\, c < 0, \tag{12}$$

for every vector c of the form

$c = c_0 + \varepsilon y$, (where ε is a small real scalar and y is an n-dimensional vector satisfying the condition $\|y\| = 1$) (13). Owing to the condition $B \neq 0$ (see (*ii*) one can find a vector c satisfying Conditions (12) and (13) and the inequality $B\,c \neq 0$. (Otherwise we would obtain $By = 0$ for every y, thus entailing $B = 0$). Choosing the vector c as specified above, consider the control function

$$\check{u}(t) = c e^{\sigma_0 t}. \tag{14}$$

Then Equ. (1) has the particular solution

$$\check{x}(t) = (\sigma_0 I - A)^{-1} B c e^{\sigma_0 t}. \tag{15}$$

The conditions Re $\sigma_0 > 0$ and $Bc \neq 0$ imply that

$$\lim_{t \to \infty} \|\check{x}(t)\| = \infty. \tag{16}$$

Replacing (14) and (15) in (2) and using (37, § 5) one obtains

$$\check{\eta}(0, t_1) = c^* H(\bar{\sigma}_0, \sigma_0)\, c \int_0^{t_1} e^{2\,\text{Re}\,\sigma_0 t}\, dt. \tag{17}$$

This and (12) yield

$$\check{\eta}(0, t_1) < 0, \quad \text{for any } t_1 > 0, \tag{18}$$

$$\lim_{t_1 \to \infty} \check{\eta}(t_0, t_1) = -\infty, \tag{19}$$

where $\check{\eta}(t_0, t_1)$ is obtained by using again (2) and replacing $\int_0^{t_1}$ by $\int_{t_0}^{t_1}$.

We now show that the foregoing conclusions contradict each of the properties (H), (h_{so}), (h_{po}) and (h_c). Since, under the conditions stated above, the system under investigation has a solution for which Conditions (18) and (16) are simultaneously satisfied, Property H_s of Definition 8, § 13 does not occur and hence Property (H) is not satisfied. It also follows that Property (h_{so}) is not satisfied. Noting that the functions in (14) and (15) satisfy the equations of the single-input system (3)—(5) (for the same value of c as in (14) and (15)), one also deduces that Property (h_c) is not satisfied. In order to show that Property (h_{po}) is also contradicted, remark that by Proposition 4, § 34 (Appendix A) there exists a continuous control function u_0 such that the corresponding solution x_0 of Eq. (1), for $x_0(0) = 0$, has at the point $t = 1$ the same value as solution (15), i.e.

$$x_0(1) = \check{x}(1) = (\sigma_0 I - A)^{-1} B\, c\, e^{\sigma_0}.$$

Hence Eq. (1) is satisfied by the functions

$$u(t) = \begin{cases} u_0(t) & \text{for } 0 \leqslant t \leqslant 1 \\ \check{u}(t) & \text{for } t > 1, \end{cases} \qquad x(t) = \begin{cases} x_0(t) & \text{for } 0 \leqslant t \leqslant 1 \\ \check{x}(t) & \text{for } t > 1 \end{cases}$$

and the corresponding integral can be written for $t_1 > 1$, as

$$\eta(0, t_1) = \eta_0(0, 1) + \check{\eta}(1, t_1), \quad (t_1 > 1).$$

Owing to Condition (19) we have $\lim_{t_1 \to \infty} \eta(0, t_1) = -\infty$ which obviously contradicts Property (h_{po}).

We have thus proved all the implications of scheme (9) which precede Property (f). The implication $(f) \to (F)$ results from the fact that the function $\sigma \to H(\bar{\sigma}, \sigma)$ is continuous at any point $\sigma = i\omega$ at which $\det(i\omega I - A) \neq 0$. It remains to prove the implications which follow after Property (F) of scheme (9).

To prove the implication $(F) \Rightarrow (h_+)$ we first establish some consequences of Condition (iv) and Property (F): Since Property 2° of Theorem 1, § 9 is satisfied, system (1)—(2) is positive. Therefore, by Property 3° of the same theorem η can be written in the form (8) (we have denoted $U = J - N$) [1]).

[1]) By Property 3° of Theorem 1, § 9 there exist in general several ways of writing η in the form (8), depending on the choice of the polynomial ψ in Relation (3, § 9). This remark will be used at a later stage of the proof where ψ will be chosen in a particular way.

Repeating the arguments used for proving the implication $(F) \Rightarrow (h_+)$ of § 14 (see also footnote 1 of § 14) one obtains a result which we state as a lemma:

Lemma 1. *If a system of the form (1), (8) is minimally stable, then U is positive semidefinite.*

A sufficient condition for the matrix U to be strictly positive definite, is given by the following lemma:

Lemma 2. *Assume that system (1), (8) is minimally stable and the following condition is satisfied: "For every pair of continuous functions u, x satisfying Eq. (1) and the conditions $Ux(t) \equiv 0$ and $Vu(t) + W^*x(t) \equiv 0$, one has $x(t) \equiv 0$." Then U is positive definite.*

Indeed, by Lemma 1 U is positive semidefinite. Therefore we have only to show that there exists no vector $x_0 \neq 0$ such that $Ux_0 = 0$. Assume that such a vector exists. Then, by the property of minimal stability, there exist two continuous functions u and x satisfying the conditions $x(0) = x_0 \neq 0$, $\lim\limits_{t\to\infty} x(t) = 0$ and $\eta(0, t_1) \leqslant 0$ (for all $t_1 \geqslant 0$). From the last inequality, from (8), and from $Ux_0 = 0$, one obtains the inequality

$$x^*(t_1)\, Ux(t_1) + \int_0^{t_1} \|Vu(t) + W^*x(t)\|^2\, dt \leqslant 0, \text{ for all } t_1 \geqslant 0.$$

Since U is positive semidefinite, the above inequality can occur only if both terms of the left hand side are identically zero. Hence, since the functions u and x are continuous, one obtains the identities $Vu(t) + W^*x(t) \equiv 0$ and $x^*(t)Ux(t) \equiv 0$. From the last identity and the fact that U is positive semidefinite, one obtains (as in footnote 3 of § 14) the identity $Ux(t) \equiv 0$. Furthermore $x(t) \not\equiv 0$ because $x(0) = x_0 \neq 0$. This contradicts the conditions of Lemma 2 and proves that U is positive definite.

To prove $(F) \Rightarrow (h_+)$ it will be therefore sufficient to show that if Conditions (*iv*) and (*v*) are satisfied, then the conditions of Lemma 2 are also satisfied. Since $\pi(-\sigma, \sigma) \not\equiv 0$ (see Condition (*v*)), the polynomial ψ of Relation (3, § 9) (see footnote 2), can be chosen such that it has no root in the semiplane $\operatorname{Re} \sigma < 0$. Then the conditions of Lemma 2 are satisfied. Indeed, by the results of § 9 we can bring system (1), (8) (under hypothesis (*v*)) to one of the forms (44, § 9) or (49, § 9)—(51, § 9). It will be sufficient to examine system (49, § 9)—(51, § 9) only, since the same reasonings are also valid (with certain simplifications) for the system (44, § 9) (whose form is obtained from equations (49, § 9)—(51, § 9) by assuming that all the vectors y_{kj} are absent. Assume therefore that system (1), (8) has the form

(49, § 9)—(51, § 9). Comparing formulas (8) and (51, § 9) one finds that the identity $Vu + Wx \equiv 0$ takes the form $u_0(t) \equiv 0$, $y_{1i}(t) \equiv 0$, $i = 1, 2, \ldots, s$. Introducing these relations in Eqs. (49, § 9) and (50, § 9) one sees that for every solution for which the identity $Vu + W^*x \equiv 0$ is satisfied, one has the following equations:

$$\left.\begin{aligned} \frac{\mathrm{d}y_{kj}}{\mathrm{d}t} &= y_{k+1,j}, \quad 0 < k < p_j, \quad (\text{if } p_j > 1) \\ \frac{\mathrm{d}y_{p_j j}}{\mathrm{d}t} &= u_j \end{aligned}\right\} \begin{array}{l} \text{for} \\ 0 < j \leqslant s \\ (\text{if } s > 0 \end{array} \tag{20}$$

$$\frac{\mathrm{d}w}{\mathrm{d}t} = A_{22}w, \quad (\text{if } n_2 > 0). \tag{21}$$

From Eqs. (20) and the identities $y_{1i}(t) \equiv 0$ we obtain, step by step, the identities $y_{kj}(t) \equiv 0$. Furthermore, from (52, § 9) and the fact that the polynomial ψ has been chosen such that it has no root in the semi-plane $\operatorname{Re} \sigma < 0$, it follows that the matrix A_{22} has no characteristic value with a strictly negative real part. Hence, the only function w which satisfies Eq. (21) and has the property $\lim_{t\to\infty} w(t) = 0$ is the identically zero function. From the foregoing it follows that the conditions of Lemma 2 are satisfied and therefore U is positive definite. This proves that $(F) \Rightarrow (h_+)$.

The implication $(h_+) \to (h)$ has been already proved in § 13 (see Proposition 4, § 13 and the proof of implication $(h_+) \to (h)$ of §14).

This concludes the proof of Theorem 1 under the assumption that all the conditions (i)—(v) are satisfied. It remains to prove the implication $(F) \to (h)$ desisting from the additional conditions (iv) and (v). This will be achieved in two steps, first assuming that only condition (iv) is satisfied and then dropping this condition.

Assume therefore that the pair (A, B) is completely controllable and Property (F) is satisfied. Then system (1), (2) is positive and can be written in the form (1), (8). By Lemma 1 the hermitian matrix U is positive semidefinite. Hence, there exists a hermitian matrix F such that $U = F^2$ (see F. R. Gantmacher [1] Chapter IX, § 12). Then the term x^*Ux of (8) can be written as $\|Fx\|^2$. Therefore we may confine ourselves to examine only the system consisting of Eq. (1) and the integral

$$\eta(0, t_1) = \Big[\|Fx\|^2\Big]_0^{t_1} + \int_0^{t_1} \|Vu + W^*x\|^2 \,\mathrm{d}t. \tag{22}$$

The condition $\pi(\lambda, \sigma) \not\equiv 0$ has a useful consequence, mentioned in the following lemma.

Lemma 3. *Consider the solutions of Eq. (1) which have the form $u(t) = ce^{\sigma t}$, $x(t) = x_0 e^{\sigma t}$, where c and x are constant vectors and σ satisfies the condition $\det(\sigma I - A) \neq 0$. If the characteristic polynomial π of the system (1), (22) is not identically zero, then there exists no vector $c \neq 0$ with the following property: For every σ for which $\det(\sigma I - A) \neq 0$, the functions of the form shown above, satisfy the identities $Fx(t) \equiv 0$ and $Vu(t) + W^* x(t) \equiv 0$.*

Proof. By the conditions of the lemma, x must be of the form $x(t) = (\sigma I - A)^{-1} Bc\, e^{\sigma t}$ and hence the identities of the lemma become $G_1(\sigma)\, c\, e^{\sigma t} \equiv 0$ and $G_2(\sigma) c\ e^{\sigma t} \equiv 0$, where $G_1(\sigma) = F(\sigma I - A)^{-1} B$ and $G_2(\sigma) = V + W^*(\sigma I - A)^{-1} B$. Since the identities obtained must hold for every σ (except a finite number of points), we deduce the identities $G_1(\sigma) c \equiv 0$ and $G_2(\sigma)\, c \equiv 0$. Furthermore, using formula (37, § 5) one finds that the characteristic function of system (1), (22) can be written as $H(\lambda, \sigma) = (\lambda + \sigma)[G_1(\bar{\lambda})]^* G_1(\sigma) + [G_2(\bar{\lambda})]^* G_2(\sigma)$. Under the conditions stated above we therefore obtain $H(\lambda, \sigma) c = 0$ whence $\det H(\lambda, \sigma) = 0$ and also $\pi(\lambda, \sigma) = 0$ (see (56, § 5)). This equality, established above with the exception of a finite number of values of λ and σ is true, by continuity, for every λ and σ.

The main step in proving the implication $(F) \rightarrow (h)$ consists in establishing the following lemma:

Lemma. 4. *Any system of the form (1), (22) can be written in the form* [1])

$$\frac{dy}{dt} = A_{11} y + B_{10} u_0 + B_{11} u_1 \tag{23}$$

$$\frac{dw}{dt} = A_{21} y + A_{22} w + B_{20} u_0 + B_{2r} u_r \tag{24}$$

$$\eta(0, t_1) = \Big[\|F_1 y\|^2 \Big]_0^{t_1} + \int_0^{t_1} (\|u_0\|^2 + \|C^* y\|^2)\, dt \tag{25}$$

and the following properties are satisfied: (α) if the continuous functions y, u_0 and u_1 satisfy equation (23) and the identities

---

[1]) Here and in the sequel the symbols used have the same meaning as in § 36. We also admit, as in § 36, that one of the vectors y and w can be zero-dimensional (in which case the corresponding equation is absent). Likewise, some of the vectors u_0, u_1 and u_r can be zero-dimensional. A more precise form of equations (23)—(25) will be given in the proof.

$u_0(t) \equiv 0$, $C^* y(t) \equiv 0$ and $F_1 y(t) \equiv 0$, then $y(t) \equiv 0$; (β) if the characteristic polynomial π of the system is not identically zero, then in (24) u_r is absent (i.e. is zero dimensional; (γ) if the system is minimally stable and satisfies, as above, the condition $\pi(\lambda, \sigma) \not\equiv 0$, then A_{22} is a Hurwitz matrix.

Proof. Consider the block

$$\frac{dx}{dt} = Ax + Bu, \quad v = Vu + W^* x, \tag{26}$$

which is of the form considered in Theorem 1, § 36 (Appendix A) for $m_1 = m_2 = m$. Applying the theorem quoted, one can write the block in the form (69, § 36)—(72, § 36), whence one deduces that system (1), (22) can be written as

$$\left.\begin{aligned} \frac{dy_{kj}}{dt} &= \sum_{i=1}^{j-1} T_{kji} y_{1i} + y_{k+1,j} + \hat{B}_{kj} u_0, \quad \text{for } 0 < k < p_j \\ &\qquad\qquad\qquad\qquad\qquad (\text{if } p_j > 1) \\ \frac{dy_{p_j j}}{dt} &= \sum_{i=1}^{j-1} T_{p_j j i}\, y_{1i} + \hat{B}_{p_j j} u_0 + \hat{u}_j \end{aligned}\right\} \begin{matrix} \text{for } 0 < j \leqslant s \\ (\text{if } s > 0) \end{matrix} \tag{27}$$

$$\frac{dz}{dt} = \sum_{i=1}^{s} F_{1i} y_{1i} + \hat{A}_{11} z + \hat{B}_{10} u_0 \qquad (\text{if } n_1 > 0) \tag{28}$$

$$\frac{d\hat{w}}{dt} = \sum_{i=1}^{s} F_{2i} y_{1i} + \hat{A}_{21} z + \hat{A}_{22} \hat{w} + B_{20} u_0 + \hat{B}_{2r} \hat{u}_r, \ (\text{if } n_2 > 0) \tag{29}$$

$$\begin{aligned} \eta(0, t_1) = {} & \left[\left\| \sum_{j=1}^{s} \sum_{k=1}^{p_j} D_{kj} y_{kj} + D_1 z + D_2 \hat{w} \right\|^2 \right]_0^{t_1} + \\ & + \int_0^{t_1} \left(\|u_0\|^2 + \sum_{i=1}^{s} \|y_{1i}\|^2 + \|C_1^* z\|^2 \right) dt, \end{aligned} \tag{30}$$

where the numbers s, p_j, n_j and n_2 have the same properties as in Theorem 1, § 36. Eqs. (27) — (29) correspond in all respects to Eqs. (69, § 36) —(71, § 36) the only difference being that the symbols which have been used before (with other meanings) in Eqs. (23) — (25) are marked with the superscript ^. Only expression (30) needs further explanation. The term between the square brackets [ ] is obtained by writing the corresponding term Fx, of (22), in the form $FR^{-1}\tilde{x}$ (see (5, § 36)) and then replacing the vector $\tilde{x}$ by its components y_{kj}, z and $\hat{w}$ (we recall that Eqs. (69, § 36)—(71, § 36) are just a more precise manner of writing the transformed Eq. (7, § 36)). Each of the matrices D_{kj}, D_1 and D_2 has n rows and a number of columns equal to the dimensions of the vector by which the respective matrix is multiplied in (30). We also point out that, by Theorem 1, § 36, if $n_1 > 0$ (i.e. if Eq. (28) is not absent), the pair $(C_1^*, \hat{A}_{11})$ is completely observable.

Assume now that $n_2 > 0$ (for $n_2 = 0$ the reasoning becomes simpler). Consider a new block of the form

$$\frac{d\hat{w}}{dt} = \hat{A}_{22}\hat{w} + \hat{B}_{2r}\hat{u}_r \tag{31}$$

$$\check{v} = D_2\hat{w}, \tag{32}$$

where we have introduced the new "output" $\check{v}$. Equation (31) is obtained from (29) for $y_{1i} = 0$, $z = 0$ and $u_0 = 0$. The block (31), (32) is of the form (1, § 36)—(2, § 36) where $D = 0$, $m_2 = r_r$, the number n from § 36 equals n_2 and the number m_1 of § 36 is equal to the number n of this paragraph.

Applying Theorem 1, § 36 one sees that this block too may be written in the form (69, § 36)—(72, § 36) (in order to avoid confusions the symbols used in the equations thus obtained will be marked by the superscript $^\vee$). Then the term in u_0 is absent since in Eq. (32), written in the form (2, § 36) we have $D = 0$. It is easy to see that if we replace Eq. (31) by Eq. (29) and if we apply to Eqs. (29), (32) the transformation applied to bring the block (31), (32) to the form given in Theorem 1, § 36, then the only difference is that some terms in y_{1i}, z and u_0 are added to the right hand members of the respective equations of the form (69, § 36) — (71, § 36)). From the foregoing it follows that Eqs. (29), (32) can be written in the form

$$\left.\begin{aligned}
\frac{d\check{y}_{kj}}{dt} &= \sum_{i=1}^{s} \hat{T}_{kji}y_{1i} + \overline{\overline{T}}_{kj}z + \sum_{i=1}^{j-1} \check{T}_{kji}\check{y}_{1i} + \\
&\quad + \check{y}_{k+1,j} + \check{B}_{kj0}\, u_0, \quad \text{for} \quad 0 < k < \check{p}_j \\
\frac{d\hat{y}_{p_j j}}{dt} &= \sum_{i=1}^{s} \hat{T}_{p_j ji}\, y_{1i} + \overline{\overline{T}}_{p_j j}\, z + \sum_{i=1}^{j-1} \check{T}_{p_j ji}\, \check{y}_{1i} + \\
&\quad \check{B}_{p_j j0}\, u_0 + \check{u}_j
\end{aligned}\right\} \quad \text{for} \quad 0 < j \leqslant \check{s} \tag{33}$$

$$\frac{d\check{z}}{dt} = \sum_{i=1}^{s} \hat{F}_{1i}y_{1i} + A_{10}z + \sum_{i=1}^{\check{s}} \check{F}_{1i}\check{y}_{1i} + \check{A}_{11}\check{z} + \check{B}_{10}u_0, \ (\text{if } \check{n}_1 > 0) \tag{34}$$

$$\begin{aligned}\frac{dw}{dt} &= \sum_{i=1}^{s} \hat{F}_{2i}y_{1i} + A_{20}z + \sum_{i=1}^{\check{s}} \check{F}_{2i}\check{y}_{1i} + \check{A}_{21}\check{z} + A_{22}w + \\
&\quad + \check{B}_{20}u_0 + \check{B}_{2r}u_r, \quad (\text{if } \check{n}_2 > 0)\end{aligned} \tag{35}$$

$$U\check{v} = \begin{bmatrix} \check{y}_{11} \\ \check{y}_{12} \\ \cdot \\ \cdot \\ \check{y}_{1\check{s}} \\ \check{C}_1^* \check{z} \end{bmatrix}, \quad (\check{C}_1, \check{A}_{11}) : \text{a completely observable pair} \quad (\text{if } \hat{n}_1 > 0), \tag{36}$$

where U is a unitary matrix ($U^* = U^{-1}$).

Using (32) one can write the integral (30) in the form

$$\eta(0, t_1) = \left[\left\| \sum_{j=1}^{s} \sum_{k=1}^{p_j} D_{kj} y_{kj} + D_1 z + \check{v} \right\|^2 \right]_0^{t_1} + \int_0^{t_1} \left(\|u_0\|^2 + \sum_{i=1}^{s} \|y_{1i}\|^2 + \|C_1^* z\|^2 \right) \mathrm{d}t. \tag{37}$$

We have found that there exists a transformation, as in § 5, by means of which from (1), (22) one obtains a system consisting of Eqs. (27)—(28), (33)—(35) and the integral (37) (where $\check{v}$ results from (36)). In order to bring this system to the form (23)—(25) it is sufficient to introduce the vector v whose components are the vectors y_{kj}, ($j = 1, 2, \ldots, s$, $k = 1, 2, \ldots, p_j$), z, $\check{y}_{kj}$, ($j = 1, 2, \ldots, \check{s}$, $k = 1, 2, \ldots, p_j$) and $\check{z}$, in which case equations (27), (28), (33) and (34) are grouped in a single equation of the form (23). (The components of the vector u_1 of (23) are $\hat{u}_j$, $j = 1, 2, \ldots, s$ and $\check{u}_j$, $j = 1, 2, \ldots, \check{s}$). We shall show that the system thus obtained has the property (α) of Lemma 4. Comparing (25) and (37) we see that the identities required in Property (α) of Lemma 4 reduce to

$$u_0(t) \equiv 0 \tag{38}$$

$$y_{1i}(t) \equiv 0, \qquad i = 1, 2, \ldots, s \tag{39}$$

$$C_1^* z(t) \equiv 0 \tag{40}$$

$$\sum_{j=1}^{s} \sum_{k=1}^{p_j} D_{kj} y_{kj}(t) + D_1 z(t) + \check{v}(t) \equiv 0. \tag{41}$$

From (38), (39) and (27) we obtain the equations $\mathrm{d}y_{kj}/\mathrm{d}t = y_{(k+1)j}$, $k = 1, 2, \ldots, p_j$, whence, using (39), we deduce step by step the identities

$$y_{kj}(t) \equiv 0, \; (j = 1, 2, \ldots, s, \; k = 1, 2, \ldots, p_j). \tag{42}$$

From (38), (39) and (28) it follows that $dz/dt = \hat{A}_{11}z$. Since the pair $(C_1^*, \hat{A}_{11})$ is completely observable, the identity (40) implies that

$$z(t) \equiv 0. \tag{43}$$

From (42), (43) and (41) it follows that $\check{v}(t) \equiv 0$ or (see (36)):

$$\check{y}_{1i}(t) \equiv 0, \qquad i = 1, 2, \ldots, \check{s} \tag{44}$$

$$\check{C}_1\check{z}(t) \equiv 0. \tag{45}$$

From (38), (42), (43) and (44) it follows that conditions (33) reduce to $d\check{y}_{kj}/dt = \check{y}_{(k+1)j}$, whence one obtains, as before,

$$\check{y}_{kj}(t) \equiv 0, \ (j = 1, 2, \ldots, \check{s}, \ \ k = 1, 2, \ldots, \check{p}_j). \tag{46}$$

Finally, from identities (38), (42), (43) and (44) it follows that Eq. (34) reduces to $d\check{z}/dt = \check{A}_{11}\check{z}$, whence, using the identity (45) and the fact that the pair $(\check{C}_1^*, \check{A}_{11})$ is completely observable, one deduces the identity

$$\check{z}(t) \equiv 0. \tag{47}$$

The identities (42), (43), (46) and (47) show that all the components of the vectors $y(t)$ are identically zero, which proves the Property (α) of Lemma 4.

As for Property (β) we remark that if the vector u_r of Eq. (24) is not absent, then expression (25) is completely independent of the component u_r of the control function. Therefore, the identities of Lemma 3 can be satisfied for a non-zero vector c, whence, by Lemma 3, one obtains the contradictory conclusion $\pi(\lambda, \sigma) \equiv 0$.

Finally, we shall prove Property (γ) of Lemma 4. Consider the initial condition $y(0) = 0$ and $w(0) = w_0$, where w_0 is arbitrary. By the property of minimal stability there exists a continuous solution of the system with the properties $\lim_{t\to\infty} y(t) = 0$, $\lim_{t\to\infty} w(t) = 0$ and $\eta(0, t_1) \leqslant 0$ for any $t_1 \geqslant 0$. Owing to the condition $y(0) = 0$ the last inequality becomes

$$\|F_{1y}(t)\|^2 + \int_0^{t_1} (\|u_0(t)\|^2 + \|C^*y(t)\|^2)\, dt \leqslant 0, \quad \text{for all } t_1 \geqslant 0$$

and hence $u_0(t) \equiv 0$, $C^*y(t) \equiv 0$ and $F_1y(t) \equiv 0$. By Property (α) of Lemma 4, $y(t) \equiv 0$. Since the vector u_r is absent (see Property (β)) we find that w satisfies the equation $dw/dt = A_{22}w$ (obtained from (24)). Since $\lim_{t\to\infty} w(t) = 0$ and the initial condition w_0 is arbitrary, A_{22} must be a Hurwitz matrix. This concludes the proof of Lemma 4 [1]).

In order to complete the proof of the implication $(F) \to (h)$ (under Hypothesis (*iv*), it is sufficient to show that system (23)—(25) (under Hypotheses (*i*)—(*iii*)), has Property (h). By Property (α) of Lemma 4, the system consisting of Eqs. (23) and (25) only, satisfies the conditions of Lemma 2. Hence, if we write the expression $\|F_1y\|^2$ in the form y^*Uy, where $U = F_1^*F_1$ we obtain, by Lemma 2, the conclusion that U is positive definite. Hence, the system consisting only of (23) and (25) has Property (h_+), whence (taking into account the implication $(h_+) \to (h)$ we conclude that the system has Property (h). We shall now show that the whole system (23)—(25) has Property (h) (i.e. Properties h_s and h_p of Definitions 3, § 13 and 4, § 13). Clearly since the system consisting only of Eqs. (23) and (25) has Property h_p, the whole system (23)—(25) has Property h_p. To establish Property h_s it is sufficient to show that the conditions of Proposition 3, § 13 are satisfied. Therefore we assume that the inequality $\eta(0, t_1) \leqslant \beta_0^2$ is satisfied (see (23, § 13)) [2]). Since, as we have seen, the system consisting only of Equs. (23) and (25) has the Property h_s, there exists a positive constant δ such that

$$\| y\ (t) \| \leqslant \delta\ (\beta_0 + \| y\ (0) \|). \tag{48}$$

From the inequality $\eta(0, t_1) \leqslant \beta_0^2$ and from (25) one obtains also the inequality

$$\int_0^{t_1} \| u_0\ (t)\|^2\ dt \leqslant \beta_0^2 + \| F_1 y\ (0)\ \|^2 \leqslant \beta_0^2 + \beta_1 \| y\ (0)\|^2, \ (\beta_1 > 0). \tag{49}$$

By Property (β) of Lemma 4 the vector u_r of Eq. (24) is absent and hence this equation has the form

$$\frac{dw}{dt} = A_{21}y + A_{22}\,w + B_{20}\,u_0. \tag{50}$$

[1]) Furthermore we have shown that system (23)—(25) can be written in the more precise form given by Eqs. (27), (28), (33)—(37). Using this form one can obtain other properties, besides those specified in Lemma 4.

[2]) In all the inequalities occurring in the proof we shall assume that t_1 takes any value of the interval $0 \leqslant t \leqslant T$. We also mention that we shall denote by α_j and β_j different positive constants the exact values of which are not needed for the evaluations that we have in view.

The solution of this equation can be written in the form

$$w(t) = w_1(t) + w_2(t) + w_3(t) \tag{51}$$

$$w_1(t) = e^{A_{22} t} w(0) \tag{52}$$

$$w_2(t) = \int_0^t e^{A_{22}(t-\tau)} A_{21} y(\tau) \, d\tau \tag{53}$$

$$w_3(t) = \int_0^t e^{A_{22}(t-\tau)} B_{20} u_0(\tau) \, d\tau. \tag{54}$$

Since A_{22} is a Hurwitz matrix (see Property (γ) of Lemma 4), there exist two constants $\alpha_1 > 0$ and $\alpha_2 > 0$ such that

$$\| e^{A_{22} t} \| \leqslant \alpha_1 e^{-\alpha_2 t}. \tag{55}$$

From this and from (52) one obtains

$$\| w_1(t) \| \leqslant \alpha_1 \| w(0) \|. \tag{56}$$

From (55) and (53) we deduce that there exists a constant $\alpha_3 \geqslant 0$ such that

$$\| w_2(t) \| \leqslant \alpha_3 \sup_{0 \leqslant \tau \leqslant t} \| y(\tau) \|,$$

whence, using (48), one obtains

$$\| w_2(t) \| \leqslant \alpha_4 (\beta_0 + \| y(0) \|), \quad (\alpha_4 \geqslant 0). \tag{57}$$

From (54) and (55) one obtains the inequality

$$\| w_3(t) \| \leqslant \alpha_5 \int_0^{t_1} e^{-\alpha_2 (t-\tau)} \| u_0(\tau) \| \, d\tau. \tag{58}$$

Using the inequality of Cauchy-Buniakowsky-Schwarz, one sees that the right hand member of the inequality is not greater than the expression

$$\alpha_5 \sqrt{\int_0^t e^{-2\alpha_2 (t-\tau)} \, d\tau} \sqrt{\int_0^t \| u_0(\tau) \|^2 \, d\tau} \cdot$$

One also has

$$\int_0^t e^{-2\alpha_2(t-\tau)}\, d\tau \leqslant 1/(2\alpha_2)$$

Using (49), one obtains from (58)

$$\| w_3(t) \| \leqslant \alpha_6 \sqrt{(\beta^2_0 + \beta_1 \| y(0) \|^2)} \leqslant \alpha_7 (\beta_0 + \| y(0) \|). \tag{59}$$

From (48), (51), (56), (57) and (59) one easily sees that the vector $x(t)$ (whose components are the vectors $y(t)$ and $w(t)$) satisfies an inequality of the form (24, § 13). Hence, by Proposition 3, §13, it follows that system (23)—(25) has Property (h). Thus the implication $(F) \to (h)$ is proved (under Hypothesis (iv)).

So far we have proved Theorem 1 assuming that the Conditions (i)—(iv) are satisfied. We shall now pass to the last step of the proof, namely to prove the implication $(F) \to (h)$ making use of the Assumptions (i)—(iii) only. To accomplish this we shall apply the following lemma.

Lemma 5. *Any system of the form (1), (2) which satisfies the conditions (i), (ii) and (iii) and has Property (F) (see Theorem 1) can be split (by using a transformation of the type studied in § 5) into a "completely controllable part" of the form*

$$\frac{dy}{dt} = A_{11} y + B_{10} u_0 + B_{11} u_1 \tag{60}$$

$$\eta_c(0, t_1) = \Big[\| F_1 y \|^2 \Big]_0^{t_1} + \int_0^{t_1} (\| u_0 \|^2 + \| C^* y \|^2)\, dt \tag{61}$$

and a "residue" of the form

$$\frac{dw}{dt} = A_{22} w \tag{62}$$

$$\eta_r(0, t_1) = \Big[y^* J_{12} w + w^* J_{21} y + w^* J_{22} w \Big]_0^{t_1} + \int_0^{t_1} (u_0^* L_2^* w + u_1^* L_3^* w + \tag{63}$$
$$+ w^* L_2 u_0 + w^* L_3 u_1 + y^* M_{12} w + w^* M_{21} y + w^* M_{22} w)\, dt,$$

while the integral $\eta(0, t_1)$ is given by the sum

$$\eta(0, t_1) = \eta_c(0, t_1) + \eta_r(0, t_1). \tag{64}$$

Furthermore the following properties are satisfied:

(a) $B_{11}^* B_{11}$ *is a nonsingular matrix; (b)* A_{22} *is a Hurwitz matrix; (c) there exist three constants* $\gamma_1 \geqslant 0$, $\gamma_2 \geqslant 0$ *and* $\gamma_3 \geqslant 0$ *such that*

$$- \eta_r (0, t_1) \leqslant \gamma_1 \| w(0) \| \sqrt{\int_0^{t_1} \| u_0 (t) \|^2 \, dt} + \gamma_2 \| w(0) \| \sup_{0 \leqslant t \leqslant t_1} \| y(t) \| + \gamma_3 \| w(0) \|^2. \tag{65}$$

Proof. Since $B \neq 0$ one can apply Proposition 3, § 34 and bring equation (1) to the form

$$\frac{d\tilde{y}}{dt} = A_{11}\tilde{y} + A_{12} w + B_1 u, \quad (A_{11}, B_1) = \text{a completely controllable pair} \tag{66}$$

$$\frac{dw}{dt} = A_{22} w. \tag{67}$$

By reasonings quite similar to those used in § 14 for bringing (23, § 14)—(24, § 14) to the form (25, § 14)—(26, § 14), one can bring (66), (67) to the form

$$\frac{dy}{dt} = A_{11} y + B_1 u, \; (A_{11}, B_1) = \text{a completely controllable pair} \tag{68}$$

$$\frac{dw}{dt} = A_{22} w. \tag{69}$$

Thus Eq. (1) is split into the two Eqs. (68) and (69). Partitioning in the corresponding way expression (2), by putting

$$x = \begin{pmatrix} y \\ w \end{pmatrix}, \; J = \begin{pmatrix} J_{11} & J_{12} \\ J_{21} & J_{22} \end{pmatrix}, \; L = \begin{pmatrix} L_1 \\ L_0 \end{pmatrix}, \; M = \begin{pmatrix} M_{11} & M_{12} \\ M_{21} & M_{22} \end{pmatrix},$$

one can write the integral (2) in the form (64), where

$$\eta_c(0, t_1) = [y^* J_{11} y]_0^{t_1} + \int_0^{t_1} (u^* K u + u^* L_1^* y + y^* L_1 u + y^* M_{11} y) \, dt \tag{70}$$

$$\eta_r(0, t_1) = [y^* J_{12} w + w^* J_{21} y + w^* J_{22} w]_0^{t_1} + \int_0^{t_1} (u^* L_0^* w + u^* L_0 u + y^* M_{12} w + w^* M_{21} y + w^* M_{22} w) \, dt. \tag{71}$$

We remark that the characteristic matrix of the whole system (68), (69), (70), (71), (64) is identical to the characteristic matrix of the "completely controllable part" consisting of (68) and (70). Consequently, if the initial system satisfies Conditions (*iii*) and (F), then its completely controllable part satisfies the same conditions. A similar statement is also valid for Conditions (*i*) and (*ii*) (for Condition (*i*) we proceed as in § 14 by applying the property of minimal stability for the initial conditions $w(0) = 0$ and $y(0) = y_0$, where y_0 is arbitrary). Hence the system (68), (70) can be written in the form (1), (22) and — after applying Lemma 4 — in the form (23)—(25); or it may be written (as we shall assume in the sequel) in the more precise form (27), (28), (33)—(37) (see footnote 1, page 177). Owing to Condition (*iii*) the vector u_r does not appear in these equations (see Property (β) of Lemma 4). Introduce the vector y whose components are the vectors y_{kj}, z, $\check{y}_{kj}$, $\check{z}$ and w, and consider also the vector u_1 whose components are the vectors $\hat{u}_j$, $j = 1, 2, \ldots, s$ and $\check{u}_j$, $j = 1, 2, \ldots, \check{s}$. Then it is easy to see that the system can be written in the form (60), (61). Furthermore, by comparing Eq. (60) with Eqs. (27), (28), (33)—(35) (where the vector u_r is absent, as we have seen) and taking into account the definition of the vector u_1, we find that the equality $\|B_{11}u_1\|^2 = 0$ can occur only if one has $\hat{u}_j = 0$, $j = 1, 2, \ldots, s$ and $\check{u}_j = 0$, $j = 1, 2, \ldots, \check{s}$, i.e. only if the vector u_1 is zero. Consequently the quadratic form $u_1^* B_{11}^* B_{11} u_1$ is positive definite and therefore $B_{11}^* B_{11}$ is nonsingular (Property (*a*) of Lemma 5).

The form of Relation (71) remains unaltered after the transformation which brings the system (68), (70) to the form (60), (61); only the values of the coefficients are different. In order to avoid overcharging the notations we shall leave unaltered the symbols of Relation (71) assuming in what follows that all the transformations described above have been already effected. If, in accordance with the partitioning of the vector u into the components u_0 and u_1 (which result from bringing the system (68), (70) to the form (60), (61)), we also partition the matrix L_0 of (71) as $L_0 = (L_2\ L_3)$ so that $L_0 u = L_2 u_0 + L_3 u_1$, then the integral (71) takes the form (63). Thus, the system under investigation is brought to the form (60)—(64) and we have also proved Property (*a*) of Lemma 5. The fact that A_{22} is a Hurwitz matrix (see Property (*b*)) results from the property of minimal stability (*i*), remarking that from Property (*i*) it follows that for any initial condition $w(0)$ Eq. (62) admits a solution with the property $\lim\limits_{t\to\infty} w(t) = 0$. We shall now prove Property (*c*) of Lemma 5. Since the matrix $B_{11}^* B_{11}$ is nonsingular (Property (*a*)), we can multiply equation (60) on the left by $(B_{11}^*B_{11})^{-1}\ B_{11}^*$ and obtain

$$u_1 = T_1 \frac{dy}{dt} + T_2 y + T_3 u_0, \tag{72}$$

where the exact values of the matrices $T_1 - T_3$ are not needed in the sequel. Consider the terms of (63) of the form

$$\eta_1 = \int_0^{t_1} w^* L_3 u_1 \, dt.$$

Replacing u_1 by (72) one obtains

$$\eta_1 = \int_0^{t_1} w^* L_3 \left(T_1 \frac{dy}{dt} + T_2 y + T_3 u_0 \right) dt.$$

Integrating by parts the term which contains dy/dt, one obtains (by proceeding as we did in §14 to obtain (30, §14)) the equality

$$\eta_1 = [w^* L_3 T_1 y]_0^{t_1} + \int_0^{t_1} w^*(T_4 y + T u_0)\, dt.$$

Obviously

$$\int_0^{t_1} u_1^* L_3^* w \, dt = \bar{\eta}_1 .$$

Substituting these expressions in (63) gives

$$\eta_r(0, t_1) = \eta_{r1}(0, t_1) + \eta_{r2}(0, t_1), \tag{73}$$

$$\eta_{r1}(0, t_1) = [y^* \tilde{J}_{12} w + w^* \tilde{J}_{21} y + w^* J_{22} w]_0^{t_1} + \int_0^{t_1} (y^* \tilde{M}_{12} w + $$

$$+ w^* \tilde{M}_{21} y + w^* M_{22} w)\, dt \tag{74}$$

$$\eta_{r2}(0, t_1) = \int_0^{t_1} (u_0^* \tilde{L}_2^* w + w^* \tilde{L}_2 u_0)\, dt, \tag{75}$$

where the exact values of the new matrices introduced (marked by the superscript $\sim$) are not needed for the evaluations which follow. Since A_{22} is a Hurwitz matrix, there exist two constants $\alpha_1 > 0$ and $\alpha_2 > 0$ such that any solution of Eq. (69) admits the evaluation

$$\| w(t)\| \leqslant \alpha_1 e^{-\alpha_2 t} \|w(0)\|. \tag{76}$$

Using Buniakowski-Schwartz inequality one obtains from (75) and (76) the inequality

$$|\eta_{r2}(0, t_1)| \leqslant \alpha_3 \| w(0)\| \sqrt{\int_0^{t_1} \|u_0\|^2 dt}, \quad (\alpha_3 \geqslant 0). \tag{77}$$

Using again (76) one obtains from (74) the inequality

$$|\eta_{r1}(0, t_1)| \leqslant \alpha_4 \|w(0)\| \sup_{0 \leqslant t \leqslant t_1} \|y(t)\| + \alpha_5 \|w(0)\|^2, \quad (\alpha_{4,5} \geqslant 0). \tag{78}$$

The inequalities (77) and (78) show that the integral $\eta_{ir}(0, t_1)$ (given by (73)) satisfies indeed an inequality of the form (65). Lemma 5 is thus proved.

To complete the proof of the implication $(F) \rightarrow (h)$ we must show that system (60)—(64) has Properties h_s and h_p (see Definitions 3, § 13 and 4, § 13). We first remark that since implication $(F) \rightarrow (h)$ has been proved before for completely controllable systems, we know that the completely controllable part of the system under investigation (i.e. system (60), (61)) has Properties h_s and h_p, a fact which will be used in the proof. We shall now show that the system has Property h_p. From (61) we obtain the inequality

$$\eta_c(0, t_1) \geqslant -\alpha_6 \|y(0)\|^2 + \int_0^{t_1} \|u_0\|^2 dt, \quad (\alpha_6 \geqslant 0).$$

Using also (65) we find that the integral $\eta(0, t_1)$ (see (64)) satisfies the inequality

$$\eta(0, t_1) \geqslant -\alpha_6 \|y(0)\|^2 + \left(\sqrt{\int_0^{t_1} \|u_0\|^2 dt} - \frac{1}{2} \gamma_1 \|w(0)\| \right)^2 -$$

$$- \gamma_2 \|w(0)\| \sup_{0 \leqslant t \leqslant t_1} \|y(t)\| - \left(\gamma_3 + \frac{1}{4} \gamma_1^2 \right) \|w(0)\|^2 \geqslant$$

$$\geqslant -\alpha_6 \|y(0)\|^2 - \gamma_2 \|w(0)\| \sup_{0 \leqslant t \leqslant t_1} \|y(t)\| - \alpha_7 \|w(0)\|^2, \quad (\alpha_7 \geqslant 0).$$

Defining the norm of the vector $x = \begin{pmatrix} y \\ w \end{pmatrix}$ by the relation $\|x\| = \sqrt{\|y\|^2 + \|w\|^2}$ we immediately see that the right hand member of the inequality obtained is greater than or equal to an expression of the form $-\beta_3\|x(0)\|^2 - \beta_4\|x(0)\|\sup_{0\leqslant t\leqslant t_1}\|x(t)\|$ and hence Property h_p is satisfied. To obtain also Property h_s we make use of Proposition 3, § 13. We assume that the inequality $\eta(0, t_1) \leqslant \beta_0^2$ (i.e. (23, § 13)) is satisfied [1]). Adding this inequality and inequality (65) and using (64) one obtains

$$\eta_c(0, t_1) \leqslant \beta_0^2 + \gamma_1\|w(0)\|\sqrt{\int_0^{t_1}\|u_0(t)\|^2\,dt} +$$

$$+ \gamma_2\|w(0)\|\sup_{0\leqslant t\leqslant t_1}\|y(t)\| + \gamma_3\|w(0)\|^2. \tag{79}$$

By (61) we have

$$-\|F_1y(0)\|^2 + \int_0^{t_1}\|u_0\|^2\,dt < \eta_c(0, t_1). \tag{80}$$

Adding (79) and (80) we may arrange the result in the form

$$\left(\sqrt{\int_0^{t_1}\|u_0\|^2\,dt} - \frac{1}{2}\gamma_1\|w(0)\|\right)^2 \leqslant \beta_0^2 + \|F_1y(0)\|^2 +$$

$$+\left(\gamma_3 + \frac{1}{4}\gamma_1^2\right)\|w(0)\|^2 + \gamma_2\|w(0)\|\sup_{0\leqslant t\leqslant t_1}\|y(t)\|. \tag{81}$$

We now remark that there exist nonnegative constants $\alpha_8 - \alpha_{10}$ such that the right hand member of (81) does not exceed

$$(\beta_0 + \alpha_8\|y(0)\| + \alpha_9\|w(0)\| + \alpha_{10}\sup_{0\leqslant t\leqslant t_1}\|y(t)\|)^2.$$

---

[1]) We assume that here and in subsequent inequalities t_1 can take any value in an interval of the form $0 \leqslant t_1 \leqslant T$.

(It is sufficient to choose the constants $\alpha_8 - \alpha_{10}$ such that

$$\alpha_8 \| y(0) \| \geqslant \| F_1 y(0) \|, \qquad \alpha_9^2 \geqslant \left(\gamma_3 + \frac{1}{4} \gamma_1^2\right) \text{ and } 2\alpha_9 \alpha_{10} \geqslant \gamma_2. \Big)$$

Then

$$\sqrt{\int_0^{t_1} \| u_0 \|^2 \, dt} \leqslant \beta_0 + \alpha_8 \| y(0) \| + \alpha_{11} \| w(0) \| + \alpha_{10} \sup_{0 \leqslant t \leqslant t_1} \| y(t) \|$$

(where $\alpha_{11} = \alpha_9 + \gamma_1/2$). Using this inequality we obtain from (79)

$$\eta_c(0, t_1) \leqslant \beta_0^2 + \gamma_1 \| w(0) \| (\beta_0 + \alpha_8 \| y(0) \| + \alpha_{11} \| w(0) \|) +$$

$$+ \alpha_{12} \| w(0) \| \sup_{0 \leqslant t \leqslant t_1} \| y(t) \| + \gamma_3 \| w(0) \|^2 \leqslant$$

$$\leqslant (\beta_0 + \alpha_{13} \| w(0) \| + \alpha_{14} \| y(0) \|)^2 + \alpha_{12} \| w(0) \| \sup_{0 \leqslant t \leqslant t_1} \| y(t) \| \quad (82)$$

(where $\alpha_{12} = \gamma_2 + \alpha_{10}$ and α_{13} and α_{14} satisfy the Conditions $2\alpha_{13} \geqslant \gamma_1$, $\alpha_{13}^2 \geqslant \gamma_1 \alpha_{11} + \gamma_3$ and $2\alpha_{13}\alpha_{14} \geqslant \alpha_8 \gamma_1$). Thus we have established that the inequality (82) — which has the form (7, § 13) for $\beta_1 = \beta_0 + \alpha_{13} \| w(0) \| + \alpha_{14} \| y(0) \|$ and $\beta_2 = \alpha_{12} \| w(0) \|$, — is satisfied for the completely controllable part of the system. Since (as shown before) the completely controllable part of the system has the Property (h_s) one obtains from (82) the inequality [1]) (8, § 13). Hence, by introducing the above values of the constants β_1 and β_2, one obtains

$$\| y(t) \| \leqslant \delta (\beta_0 + \alpha_{15} \| w(0) \| + \alpha_{16} \| y(0) \|),$$

(where $\alpha_{15} = \alpha_{12} + \alpha_{13}$ and $\alpha_{16} = 1 + \alpha_{14}$). Introducing the vector $x = \begin{pmatrix} y \\ w \end{pmatrix}$ and taking into account the inequality $\| w(t) \| \leqslant \alpha_1 \| w(0) \|$ (resulting from (76)), we find that $\| x(t) \|$ satisfies an inequality of the form (24, § 13). Applying Proposition 3, § 13 we see that the system under investigation has Property h_s. This completes the general proof of the implication $(F) \rightarrow (h)$ and the proof of Theorem 1.

We conclude with a remark (already mentioned in Section 2 of § 13) concerning Condition *(iii)* ($\pi(\lambda, \sigma) \not\equiv 0$). In the above proof this condition has been only used to ensure that the vector

---

[1]) Naturally, the part played by x in the inequalities (7, § 13) and (8, § 13) is taken in this case by y.

u_r of (24) is absent (is zero-dimensional). However, from Eqs. (23)—(25) it follows that if the vector u_r had at least one component, it would not affect in any way the values of the integral $\eta(0, t_1)$. Then, if we have also the equality $B_{2r}u_r = 0$, the vector u_r can be eliminated without changing in fact our problem. If however $B_{2r}u_r \neq 0$ the function $u_r(t)$ can be chosen such that Eq. (24) admits a solution with the property $\lim_{t\to\infty} |w(t)| = \infty$ and the integral $\eta(0, t_1)$ is identically zero. (It is sufficient to take the solution of Eq. (23) for $y(0) = 0$, $u_0(t) \equiv 0$ and $u_1(t) \equiv 0$). This contradicts Property h_s and hence the system cannot be hyperstable. The same remarks remain valid for the more general system (60)—(64) since, as we know, the completely controllable part (60), (61) can be written in the form (23)—(25). Thus, completing the remarks of Section 2, § 13, we obtain the conclusion that no effective restriction concerning our problem is introduced by admitting the condition $\pi(\lambda, \sigma) \not\equiv 0$.

§ 17. Multi-input hyperstable blocks

The results obtained in the foregoing paragraph may be easily extended to the case of multi-input blocks described by the equations

$$\frac{\mathrm{d}x}{\mathrm{d}t} = Ax + Bu \tag{1}$$

$$v = C^*x + Du, \tag{2}$$

where the symbols x, A, B and u have the same meaning as in § 16, the vector v has m components, C is an $n \times m$ matrix and D is an $m \times m$ matrix.

Besides the block (1)—(2) we consider the associated multi-input system (see Section 9, § 13) of the form

$$\left.\begin{aligned} &\frac{\mathrm{d}x}{\mathrm{d}t} = Ax + Bu \\ &\eta(0, t_1) = \mathrm{Re}\left(\int_0^{t_1} u^* v \,\mathrm{d}t\right) = \int_0^{t_1}\left(u^*(D+D^*)u + \frac{1}{2}u^*C^*x + \frac{1}{2}x^*Cu\right)\mathrm{d}t \end{aligned}\right\} \tag{3}$$

which has the form studied in § 16 for

$$K = D + D^*,\ \ L = \frac{1}{2}C,\ J = M = 0. \tag{4}$$

In accordance with the definitions given in Section 9, § 13 block (1)—(2) is said to be hyperstable if the associated system (3) is hyperstable. Likewise, block (1)—(2) is said to be minimally stable if the associated system is minimally stable.

By Relation (37, § 5) it follows that the characteristic function of the associated system (3) is expressed by

$$H(\lambda, \sigma) = \frac{1}{2}(\bar{G}'(\lambda) + G(\sigma)), \tag{5}$$

where

$$G(\sigma) = D + C^*(\sigma I - A)^{-1} B \tag{6}$$

is the transfer matrix of block (1)—(2).

We also introduce a family of single-input blocks of the form (studied in § 15)

$$\frac{dx}{dt} = Ax + b\mu, \quad \nu = a^* x + \alpha\mu, \tag{7}$$

where

$$b = Bc, \quad a = Cc, \quad \alpha = c^* Dc \tag{8}$$

and where c is a vector satisfying the condition $\|c\| = 1$. The blocks of the form (7)—(8) are obtained from the block (1)—(2) by introducing the relations

$$u(t) = c\mu(t), \quad \nu(t) = c^* v(t). \tag{9}$$

From Relations (8) it follows that the transfer function γ of the block (7)—(8) is related to the transfer function G of block (1)—(2) by the relation

$$\gamma(\sigma) = c^* G(\sigma) c. \tag{10}$$

Corresponding to each block of the form (7)—(8), we define also an associated system of the form (3, § 15) described by the equations

$$\frac{dx}{dt} = Ax + Bc\mu, \quad \eta(0, t_1) = $$
$$= \int_0^{t_1} \left(\operatorname{Re} c^* Dc |\mu|^2 + \frac{1}{2}\mu^* c^* C^* x + \frac{1}{2} x^* Cc\mu\right) dt. \tag{11}$$

One immediately sees that the simple system (11) (associated to block (7)—(8)) is related to system (3) (associated to block (1)—

(2)) in the same way as the single-input system (3, § 16)—(4, § 16) is related to system (1, § 16)—(2, § 16). Furthermore the characteristic function χ of system (11) is related to the characteristic function H, (4), and to the transfer function γ, (10) by the relations

$$\chi(\lambda, \sigma) = c^* H(\lambda, \sigma)\, c = \frac{1}{2}(\overline{\gamma}(\lambda) + \gamma(\sigma)). \tag{12}$$

The principal properties of the multi-input hyperstable blocks are stated in

Theorem 1. *Suppose that block (1)—(2) is minimally stable and the conditions $B \neq 0$ and*

$$\det(D^* + D + C^*(\sigma I - A)^{-1} B) \not\equiv 0. \tag{13}$$

are satisfied. Then the following properties are equivalent

(H_0): "Block (1)—(2) is hyperstable,"

(F_0): "The matrix $\overline{G}'(-i\omega) + G(i\omega)$ is positive semidefinite for every real number ω satisfying the condition $\det(i\omega I - A) \neq 0$,"

(H'_0): "System (3) satisfies one of the conditions (h), (h_{so}), (h_{po}), (h_c) and (f) of Theorem 1, § 16".

$(h_{c\lambda})$: "All the feedback systems defined by the relations

$$\frac{dx}{dt} = Ax + Bc\mu, \quad \nu = c^* C^* x + c^* D c\mu, \quad \frac{d\nu}{dt} = -\lambda\mu \tag{14}$$

and

$$\frac{dx}{dt} = Ax + Bc\mu, \quad \nu = c^* C^* x + c^* Dc\mu, \quad \frac{d\mu}{dt} = -\lambda\nu \tag{15}$$

have stable trivial solutions for every vector c satisfying the conditions $\|c\| = 1$ and $Bc \neq 0$ and for every real and strictly positive number λ, except, in the case of system (14) for $c^* Dc = 0$, the value $\lambda_0 = -c^* C^* Bc$".

The greatest part of the proof of this theorem is just an application of Theorem 1, § 16. We first show that from the conditions stated in the theorem it follows that Conditions (*i*)—(*iii*) of Theorem 1, § 16 are satisfied. This is obvious for Conditions (*i*) and (*ii*). As for Condition (*iii*), observe that if $\pi(\lambda, \sigma) \equiv 0$ then, by (56, § 5), one has $\det H(\lambda, \sigma) \equiv 0$. Considering Relations (5) and (6) and taking the limit as λ tends to ∞ in the above identity, one obtains $\det(D^* + G(\sigma)) \equiv 0$, which contradicts Condition (13) (see also (6)).

Hence one can apply Theorem 1, § 16 and deduce that Property (H_0) of Theorem 1 is equivalent to any of the properties (F), (h), (h_{so}), (h_{po}), (h_c) and (f) stated in Theorem 1, § 16, written for system (3). Using (5) we immediately see that Properties (H_0), (F_0) and (H_0') are equivalent. We must show that these properties are also equivalent to Property $(h_{c\lambda})$; this will result by proving the implications $(H_0') \to (h_{c\lambda}) \to (H_0')$.

To prove the implication $(H_0') \to (h_{c\lambda})$ we remark that from Property (H_0') it follows in particular, using Property (h_c) that system (11) is hyperstable and hence block (7)—(8) is hyperstable for every vector c with the property $\|c\| = 1$. Hence Property $(h_{c\lambda})$ is obtained by the arguments used in § 15 in the proof of the implication $(H) \to (h_\lambda)$. (Notice that in the proof of the implication $(H) \to (h_\lambda)$ of § 15 no use was made of the assumptions required in Theorem 1, § 15).

We now prove the implication $(h_{c\lambda}) \to (H_0')$. If Property (H_0') is false then so is also Property (f) of Theorem 1, § 16 (written for system (3)). As in § 16, it follows that there exists a vector c which satisfies the condition $Bc \neq 0$ and for which the inequality (12, § 16) is satisfied. Obviously, the vector c can be chosen such that $\|c\| = 1$. We now consider the block (7)—(8) for that value of c for which the inequality (12, § 16) is satisfied. Using (12) we find that the transfer function of this block is not positive real. Notice that in the proof of implication $(h_\lambda) \to (f')$ of § 15 no use has been made of the assumptions of Theorem 1, § 15. Using the same reasonings, from the fact that the transfer function of block (7), (8) determined above is not positive real one deduces that there exists a feedback system of the form (14) or (15) whose trivial solution is not stable, for a value of λ which satisfies the conditions stated in property (h_c). Hence, property $(h_{c\lambda})$ is contradicted. This proves the implication $(h_{c\lambda}) \to (f')$ and concludes the proof of Theorem 1.

§ 18. Discrete hyperstable systems and blocks

The results obtained in the foregoing chapters can be easily extended to the case of discrete systems. The proofs required to that end are completely similar to those used in the case of continuous systems and are often even simpler. This paragraph will be therefore mostly confined to the statement of the definitions and of the principal results.

In the case of the discrete systems instead of systems of the general form of (10, § 13)—(11, § 13) we have to deal with systems of the form

$$x_{k+1} = f(x_k, u_k, k), \quad k = k_0, k_0 + 1, \ldots \tag{1}$$

$$\eta(k_0, k_1) = [\varphi(x_k, k)]_{k_0}^{k_1+1} + \sum_{k=k_0}^{k_1} \psi(x_k, u_k, k), \; k_1 \geqslant k_0, \tag{2}$$

where the vector-valued function $(y, v, k) \mapsto f(y, v, k)$ and the scalar-valued functions $(y, k) \mapsto \varphi(y, k)$ and $(y, v, k) \mapsto \psi(y, v, k)$ assign finite values (in general complex values) for every complex n-vector y, every complex m-vector v and every integer $k \geqslant k_0$. The first term of the right-hand member of Relation (2) has, as usual, the meaning

$$[\varphi(x_k, k)]_{k_0}^{k_1+1} = \varphi(x_{k_1+1}, k_1 + 1) - \varphi(x_{k_0}, k_0). \tag{3}$$

By "solution" of a system of the form (1)—(2) in an interval $[k_0, K]$, $(K \geqslant k_0)$ we mean (cf. Definition 7, § 13) a set consisting of: (a) a sequence $k \mapsto u_k$, defined for $k = k_0, k_0 + 1, \ldots, K$; (b) a sequence $k \mapsto x_k$ defined for $k = k_0, k_0 + 1, \ldots, K + 1$ and satisfying together with u_k Eq. (1), and (c) a sequence $k_1 \mapsto \eta(k_0, k_1)$ defined for $k_1 = k_0, k_0 + 1, \ldots, K$ and related to u_k and x_k by (2).

The concept of a discrete hyperstable system is defined by a direct extension of Definition 8, § 13:

Definition 1. *The system (1) — (2) is said to be hyperstable if there exists a function α of class M_i and three functions β, γ and δ of class M_s, such that the following two properties are satisfied:*

Property $\mathbf{H}_s$: *For every solution of system (1)—(2) the following statement is true: For every constant $\beta_0 \geqslant 0$ for which one has the inequality*

$$\eta(k_0, k_1) \leqslant \beta_0^2, \quad \textit{for } k_1 = k_0, k_0 + 1, \ldots, K \tag{4}$$

one also has the inequality

$$\alpha(x_k) \leqslant \beta_0 + \beta(x_{k_0}), \textit{ for } k = k_0, k_0 + 1, \ldots, K + 1. \tag{5}$$

Property $\mathbf{H}_p$: *The inequality*

$$\eta(k_0, k_1) \geqslant -\gamma^2(x_{k_0}) - \delta(x_{k_0}) \max_{k_0 \leqslant k \leqslant k_1} \alpha(x_k), \textit{ for } k_1 = k_0, k_0 + k_1, \ldots K, \tag{6}$$

is satisfied for every solution of system (1)—(2).

We shall examine in some detail the discrete systems described by the equations

$$x_{k+1} = Ax_k + Bu_k \tag{7}$$

$$\eta(0, k_1) = [x_k^* J x_k]_0^{k_1+1} + \sum_{k=0}^{k_1} (u_k^* K u_k + u_k^* L^* x_k + x_k^* L u_k + x_k^* M x_k), \tag{8}$$

where the notations have the same meaning as in § 6.

In accordance with Definition 2, § 13, system (7)—(8) is said to be minimally stable, if for every initial condition x_0 there exists a solution of the system, defined for all the integers $k \geqslant 0$ and having the properties

$$\eta(0, k_1) \leqslant 0, \text{ for every integer } k_1 \geqslant 0 \tag{9}$$

$$\lim_{k\to\infty} x_k = 0. \tag{10}$$

In the case of the systems of the form (7)—(8) one can introduce Properties h_s and h_p (cf. Definitions 3, § 13 and 4, § 13):

Property $\mathbf{h}_s$: *There exist a constant $\delta \geqslant 0$ such that for every solution of system (7), (8) the following statement is true: For every pair of constants, $\beta_1 \geqslant 0$ and $\beta_2 \geqslant 0$ for which one has the inequality*

$$\eta(0, k_1) \leqslant \beta_1^2 + \beta_2 \max_{0\leqslant k\leqslant k_1+1} \|x_k\|, \textit{ for } k_1 = 0, 1, \ldots, K \tag{11}$$

one also has the inequality

$$\|x_k\| \leqslant \delta(\beta_1 + \beta_2 + \|x_0\|), \textit{ for } k = 0, 1, \ldots; K+1 \tag{12}$$

Property $\mathbf{h}_p$: *There exists a pair of constants $\beta_3 \geqslant 0$ and $\beta_4 \geqslant 0$ such that for every solution of the system (7)—(8) one has*

$$\eta(0, k_1) \geqslant -\beta_3 \|x_0\|^2 - \beta_4 \|x_0\| \max_{0\leqslant k\leqslant k_1+1} \|x_k\|, \textit{ for } k=0, 1, \ldots. \tag{13}$$

It can be easily proved that if system (7)—(8) has properties h_s and h_p, then the system is hyperstable (cf. Proposition 1, § 13).

Propositions 2, § 13—4, § 13 remain valid (with the necessary changes) in the case of discrete systems. The definition of the sum of two hyperstable systems and the corresponding properties stated in Propositions 5, § 13—11, § 13 can also be immediately extended to discrete systems.

The special properties of the hyperstable systems of the form (7)—(8) are given by the following theorem (cf. Theorem 1, § 16):

Theorem 1. *Assume that the following conditions are satisfied: (i) system (7)—(8) is minimally stable; (ii) $B \neq 0$,* *and (iii) $\pi(\lambda, \sigma) \equiv 0$. Then the following properties are equivalent*

(H): "The system (7)—(8) is hyperstable"

(F): "The matrix $H(\bar{\sigma}, \sigma)$ is positive semidefinite for any number σ which satisfies the conditions $|\sigma| = 1$ and $\det(\sigma I - A) \neq 0$",

(h): "The properties (h_s) and (h_p) hold"

(h_{so}): "The property (h_s) holds for the particular case $\beta_1 = 0$, $\beta_2 = 0$".

(h_{po}): "The property (h_p) holds for the particular case $x_0 = 0$".

(h_c): "All the single-input systems obtained from the system (7)—(8) for $u = c\mu(t)$ (where $\|c\| = 1$), are hyperstable".

(f): "The matrix $H(\bar{\sigma}, \sigma)$ is positive semidefinite for any number σ which satisfies the conditions $|\sigma| > 1$ and $\det(\sigma I - A) \neq 0$".

Moreover, if, besides conditions (i), (ii) and (iii) one also has the Conditions (iv) "the pair (A, B) is completely controllable" and (v): $\pi(\bar{\sigma}, \sigma) \not\equiv 0$, then each of the above properties is also equivalent to the property

(h_+): "The expression $\eta(0, k_1)$ can be written in the form (similar to 8, § 10):

$$\eta(0, k_1) = [x_k^* U x_k]_0^{k_1+1} + \sum_{k=0}^{k_1} \| V u_k + W^* x_k \|^2,$$

where U is positive definite".

The proof of Theorem 1 is in every way similar to that of Theorem 1, § 16. The proof requires the results of § 36 which actually are also valid in the case of discrete systems (making everywhere the same changes that were required to pass from (1, § 16)—(2, § 16) to (7)—(8)). In the above theorem Condition (v) $\pi(-\sigma, \sigma) \not\equiv 0$ of Theorem 1, § 16, is replaced by $\pi(1/\sigma, \sigma) \not\equiv 0$ (since in the case of our systems Relation (3, § 9) is replaced by (3, § 10)).

Consider now the single-input discrete blocks of the form

$$x_{k+1} = Ax_k + b\mu_k \tag{14}$$

$$\nu_k = a^* x_k + \alpha\mu_k \tag{15}$$

(cf. (1, § 15)—(2, § 15)). The system associated to (14)—(15) has the form

$$x_{k+1} = Ax_k + b\,\mu_k \tag{16}$$

$$\eta(0, k_1) = \mathrm{Re}\left(\sum_{k=0}^{k_1} \mu_k^* \nu_k\right) = \sum_{k=0}^{k_1}\left(\mathrm{Re}\,\alpha|\mu_k|^2 + \frac{1}{2}\mu_k^* a^* x_k + \frac{1}{2} x_k^* a\mu_k\right).$$

The principal properties of the hyperstable blocks of the form (14)—(15) are stated in the following theorem (cf. Theorem (1, § 15) [1]) :

Theorem 2. *Assume that block (14)—(15) is minimally stable (i.e. the associated system (16) is minimally stable) and that the conditions* $b \neq 0$ *and* $2\mathrm{Re}\,\alpha + a^*(\sigma I - A)^{-1} b \not\equiv 0$ *are also satisfied. Then the following properties are equivalent:*

(H): "Block (14)—(15) is hyperstable",

(F): "The transfer function

$$\gamma(\sigma) = \alpha + a^* (\sigma I - A)^{-1} b \tag{17}$$

satisfies the inequality $\mathrm{Re}\,\gamma(\sigma) \geqslant 0$ *for every number* σ *which satisfies the conditions* $|\sigma| = 1$ *and* $\det(\sigma I - A) \neq 0$*",*

(H'): "The associated system (16) satisfies one of the properties (h), (h_{so}), (h_{po}) *and* (f) *of Theorem 1 (particularized for the case of single-input systems)".*

(h_λ): *"All the feedback systems*

$$x_{k+1} = Ax_k + b\mu_k, \quad \nu_k = a^* x_k + \alpha\mu_k, \quad \eta_{k+1} = \eta_k + \nu_k,$$

$$\mu_k = -\lambda(2\eta_k + \nu_k) \tag{18}$$

$$x_{k+1} = Ax_k + b\mu_k, \quad \nu_k = a^* x_k + \alpha\mu_k, \quad \eta_{k+1} = -\eta_k + \nu_k,$$

$$\mu_k = -\lambda(-2\eta_k + \nu_k), \tag{19}$$

have stable trivial solutions for every real, strictly positive number λ, *except the case* $1 + \alpha\lambda = 0$*."*

(f'): "The function $\sigma \mapsto \gamma\left(\dfrac{\sigma+1}{\sigma-1}\right) = \tilde{\gamma}(\sigma)$ *is positive real."*

The proof differs from the proof from § 15 only in some details. Notice that Properties (h_λ) and (f') have a different form. As in § 15, one shows that if the conditions of Theorem 2 are satisfied then the associated system (16) satisfies Conditions (i)—(iii) of Theorem 1. Making also use of the fact that the characteristic function χ of system (16) (see (31, § 6)) is related to the transfer function (17) by the formula

$$\chi(\lambda, \sigma) = \frac{1}{2}(\bar{\gamma}(\lambda) + \gamma(\sigma)), \tag{20}$$

---

[1]) For simplification we state only the analogues of the statements 1° and 2° of Theorem 1, § 15.

we see that Properties (H), (F) and (H') are equivalent. These properties are also equivalent to Property (f'). Indeed, from (20) it follows that Property (f) of Theorem 1, written for system (16), reduces to the condition $\operatorname{Re} \gamma(\sigma) \geqslant 0$ for every σ which satisfies the conditions $|\sigma| > 1$ and for which the function $\gamma(\sigma)$ is defined. Obviously, this condition can be also expressed in the form (f'). We have now to prove that Properties (H), (F), (H') and (f') are also equivalent to Property (h_λ). As in § 15, we use the scheme of implications:

$$(H) \rightarrow (h_\lambda) \rightarrow (f').$$

To prove $(H) \rightarrow (h)$ we show that if $1 + \alpha\lambda \neq 0$, then Eqs. (18) and (19) can be written in the usual form (i.e. the matrix-vectorial form $y_{k+1} = Cy_k$). Substituting the value of μ_k given by the last relation of (18), in the first two equations of (18), one obtains

$$x_{k+1} = A\, x_k - b\, \lambda\, (2\, \eta_k + \nu_k) \tag{21}$$

$$\nu_k = a^*\, x_k - \alpha\, \lambda\, (2\, \eta_k + \nu_k). \tag{22}$$

Since $1 + \alpha\lambda \neq 0$ one deduces from (22) the relation $\nu_k = (a^*x_k - 2\alpha\lambda\eta_k)/(1 + \alpha\lambda)$. Introducing this value in Eq. (21) and in the equation $\eta_{k+1} = \eta_k + \nu_k$ (see (18)), one obtains a system of the usual form. The same operations bring system (19) (for $1 + \alpha\lambda \neq 0$) to the usual form. Notice now that the systems (18) and (19) have the structure in Fig. 4.1.c (particularized for single-input systems), where B_1 is described by

$$B_1 : x_{k+1} = Ax_k + b\mu_k\,, \qquad \nu_k = a^*\, x_k + \alpha\, \mu_k$$

while B_2, in the case of system (18,) is given by

$$B_2' : \eta_{k+1} = \eta_k + \tilde{\mu}_k, \qquad \tilde{\nu}_k = \lambda\, (\, 2\, \eta_k + \tilde{\mu}_k)$$

and in the case of system (19), by

$$B_2'' : \eta_{k+1} = -\, \eta_k + \tilde{\mu}_k, \qquad \tilde{\nu}_k = \lambda\, (-\, 2\, \eta_k + \tilde{\mu}_k)$$

(the quantities μ_k, ν_k, $\tilde{\mu}_k$ and $\tilde{\nu}_k$ satisfy the relations $\tilde{\mu}_k = \nu_k$ and $\tilde{\nu}_k = -\,\mu_k$). Hence, to conclude the proof of the implication $(H) \rightarrow (h_\lambda)$ it remains only to show that blocks B_2' and B_2'' are hyperstable for every $\lambda > 0$. In the case of block B_2' we obtain from the equations of the block the equalities

$$\lambda\, |\, \eta_{k+1}\, |^2 - \lambda |\, \eta_k |^2 = \lambda\, |\, \eta_k + \tilde{\mu}_k\, |^2 - \lambda |\, \eta_k |^2 = 2\lambda\, \operatorname{Re}(\tilde{\mu}_k^*\, \eta_k) +$$

$$+ \lambda\, |\tilde{\mu}_k|^2 = \lambda\, \operatorname{Re}(\tilde{\mu}_k^*\, (2\, \eta_k + \tilde{\mu}_k)) = \operatorname{Re}(\tilde{\mu}_k^*\, \tilde{\nu}_k)$$

and hence

$$\eta_2(0, k_1) = \mathrm{Re}\left(\sum_{k=0}^{k_1} \widetilde{\mu}_k^* \widetilde{\nu}_k\right) = \lambda |\eta_{k_1+1}|^2 - \lambda |\eta_0|^2$$

which shows that B_2' is hyperstable, for every $\lambda > 0$. Proceeding similarly with the block B_2'' one obtains again the relation $\lambda |\eta_{k+1}|^2 - \lambda |\eta_k|^2 = \mathrm{Re}(\widetilde{\mu}_k^* \widetilde{\nu}_k)$, and thus we find that the block is hyperstable for every $\lambda > 0$. Thus the implication $(H) \to (h_\lambda)$ is proved.

Concerning the implication $(h_\lambda) \to (f')$ we remark that if Property (f') is false, then so is also Property (B) of Theorem 1, Appendix C. Hence, there exists a number σ_0 satisfying the condition $\mathrm{Re}\,\sigma_0 > 0$ and for which one of the numbers $\widetilde{\gamma}(\sigma_0)\sigma_0$ or $\widetilde{\gamma}(\sigma_0)/\sigma_0$ is finite, real and strictly negative. Each of the functions $\widetilde{\gamma}(\sigma)\sigma$ and $\widetilde{\gamma}(\sigma_0)/\sigma_0$ is either constant or else analytic and non constant in an open connected set which contains the point σ_0. In the latter case the image of that open connected set is also an open connected set. Therefore the number σ_0 having the properties stated above can be always chosen such that $\sigma_0 - 1 \neq 0$, $1 - \alpha/(\widetilde{\gamma}(\sigma_0)\sigma_0) \neq 0$ and $1 - \alpha\sigma_0/\widetilde{\gamma}(\sigma_0) \neq 0$. These conditions make it possible to perform the operations which follow.

Introducing the number $\sigma_1 = (\sigma_0 + 1)/(\sigma_0 - 1)$ (which satisfies the condition $|\sigma_1| > 1$) and using the difinition of the function $\widetilde{\gamma}$ (see Property (f')) we find that one of the numbers $\rho_1 = \gamma(\sigma_1)(\sigma_1 + 1)/(\sigma_1 - 1)$ or $\rho_2 = \gamma(\sigma_1)(\sigma_1 - 1)/(\sigma_1 + 1)$ is real, finite and strictly negative. If we have $\det(\sigma_1 I - A) = 0$ then, since $|\sigma_1| > 1$, the trivial solution of system (18) is unstable for $\lambda = 0$ and hence it is also unstable for a positive and sufficiently small λ; this contradicts Property (h_λ). It is therefore sufficient to assume that $\det(\sigma_1 I - A) \neq 0$.

If $\rho_1 < 0$, we consider Eqs. (18) for $\lambda = -1/\rho_1$ and we find that they are satisfied by the sequences $\mu_k = \sigma_1^k$, $x_k = (\sigma_1 I - A)^{-1} b \sigma_1^k$, $\nu_k = \gamma(\sigma_1)\sigma_1^k$ and $\eta_k = \gamma(\sigma_1)\sigma_1^k/(\sigma_1 - 1)$.

Since $|\sigma_1| > 1$ the trivial solution of the system (18) is unstable although the value $\lambda = -1/\rho_1$ is strictly positive, thus contradicting Property (h_λ).

Similarly, if $\rho_2 < 0$ the trivial solution of system (19) is unstable for $\lambda = -1/\rho_2$; as before, this contradicts Property (h_λ)[1]. This concludes the proof of Theorem 2.

The case of multi-input blocks can be treated in a similar way thus establishing the discrete analogue of Theorem 1, § 17.

---

[1]) We make use of the fact (proved before) that in the neighbourhood of the point $\lambda = 0$, systems (18) and (19) can be brought into the usual form, in which, moreover, the coefficients are continuous functions of λ.

§ 19. Hyperstability of more general systems

In § 13 we have defined the property of hyperstability only for systems described by an ordinary differential equation of the form (10, § 13) and an integral of the form (11, § 13). The detailed study of the property of hyperstability, presented in the foregoing chapters, dealt mainly with systems of that particular form. However, in applications other forms of describing the objects under investigation are also encountered [1]). Furthermore, as we have remarked at the beginning of Section 7, § 13, the general results presented in § 13 use only to very slight extent the assumption that the system under investigation is described by equations of the form (10, § 13)—(11, § 13), and hence these results can be easily extended to more general systems. Without the intention of treating this generalization as fully as we did in the case of the systems studied in the foregoing paragraphs, we shall indicate here some slight modifications of the exposition presented in § 13 whereby the field of applications is considerably extended. In order to avoid the repetition of similar reasonings we shall define the new concepts introduced in the present paragraph in such a manner as to keep unchanged most of the definitions and the main results of § 13; we shall only give them a new meaning.

We first introduce a more general form of systems. Consider a set of input functions u with certain properties. For instance, we may assume that u is a vector-valued function of t defined and piecewise continuous for $t \geqslant t_0$. We also consider a set of output functions v, with similar properties.

Let R be a given relation between the input functions and the output functions. Then, for every input-output pair of functions, (u, v), one can establish whether or not relation R is satisfied. We assume that for every input-output pair of functions, (u, v), for which relation R is satisfied there exists a real valued scalar function $t \mapsto \eta(t)$, defined for all $t \geqslant t_0$, called the "index" of the respective pair.

A "system" is defined if we know the sets of the input and output functions, the relation R and the index η. The systems of the form (10, § 13)—(11, § 13) are obviously included in the more general concept of system introduced above. Then the output v is defined by the relation $v = x$. The relation R is defined by the condition "the pair (u, v) satisfies relation R if the functions u and $x = v$ satisfy the differential equation

[1]) Such an example will be treated in the next paragraph.

(10, § 13)". For every input-output pair of functions which satisfy this equation, the corresponding index is defined by Relation (11, § 13) (we take $\eta(t) = \eta(t_0, t)$).

The main object of § 13 — generalized in the present paragraph — is to obtain some information about the output v (namely, certain evaluations of $v(t)$) when we possess some information concerning the index η.

There are several ways of generalizing the property of hyperstability to the abstract systems defined above. We give here the most direct one. For every input-output pair (u, v) satisfying relation R we define a vector-valued function $t \mapsto x(t)$ and a vector x_0. Thus our initial system is *extended* by the addition of the above quantities.

By "solution" of the abstract system defined above we mean a set formed by : 1° an input-output pair (u, v) satisfying relation R and 2° the index η, the function x and the vector x_0 which correspond to the considered pair (u, v). Obviously, a solution can be defined in an interval of the form $[t_0, T]$, by considering the restrictions to this interval of all the functions involved.

Now one can define the property of hyperstability for an abstract system S by using the same definition as before (see Def. 8, § 13) with the following changes : the words "system (10)—(11)" are replaced by "system S", the word "solution" is interpreted in the explained above sense, $\eta(t_0, t)$ is replaced by $\eta(t)$ and $x(t_0)$ is replaced by x_0.

Obviously, the definition thus obtained depends on the choice of η, x and x_0. Since the reason for introducing the definition of hyperstability is to use it in stability problems, we must require that the inequality $\alpha(x(t)) \leqslant \beta(x_0)$, which is obtained from (13, § 13) for $\beta_0 = 0$, should constitute an acceptable definition of the stability of the trivial solution of the system under investigation and the system should admit the trivial solution. Thus, every choice of x and x_0 entails the acceptance of a certain definition of stability for the system under investigation. The main criterion to be considered in choosing x and x_0 is that the definition of stability obtained as indicated above, should be satisfactory. This is a question that is to be solved for each system in particular.

If we add the condition that the function x be continuous in t, we find that Propositions 2, § 13 and 4, § 13 remain also valid in the case of the systems considered in this paragraph. We can also define the sum of two systems of that form by a direct extension of the definition given in Section 8, § 13. We find that if we adequately define x and x_0 for the sum system, then the sum of two hyperstable systems is always a hyperstable system (cf. Proposition 5, § 13).

The introduction of the concept of a hyperstable block is easy since from the beginning we have defined the functions u and v as input and output quantities. In the case of the blocks, however, we shall assume that the vectors $u(t)$ and $v(t)$ have the same number of components. A block is completely defined when one knows the relation R between u and v as described above. The system associated with a block is obtained by introducing beside each pair u, v which satisfies the relation R an index of the particular form

$$\eta(t_1) = \operatorname{Re}\left(\int_{t_0}^{t_1} u^*(t)v(t)\,\mathrm{d}t\right).$$

Let B_1 and B_2 be two blocks characterized by the relations R_1 and R_2 which connect their input and output quantities. We can define a new block B_p resulting from the "parallel connection of the blocks B_1 and B_2, by the following condition: the pair of functions (u, v) is a solution of the block B_p if there exist two functions u_1 and v_1 related by R_1, and two functions u_2 and v_2 related by R_2 such that the relations $u_1 = u_2 = u$ and $v = v_1 + v_2$ are satisfied. Likewise, the pair of functions (u, v) is said to be a solution of the block "with negative feedback" B_f, made up of the blocks B_1 and B_2, if there exists a pair of functions (u_1, v_1) related by R_1, as well as the pair (u_2, v_2) related by R_2 such that the relations $u_1 = u - v_2$, $u_2 = v_1$ and $v = v_1$, are satisfied.

It is easy to show, as in § 13, that if blocks B_1 and B_2 are hyperstable, then blocks B_p and B_f defined above, are also hyperstable. This property remains valid if we consider an arbitrary association of hyperstable blocks with the property that from the connection laws of the blocks we can obtain the equality $\operatorname{Re}(\Sigma u_j^* v_j) = 0$. Also, among the blocks used there may be, as in § 13, some memoryless hyperstable blocks or the inverses of some hyperstable blocks. Hence one obtains a result similar to the statement of Proposition 9, § 13 namely that if the resultant block — with zero input — admits a family of solutions which contains the trivial solution, then the trivial solution is stable (in a sense which depends on the choice of x and x_0 in the definition of the property of hyperstability for the component blocks). If the concept of stability resulting from the choice of the functions x and x_0 is acceptable for each block separately, then the concept of stability obtained for the resultant block should also be satisfactory).

We further remark that all the assertions resulting from the property of hyperstability are true "for every solution of the system". The problem of the existence of these solutions is deliberately left out of the general problem of hyperstability because it must be solved from case to case by methods specific to the equations involved. It may happen that the systems introduced have no solution. Then all the assertions resulting from the property of hyperstability, without becoming false, loose their interest (since the condition "for any solution of the system" leads to an empty set). Sometimes from the property of hyperstability we can deduce that the solution, if it exists, is bounded, a fact which may be useful for proving the existence of a solution (see C. Corduneanu [2] in this connection).

Let us now assume that we have combined several hyperstable blocks of the type considered in § 13 and in this paragraph, using only "parallel" or "negative feedback" connections or other types of connections which imply the condition $\operatorname{Re}(\sum u_j^* v_j) = 0$. Considering the system obtained when the input of the resultant block vanishes, it is easy to see that the state of all the component blocks of the form given in § 13 is bounded by a constant (which depends on the choice of x and x_0 for the blocks of the form described in this paragraph).

Sometimes, even in the case of some systems of the form given in § 13, it is preferable to use the generalized definition of hyperstability introduced in this paragraph. Then, for example, if the state of the system contains certain components whose boundedness is unimportant for the stability problem under investigation, the generalized definition allows one to ignore these components. Thus we ensure the fulfilment of a property of conditional stability concerning only some components of the state of the investigated object.

§ 20. Integral hyperstable blocks

1. Description of completely controllable integral blocks

In this paragraph we shall study, — without making use of the results obtained in the foregoing paragraphs —, a special type of integral blocks described by relations of the form

$$\nu (t) = \rho (t) + \int_0^t \varkappa (t - \tau) \mu_+ (\tau) \, d\tau, \quad t \geqslant 0, \tag{1}$$

where μ, $\varkappa$ and ρ are functions defined for $t \geqslant 0$, piecewise continuous and which may be also complex-valued.

The blocks described by a differential equation of the form

$$\frac{dx}{dt} = Ax + b\,\mu, \qquad \nu = c^*x \tag{2}$$

(see (1, § 15), (2, § 15)), for $\alpha = 0$) can, of course, be also described by Relation (1) where

$$\varkappa\,(t) = c^* e^{At} b, \tag{3}$$

$$\rho\,(t) = c^* e^{At}\, x\,(0). \tag{4}$$

Relation (1) is more general since the kernel $\varkappa$ is not required to have the form (3).

We assume that the function ρ belongs to a class of functions characterized by the following condition : "For any function ρ of the class considered we can find a function μ, piecewise continuous and defined in an interval of the form $-\infty < t_0 \leqslant \leqslant t < 0$ [1], such that for any $t > 0$ the equality

$$\rho(t) = \int_{t_0}^{0} \varkappa\,(t-\tau)\,\mu_-\,(\tau)\,d\tau \tag{5}$$

is satisfied".

The above condition involves a restriction even if the integral block (1) is equivalent to the differential block (2) since, in general, given a pair of functions $\varkappa$ and ρ of the form (3), (4) it is not always possible to satisfy a relation of the form (5). However, by Definition 1, § 31 (Appendix A), this condition is satisfied if (A, b) is completely controllable. Therefore the condition that the functions ρ may be written in the form (5) can be considered as a generalization of the property of complete controllability, for integral blocks.

Using (5) one can write (1) in the form

$$\nu\,(t) = \int_{t_0}^{t} \varkappa\,(t-\tau)\,\mu\,(\tau)\,d\tau, \tag{6}$$

where

$$\mu\,(t) = \begin{cases} \mu_+\,(t) \text{ for } t \geqslant 0 \\ \mu_-\,(t) \text{ for } t_0 \leqslant t < 0. \end{cases} \tag{7}$$

---

[1]) The results remain also valid when $t_0 = -\infty$, if $\mu(t)$ satisfies a condition of the form $|\mu(t)| \leqslant \delta_1 e^{\delta_2 t}$, where δ_1 and δ_2 are positive constants. See the author's paper [12] (which, however, contains a few misprints).

The function $\varkappa$ is assumed to be continuous and defined for every real t and to satisfy the conditions

$$\varkappa(t) = \overline{\varkappa(-t)}, \tag{8}$$

$$\varkappa(0) \neq 0, \tag{9}$$

$$|\varkappa(t)| \leqslant \delta_1, \tag{10}$$

where δ_1 is a constant. Notice that in (6) only the values of $\varkappa$ for $t \geqslant 0$ are occurring (this is why at the beginning of this paragraph we required that the function $\varkappa$ be defined for $t \geqslant 0$). Relation (8) defines in fact an "extension" of the function $\varkappa$ for the negative values of t. The only restriction introduced by this relation is that the value of $\varkappa(0)$ must be real.

$$\eta(t_0, t_1) = \operatorname{Re} \int_{t_0}^{t_1} \overline{\mu(t)}\, \nu(t)\, dt. \tag{11}$$

2. Definition of the hyperstable integral blocks

Definition 1. *The block (6) is said to be hyperstable if there exists a number $\delta_0 \geqslant 0$ such that for every function μ with the properties specified above the corresponding function ν satisfies the inequality*

$$\delta_0 |\nu(t_1)|^2 \leqslant \eta(t_0, t_1), \tag{12}$$

for all $t_1 \geqslant t_0$ [1].

The right side of inequality (12) can be written in a different form which is useful for the next developments. Replacing in (11) the function ν by its expression in (6), one obtains

$$\eta(t_0, t_1) = \int_{t_0}^{t_1} \operatorname{Re} \left(\overline{\mu(t)} \int_{t_0}^{t} \varkappa(t-\tau)\, \mu(\tau)\, d\tau \right) dt =$$

$$= \frac{1}{2} \int_{t_0}^{t_1} \mu(t)\, dt \int_{t_0}^{t} \overline{\varkappa(t-\tau)}\, \overline{\mu(\tau)}\, d\tau +$$

$$+ \frac{1}{2} \int_{t_0}^{t_1} \overline{\mu(t)}\, dt \int_{t_0}^{t_1} \varkappa(t-\tau)\, \mu(\tau)\, d\tau. \tag{13}$$

[1]) From Definition 1 it follows that the block (6) is hyperstable in the sense stated in § 19 (for $t_0 = 0$), if we define x by the relation $x(t) = \nu(t)$ and x_0 by the relation $x_0 = \eta(t_0, 0)$ (we take the number t_0 of § 19 equal to zero).

Changing the order of integration in the last term of the right hand member of (13) and using (8) one obtains

$$\frac{1}{2}\int_{t_0}^{t_1}\overline{\mu(t)}\,dt\int_{t_0}^{t}\varkappa(t-\tau)\,\mu(\tau)\,d\tau =$$

$$=\frac{1}{2}\int_{t_0}^{t_1}\mu(\tau)\,d\tau\int_{\tau}^{t_1}\overline{\varkappa(\tau-t)}\,\overline{\mu(t)}dt. \tag{14}$$

Replacing in the last integral τ by t and t by τ and introducing the expression thus obtained in (13) one has

$$\eta(t_0, t_1) = \frac{1}{2}\int_{t_0}^{t_1}\mu(t)\,dt\int_{t_0}^{t_1}\overline{\varkappa(t-\tau)}\,\overline{\mu(\tau)}\,d\tau =$$

$$\frac{1}{2}\int_{t_0}^{t_1}\overline{\mu(t)}\,dt\int_{t_0}^{t_1}\varkappa(t-\tau)\,\mu(\tau)\,d\tau, \tag{15}$$

the last equality resulting from the fact that $\eta(t_0, t_1)$ is real (see (11)).

3. A method of obtaining the desired inequalities

In establishing inequalities of the form required by Definition 1 we shall repeatedly make use of the following well-known method: we introduce certain vector spaces (where the addition of elements and the multiplication by scalars are defined in the usual way), and we define for each pair of elements x, y of the vector space a complex number $[x, y]$ having the following properties [1]:

$$[\alpha x + \beta y, z] = \alpha [x, z] + \beta [y, z], \tag{16}$$

$$[x, y] = \overline{[y, x]}, \tag{17}$$

$$[x, x] \geqslant 0, \tag{18}$$

where x, y and z are arbitrary elements of the vector space considered and α and β are arbitrary complex numbers. From (16)—(18) one obtains the inequality of Cauchy-Buniakowski-Schwarz

[1]) Relations (16)—(18) are similar to the relations which characterize "scalar products" with the only difference that the inequality $[x, x] = 0$ does not entail here the equality $x = 0$.

$$|[x, y]|^2 \leqslant [x, x]\,[y, y], \tag{19}$$

for every elements x and y of the vector space considered. In applications, starting from a certain inequality which has to be proved, we shall bring it to the form (19) by a convenient definition of the product $[x, y]$ and an adequate choice of the elements x and y. Then, in order to prove the inequality we have only to check that the "product" $[x, y]$ satisfies Conditions (16)—(18).

4. Hyperstability theorem for integral blocks

Theorem 1. *The following properties are equivalent:*

1° Block (6) is hyperstable (Definition 1).

2° There exists a constant $\delta > 0$ such that, for every pair of functions μ and ν and every constant $\gamma \geqslant 0$ satisfying (6) and the inequality

$$\eta\,(0,t) \leqslant \gamma^2 + \gamma \sup_{0 \leqslant \tau \leqslant t} |\nu(\tau)|, \quad \textit{for every } t \geqslant 0 \tag{20}$$

one also has the inequality

$$|\nu\,(t)| \leqslant \delta\,(\gamma + \sqrt{|\,\eta\,(t_0, 0)\,|})\,, \quad \textit{for every } t \geqslant 0. \tag{21}$$

3° *The Laplace transform of the kernel* [1])

$$\gamma\,(\sigma) = \int_0^\infty \mathrm{e}^{-\sigma t}\,\varkappa(t)\,\mathrm{d}t \tag{22}$$

satisfies the condition

$$\text{"}\mathrm{Re}\ \gamma(\sigma) \geqslant 0, \textit{ for every } \sigma \textit{ with } \mathrm{Re}\ \sigma > 0\text{".} \tag{23}$$

4° *The kernel $\varkappa(t)$ is a positive function* [2]). *In orher words for every finite sequence of real numbers $t_1, t_2, \ldots, t_N$ (where N is an arbitrary positive integer) and for every finite sequence of complex numbers $\alpha_1, \alpha_2, \ldots, \alpha_N$, we have*

$$\sum_{m=1}^{N} \sum_{n=1}^{N} \alpha_m\, \varkappa\,(t_m - t_n)\, \overline{\alpha}_n \geqslant 0. \tag{24}$$

---

[1]) This transform exists at least for $\mathrm{Re}\ \sigma > 0$ since by hypothesis the kernel $\varkappa(t)$ satisfies Condition (10).

[2]) See, for instance, M. Onicescu, G. Mihoc and C. T. Ionescu-Tulcea [1] Chapter VIII, § 1, Section 5. The continuity of the function $\varkappa(t)$ has been required above, immediately after formula (7).

5° *The inequality*

$$\eta(t_0, t_1) \geqslant 0$$

is satisfied for every $t_1 \geqslant t_0$ *and every pair of functions* μ *and* ν *which satisfy (6).*

The proof of the theorem follows the scheme

$$\begin{array}{c} 1^\circ \rightarrow 2^\circ \rightarrow 3^\circ \rightarrow 4^\circ \rightarrow 1^\circ. \\ \searrow 5^\circ \nearrow \end{array} \tag{25}$$

The implication $1^\circ \rightarrow 2^\circ$ is proved by writing the right hand member of inequality (12) in the form $\eta(t_0, t_1) = \eta(t_0, 0) + \eta(0, t)$ and then using the same arguments as in the proof of Proposition 4, § 13. The implication $1^\circ \rightarrow 5^\circ$ is obvious. The implications $2^\circ \rightarrow 3^\circ$ and $5^\circ \rightarrow 3^\circ$ can be proved simultaneously by showing that the hypothesis : "Property 3° is not satisfied", contradicts both Property 2° and Property 5°. Indeed, if Property 3° is not satisfied, there exists a complex number σ_0 satisfying the conditions $\mathrm{Re}\ \sigma_0 > 0$ and $\mathrm{Re}\ \gamma(\sigma_0) < 0$. Consider the input $\mu(t) = e^{\sigma_0 t}$. The corresponding function ν given by (6) can be written in the form $\nu = \nu_1 - \nu_2$ where

$$\nu_1(t) = \int_{-\infty}^{t} \varkappa(t - \tau)\, e^{\sigma_0 \tau}\, d\tau = e^{\sigma_0 t} \int_0^{\infty} e^{-\sigma_0 \rho} \varkappa(\rho)\, d\rho = \gamma(\sigma_0)\, e^{\sigma_0 t},$$

$$\nu_2(t) = \int_{-\infty}^{t_0} \varkappa(t - \tau)\, e^{\sigma_0 \tau}\, d\tau.$$

The above expression of function ν_1 is obtained by effecting the change of variable $\rho = t - \tau$ and using then Relation (22). Using the relations $\mathrm{Re}\ \sigma_0 > 0$ and $\mathrm{Re}\ \gamma(\sigma_0) < 0$ we conclude that $\lim\limits_{t \to \infty} |\nu_1(t)| = \infty$. Furthermore by conditions $\mathrm{Re}\ \sigma_0 > 0$ and (10) the function $\nu_2(t)$ is bounded. Hence, the function $\nu = \nu_1 - \nu_2$ has the property

$$\lim_{t \to \infty} |\nu(t)| = \infty. \tag{26}$$

Introducing the functions μ and ν in (11) we obtain

$$\eta(t_i, t) = \eta_1(t_i, t) + \eta_2(t_i, t),$$

where

$$\eta_j(t_i, t) = \mathrm{Re}\left(\int_{t_i}^{t} \overline{\mu(\tau)}\, \nu_j(\tau)\, d\tau\right), \quad j = 1, 2.$$

Computing the integral $\eta_1(t_1, t)$ we have

$$\eta_1(t_i, t) = \frac{\mathrm{Re}\,\gamma(\sigma_0)}{2\,\mathrm{Re}\,\sigma_0}\left(e^{2\,\mathrm{Re}\,\sigma_0 t} - e^{2\,\mathrm{Re}\sigma_0 t_i}\right).$$

Since, as we have seen, the function ν_2 is bounded, the function $t \mapsto \eta_2(t_i, t)$, for a fixed t_i, satisfies an inequality of the form

$$|\eta_2(t_i, t)| \leqslant \text{const.}\, e^{\mathrm{Re}\,\sigma_0 t}.$$

These relations together with the condition $\mathrm{Re}\,\sigma_0 > 0$ lead to the conclusion

$$\lim_{\to\infty} \eta(t_i, t) = -\infty, \tag{27}$$

for any value of t_i. Thus $\eta(0, t)$ is bounded from above and hence there exists a constant γ such that (20) is satisfied; but Condition (26) is incompatible with any relation of the form (21) and this contradicts Property 2° and proves the implication 2° → 3°. Furthermore, Condition (27) written for $t_i = t_0$ contradicts also Property 5°, thus proving also the implication 5° → 3°.

Proof of implication 3° → 4°. If Property 4° is not satisfied, there exists a positive integer N, real numbers $t_1, t_2, \ldots, t_N$ and complex numbers $\alpha_1, \alpha_2, \ldots, \alpha_N$ such that the double sum in the left hand member of (24) is negative. Therefore introducing the new function

$$\tilde{\varkappa}(t) = e^{-\alpha|t|}\, \varkappa(t), \tag{28}$$

the inequality

$$\tilde{\eta}(N) = \sum_{m=1}^{N} \sum_{n=1}^{N} \alpha_m\, \tilde{\varkappa}(t_m - t_n)\, \bar{\alpha}_n < 0 \tag{29}$$

is satisfied for $\alpha = 0$ and also (by continuity) for a small enough $\alpha > 0$. Take such an α. Using Condition (8) it is easy to see that the Fourier transform of the function $\tilde{\varkappa}$, (28), is related to the Laplace transform γ, (22), by the equality

$$\int_{-\infty}^{\infty} \tilde{\varkappa}(t)\, e^{-i\omega t}\, dt = 2 \operatorname{Re} \gamma(\alpha + i\omega). \tag{30}$$

By Property 3°, since $\alpha > 0$, the Fourier transform (30) is non-negative, for every real ω. From the continuity of the function (28) and from (29) it follows that

$$\sum_{m=1}^{N} \sum_{n=1}^{N} \alpha_m \tilde{\varkappa}(t_m - t_n + \tau)\, \bar{\alpha}_n < 0$$

in an interval $-R < \tau < R$, if R is sufficiently small. Therefore we can introduce the negative number

$$\xi = \int_{-R}^{R} \left(1 - \frac{|\tau|}{R}\right)\left(\sum_{m=1}^{N} \sum_{n=1}^{N} \alpha_m \tilde{\varkappa}(t_m - t_n + \tau)\right) d\tau < 0. \tag{31}$$

It is easily seen (for example using Parseval's formula), that the integral of (31) can be also written as

$$\xi = \frac{1}{\pi}\int_{-\infty}^{\infty} \frac{1 - \cos(R\omega)}{R\omega^2} (\operatorname{Re}\gamma(\alpha + i\omega)) \left| \sum_{m=1}^{N} \alpha_m e^{i\omega t_m} \right|^2 d\omega. \tag{32}$$

Thus $\xi \geqslant 0$ (since $\operatorname{Re} \gamma(\alpha + i\omega) \geqslant 0$), and this contradicts (31).

Proof of implication 4° → 1°. We apply the method described in Section 3 of this paragraph. Given an arbitrary positive number t, introduce the vector space whose elements are of the form $x = \{\alpha ; \mu\}$, where α is a complex number and μ is a function $\tau \mapsto \mu(\tau)$ defined for $t_0 \leqslant \tau \leqslant t$ and satisfying the conditions required to the input functions of block (6). The addition and multiplication by a complex number are defined in the usual way. We shall show that, given two arbitrary elements $x_1 = \{\alpha_1 ; \mu_1\}$ and $x_2 = \{\alpha_2 ; \mu_2\}$ the expression $[x_1, x_2]$, defined as

$$[x_1, x_2] = \alpha_1 \bar{\alpha}_2 \varkappa(0) + \alpha_1 \int_{t_0}^{t} \varkappa(t - \tau)\, \overline{\mu_2(\tau)}\, d\tau +$$

$$+\left(\int_{t_0}^{t} \mu_1(\tau)\, \varkappa(\tau - t)\, d\tau\right) \bar{\alpha}_2 + \int_{t_0}^{t} \mu_1(\tau)\, d\tau \int_{t_0}^{t} \varkappa(\tau - \rho)\, \overline{\mu_2(\rho)}\, d\rho, \tag{33}$$

satisfies Conditions (16)—(18).

Assuming that the above assertion has been proved we can write (19) for the particular values $x = \{1, 0\}$ and $y = \{0 ; \bar{\mu}\}$, where μ is an arbitrary function belonging to the class of input functions. For these values of x and y we obtain from (33) the equalities

$$[x,y] = \nu(t), \quad [x, x] = \varkappa(0); \quad [y, y] = 2\eta(t_0, t) \tag{34}$$

(using also (6) and (15)). From $[x, x] \geqslant 0$ (see (18)) and Condition (9) it follows that $\varkappa(0) > 0$. Thus, Relations (34) and (19) allow to write the equality (12) of Definition 1 where the number δ_0 can be taken equal to $1/2\varkappa(0)$. Hence, in order to conclude the proof of implication 4° → 1°, we have to show that if Property 4° is satisfied, then (33) possesses Properties (16)—(18).

Since Properties (16) and (17) are directly verified by using also (8) it remains to consider Relation (18). If this relation is not satisfied, then we can find an element $x = \{\beta, \mu\}$, such that $[x, x] < 0$, or explicitly (see (33)):

$$|\beta|^2 \varkappa(0) + 2 \operatorname{Re}\left(\beta \int_{t_0} \varkappa(t - \tau) \overline{\mu(\tau)} \, d\tau\right) + \\ + \int_{t_0}^{t} \mu(\tau) \, d\tau \int_{t_0}^{t} \varkappa(\tau - \rho) \overline{\mu(\rho)} \, d\rho = -\gamma_0 < 0, \tag{35}$$

where γ_0 is a positive number. We approximate the integrals by Riemann sums by finding a positive integer N and real equidistant numbers $t_1, t_2, \ldots, t_N$, in the interval $[t_0, t]$ such that

$$2 \frac{(t - t_0)}{N} \operatorname{Re}\left(\beta \sum_{k=1}^{N} \varkappa(t - t_k) \overline{\mu(t_k)}\right) - \\ - 2 \operatorname{Re}\left(\beta \int_{t_0}^{t} \varkappa(t - \tau) \overline{\mu(\tau)} d\tau\right) < \frac{\gamma_0}{4}, \tag{36}$$

$$\left(\frac{t - t_0}{N}\right)^2 \sum_{m=1}^{N} \sum_{n=1}^{N} \mu(t_m) \varkappa(t_m - t_n) \overline{\mu(t_n)} - \\ - \int_{t_0}^{t} \mu(\tau) \, d\tau \int_{t_0}^{t} \varkappa(\tau - \rho) \overline{\mu(\rho)} \, d\rho < \frac{\gamma_0}{4}. \tag{37}$$

Adding Relations (35)—(37) we obtain the inequality

$$|\beta|^2 \varkappa(0) + 2 \operatorname{Re}\left(\beta \frac{t - t_0}{N} \sum_{n=1}^{N} \varkappa(t - t_n) \overline{\mu(t_n)}\right) + \\ + \left(\frac{t - t_0}{N}\right)^2 \sum_{m=1}^{N} \sum_{n=1}^{N} \mu(t_m) \varkappa(t_m - t_n) \overline{\mu(t_n)} < -\frac{\gamma_0}{2} < 0. \tag{38}$$

Using the notations $t_{N+1} = t$, $\alpha_j = \frac{t - t_0}{N} \mu(t_j)$, $j = 1, 2, \ldots, N$ and $\alpha_{N+1} = \beta$ the left-hand member of (38) can be written as the double sum in (24) (where N is replaced by $N + 1$) and therefore (38) contradicts (24) (and Property 4°). This concludes the proof of implication 4° and of Theorem 1.

5. Multi-input integral blocks

The principal contents of Theorem 1 published by the author in [12] have been subsequently extended by A. Halanay [4], [5] to discrete and multi-input blocks.

We shall briefly examine the case of multi-input blocks described by an integral equation of the form

$$v(t) = \int_{t_0}^{t} K(t - \tau)\, u(\tau)\, d\tau, \quad t_0 < 0, \tag{39}$$

where u is a vector-valued function with m piecewise continuous components. The "kernel" K is an $m \times m$-matrix whose entries are continuous and bounded functions of time, defined for every real t and having the properties

$$K(t) = K^*(-t), \tag{40}$$

$$\det K(0) \neq 0. \tag{41}$$

We introduce the expression (cf. (11))

$$\eta(t_0, t) = \mathrm{Re}\left(\int_{t_0}^{t} u^*(\tau)\, v(\tau)\, d\tau\right). \tag{42}$$

Definition 2. *A block described by (39) is said to be hyperstable if there exists a number $\delta_0 \geqslant 0$ such that for every pair of functions (u, v) satisfying (39), the inequality*

$$\delta_0 \,\| v(t) \|^2 \leqslant \eta(t_0, t) \tag{43}$$

is satisfied for every $t \geqslant t_0$.

Theorem 2. *The following properties are equivalent:*

1° Block (39) is hyperstable (Definition 2).

2° There exists a number $\delta \geqslant 0$ such that for every pair of functions (u, v) and every constant $\gamma \geqslant 0$ satisfying (39) *and the inequality*

$$\eta(t_0, t) \leqslant \gamma + \gamma^2 \sup_{0\leqslant\tau\leqslant t} \| v(\tau) \| \text{ for every } t \geqslant 0, \tag{44}$$

one also has the inequality

$$\| v(t) \| \leqslant \delta (\gamma + \sqrt{|\eta(t_0, 0)|}) \text{ for every } t \geqslant 0. \tag{45}$$

3° *The Laplace transform of the kernel*

$$G(\sigma) = \int_0^\infty e^{-\sigma t} K(t)\, dt \tag{46}$$

has the property that the hermitian matrix $G(\sigma) + \overline{G'(\sigma)}$ *is positive semidefinite for any complex number* σ *with* $\operatorname{Re} \sigma > 0$.

4° *For every finite sequence of real numbers* $t_1, t_2, \ldots, t_N$ *(where* N *is an arbitrary positive integer) and every finite sequence of complex numbers* $\alpha_1, \alpha_2, \ldots, \alpha_N$, *the matrix*

$$S(N) = \sum_{m=1}^{N} \sum_{n=1}^{N} \alpha_m K(t_m - t_n) \bar{\alpha}_n \tag{47}$$

is positive semidefinite.

5° *The inequality*

$$\eta(t_0, t_1) \geqslant 0 \tag{48}$$

is satisfied for every $t_1 \geqslant t_0$ *and every pair of functions* (u, v) *which satisfy Relation (39).*

The proof is similar to the previous one. The implication 1° → 2° is obtained as in Proposition 4, § 13 and the implication 1° → 5° is obvious. The implications 2° → 3° → 4° and the implication 5° → 3° can be proved by examining the input u of the particular form $u(t) = c\mu(t)$ (where μ is a scalar function and c is an arbitrary complex m-vector satisfying the condition $\| c \| = 1$), and by defining the scalar output $\nu(t) = c^* v(t)$. Then, instead of $G(\sigma)$ we have the scalar $\gamma(\sigma) = c^* G(\sigma) c$, instead of the matrix sum (47) we have the scalar sum $c^* S(N) c$, and the proofs follow the corresponding proofs of Theorem 1.

Some changes occur in the proof of implication 4° → 1° but the main ideas remain the same. We consider a vector space whose elements are of the form $\{a; u\}$, where a is a constant m-vector and u is a piecewise continuous function $\tau \mapsto u(\tau)$ defined for $t_0 \leqslant \tau \leqslant t$. Let $x_1 = \{a_1; u_1\}$ and $x_2 =$

$= \{a_2\,;\ u_2\}$ be two arbitrary elements ; we define their "product" by the expression (cf. (33))

$$[x_1, x_2] = a_1' K(0)\,\bar{a}_2 + a_1' \int_{t_0}^{t} K(t-\tau)\,\overline{u_2(\tau)}\,d\tau +$$

$$+ \left(\int_{t_0}^{t} u_1(\tau)\,K(\tau - t)\,d\tau\right)\bar{a}_2 + \int_{t_0}^{t} u_1'(\tau)\,d\tau \int_{t_0}^{t} K(\tau - \rho)\,\overline{u_2(\rho)}\,d\rho. \tag{49}$$

As in the case of (33) we prove that the "product" (49) satisfies Conditions (16)—(18) and hence that (19) is valid. If $v(t) \neq 0$, by writing (49) for $x = \left\{\dfrac{\bar{v}(t)}{\|v(t)\|},\ 0\right\}$ and $y = \{0, \bar{u}\}$, we obtain

$$[x, y] = \frac{\bar{v}'(t)\,v(t)}{\|v(t)\|} = \|v(t)\|,$$

$$[x, x] = \frac{\bar{v}'(t)\,K(0)\,v(t)}{\|v(t)\|^2} \leqslant \sup_{\|a\|=1} a'K(0)\,\bar{a},$$

$$[y, y] = 2\eta(t_0, t).$$

Therefore, the inequality (19) can be written in the form (43), where $\delta_0 = 1/(2 \sup_{\|a\|=1} a'K(0)\bar{a})$.

If $v(t) = 0$ we repeat the reasoning for $x = \{e\,;\ 0\}$ and $y = \{0\,;\ \bar{u}\}$, where e is an arbitrary vector with the norm equal to unity. Then inequality (19) can be written in the form $\eta(t_0, t) \geqslant 0$, which shows that in this case too, inequality (43) is satisfied.

§ 21. Lemma of I. Barbălat and its use in the study of asymptotic stability

We have seen in § 13 that if the inequality (15, § 13) is satisfied for a hyperstable system of the form (10, § 13), (11, § 13), then the solutions of the blocks are bounded (see (16, § 13)).

We shall now find additional conditions ensuring also the relation $\lim_{t\to\infty} x(t) = 0$ which characterizes asymptotic stability. A natural way of treating this problem results from the remark that if (16, § 13) is satisfied, then the left hand member of (15, § 3) is bounded from above. Thus we must find the consequences of the fact that a certain integral is bounded from above when the upper limit of integration tends to infinity. An answer to this problem is given by the lemma of I. Barbălat [1] which can be stated (in a slightly modified form which better suits our purpose) as follows:

Lemma 1. (I. Barbălat) *If φ is a real function of the real variable t, defined and uniformly continuous for $t > 0$ and if the limit of the integral $\int_0^t \varphi(\tau)d\tau$ as t tends to infinity exists and is a finite number, then*

$$\lim_{t\to\infty} \varphi(t) = 0. \tag{1}$$

Indeed, if (1) is not satisfied then there exists a positive number ϑ_0 such that for every positive number T one can find a number $t(T) \geqslant T$ with $|\varphi(t(T))| \geqslant \vartheta_0$. Since φ is uniformly continuous there exists a positive number ϑ_1 such that for every $\tilde{t} > 0$ and every τ in the interval $0 \leqslant \tau < \vartheta_1$ one has $|\varphi(\tilde{t}) - \varphi(\tilde{t} + \tau)| \leqslant \vartheta_0/2$. Hence, the inequality $|\varphi(t)| - |\varphi(t(T))| \geqslant -\vartheta_0/2$ is satisfied for $t \in [t(T), t(T) + \vartheta_1]$; adding this inequality and the previous inequality $|\varphi(t(T))| > \vartheta_0$ we get $|\varphi(t)| > \vartheta_0/2$ for every t in the mentioned interval. We thus obtain

$$\left|\int_{t(T)}^{t(T)+\vartheta_1} \varphi(t)\,dt\right| = \int_{t(T)}^{t(T)+\vartheta_1} |\varphi(t)|\,dt \geqslant \frac{1}{2}\vartheta_0\vartheta_1,$$

where the first equality holds since $\varphi(t)$ retains the same sign for $t(T) \leqslant t \leqslant t(T) + \vartheta_1$. Thus the integral $\int_0^t \varphi(\tau)d\tau$ cannot tend to a finite limit when time tends to infinity: a contradiction.

From the above proof one derives the following more general statement:

Lemma 1′. *If φ is a real function of the real variable t, defined and uniformly continuous for $t \geqslant 0$, and if, for every $\Delta > 0$ there exists $\vartheta \in]0, \Delta]$ such that $\lim_{t\to\infty} \int_t^{t+\vartheta} \varphi(\tau)\,d\tau = 0$, then $\lim_{t\to\infty} \varphi(t) = 0$.*

Now we can obtain some conditions for asymptotic stability, using

Lemma 2. *Assume that system (10, § 13)—(11, § 13) is hyperstable and that there exists a function of class M_i (see definition 5, § 13) such that for every solution of the system one has the inequality*

$$\eta(t_0, t_1) \geqslant \int_{t_0}^{t_1} \rho(x(t))\,dt, \quad \textit{for every } t_1 \geqslant t_0. \tag{2}$$

Then, for every solution of the system for which u is bounded and one has

$$\eta(t_0, t_1) \leqslant \textit{constant}, \quad \textit{for every } t_1 \geqslant t_0 \tag{3}$$

one also has

$$\lim_{t\to\infty} x(t) = 0. \tag{4}$$

Indeed, from (3) and the fact that the system is hyperstable it follows that the function x is bounded. Since the function u is also bounded and the function f of (10, § 13) is piecewise continuous, the derivative dx/dt is bounded and hence the function x is uniformly continuous. From the fact that the function ρ belongs to class M_i it follows that the function $t \mapsto \rho(x(t))$ is uniformly continuous. By inequalities (2), (3) and $\rho(x(t)) \geqslant 0$ (see Definition 5, § 13), the integral in the right hand member of (2) tends to a finite limit when t_1 tends to infinity. Using the Lemma of Barbălat one obtains

$$\lim_{t\to\infty} \rho(x(t)) = 0. \tag{5}$$

Using again the fact that the function ρ belongs to class M_i one obtains (4).

Notice that if the function ρ belongs to class M_s (Definition 6, § 13) then one obtains as before (5), but no longer (4). Then we can at most obtain the conclusion that some components of $x(t)$ tend to zero when t tends to infinity. Sometimes, using this conclusion and resorting to the direct study of Eq. (10, § 13) one can also obtain the general conclusion (4).

Lemma 2 is useful in obtaining the conditions of asymptotic stability in the case of arbitrary hyperstable systems (linear or nonlinear). For linear systems of the form (1, § 13)—(2, § 13) a more exact study of asymptotic stability is given in the next two paragraphs.

§ 22. Other methods for studying asymptotic stability

Under the hypothesis of complete controllability the hyperstable systems examined in §§ 14—17 are, at the same time, positive systems and can be brought to the special forms from Section 6, § 8. For example, the integral $\eta(0, t_1)$ can be written in the form (51, § 8) and Conditions (7, § 13)—(8, § 13) give the inequality

$$\int_0^{t_1} |\mu_c(t)|^2 \, dt \leqslant \tilde{\delta}, \textit{ for every } t_1 > 0, \tag{1}$$

(see also (3, § 13)), where $\tilde{\delta}$ is a new positive constant. We thus come to the study of the solutions of a differential equation of the form

$$\frac{dx_c}{dt} = A_c x_c + b_c \mu_c, \tag{2}$$

(see (51, § 8)) when the control function μ_c satisfies an inequality of the form (1). To this purpose we state the following lemma (which also holds for multi-input systems):

Lemma 1. *Let A_c be an $n \times m$-matrix, B_c, an $n \times m$-matrix and u_c a vector-valued function with n piecewise continuous components. Define $y(t_2, t_1)$ as*

$$y(t_2, t_1) = \int_{t_1}^{t_2} e^{A_c(t_2-t)} B_c u_c(t) \, dt, \quad 0 \leqslant t_1 < t_2. \tag{3}$$

Then

$$\| y(t_2, t_1) \| \leqslant \sqrt{\| X_0(t_2, t_1) \|} \sqrt{X_1(t_2, t_1)}, \tag{4}$$

where

$$X_1(t_2, t_1) = \int_{t_1}^{t_2} \| u_c(t) \|^2 \, dt, \tag{5}$$

$$X_0(t_2, t_1) = \int_0^{t_2-t_1} e^{A_c(t_2-t_1-t)} B_c B_c^* e^{A_c^*(t_2-t_1-t)} \, dt, \tag{6}$$

$$\| X_0(t_2, t_1) \| = \sup_{\|p\|=1} | p^* X_0(t_2, t_1) p |, \tag{7}$$

p denoting an arbitrary m-vector.

Proof. We apply the method of Section 3, § 20 and introduce a vector space whose elements are of the form $x = \{u\,;\,p\}$ where p is a constant m-vector and u satisfies the conditions required in the lemma. If $\tilde{x} = \{\tilde{u}\,;\,\tilde{p}\}$ is another element of the same space, we introduce the "product"

$$[\tilde{x}, x] = \int_{t_1}^{t_2} (\tilde{u}'(t) + \tilde{p}'\, e^{A_c(t_2-t)}\, B_c)\, \overline{(u(t) + B_c^*\, e^{A_c^*(t_2-t)}\, \overline{p})}\, \mathrm{d}t, \qquad (8)$$

which satisfies all the conditions (16, § 20)—(18, § 20), as one can easily see. Hence, one can write the inequality (19, § 20) for the particular elements $x = \{0\,;\,\overline{p}\}$, $y = \{\overline{u}_c\,;\,0\}$, thus obtaining (with the notations (3)—(6)):

$$|\, p^* y\,(t_2, t_1)\,|^2 \leqslant X_1\,(t_2, t_1)\, p^* X_0\,(t_2, t_1)\, p.$$

Writing this inequality for the particular value $p = y(t_2, t_1)/\|y(t_2, t_1)\|$ (assuming $\|y(t_2, t_1)\| \neq 0$), the left hand member becomes equal to $\|y(t_2, t_1)\|^2$ and the right hand member does not exceed $\|X_0\| X_1$ (see (7)), whence (4). If $y(t_2, t_1) = 0$, (4) is obvious.

From Lemma 1 we derive

Lemma 2. *If for every* $\lambda > 0$ *we have*

$$\lim_{t\to\infty} X_1\,(t+\lambda, t) = \lim_{t\to\infty} \int_t^{t+\lambda} \|u\,(t)\|^2\, \mathrm{d}t = 0, \qquad (9)$$

then for every pair of numbers $\varepsilon > 0$ *and* $\lambda > 0$ *there exists a* $T_0 > 0$ *such that in every interval* $t_1 \leqslant t \leqslant t_1 + \lambda$ *(for every* $t_1 \geqslant T_0$*) the solutions of the equation* $\mathrm{d}x/\mathrm{d}t = Ax + Bu$ *satisfy the inequality*

$$\|x\,(t) - e^{A(t-t_1)}\, x\,(t_1)\| < \varepsilon\,;\ (T_0 \leqslant t_1 \leqslant t \leqslant t_1 + \lambda). \qquad (10)$$

Condition (9) is satisfied if the function u *satisfies one of the conditions*

$$\lim_{t\to\infty} u\,(t) = 0, \qquad (11)$$

$$\int_0^\infty \|u\,(t)\|^2\, \mathrm{d}t < \infty. \qquad (12)$$

Indeed, all the solutions of the equation $\mathrm{d}x/\mathrm{d}t = Ax + Bu$ can be written in the form

$$x\,(t) = e^{A(t-t_1)}\, x\,(t_1) + y\,(t, t_1), \qquad (13)$$

where $y(t, t_1)$ has the expression (3) for $t_2 = t$, $A_c = A$, $B_c = B$ and $u_c = u$. From (5)—(7) one obtains the inequalities

$$\|X_0(t, t_1)\| \leqslant \|X_0(t_1 + \lambda, t_1)\|; \; X_1(t, t_1) \leqslant X_1(t_1 + \lambda, t_1) \quad (14)$$

(for $t_1 \leqslant t \leqslant t_1 + \lambda$).
Notice that the integral $X_0(t_1 + \lambda, t_1)$ does not depend on t_1 but only on λ (see (6)) and hence one can define

$$\rho(\lambda) = \|X_0(t_1 + \lambda, t_1)\|. \quad (15)$$

By (6) and (7), we have $\rho(\lambda) = 0$ for a number $\lambda \neq 0$ only if X_0, (6), is identically zero. Then from (4) it follows that the left hand member of (10) is identically zero. In the case $\rho(\lambda) \neq 0$, it follows from Condition (9), that for every positive number ε there exists a number t_1 such that for every $t > t_1$ we have

$$X_1(t + \lambda, t) \leqslant \frac{\varepsilon^2}{\rho(\lambda)}.$$

Then from (4), (14) and (15) we deduce the inequality $\|y(t, t_1)\| < \varepsilon$ in the interval $T_0 \leqslant t_1 \leqslant t \leqslant t_1 + \lambda$, i.e. (10) (see (13)).

If Condition (11) is satisfied we can find a number $T_0 > 0$ such that $\|u(t)\|^2 < \varepsilon/\lambda$, for every $t > T_0$. This implies

$$\int_t^{t+\lambda} \|u(t)\|^2 \, dt \leqslant \varepsilon$$

and hence (9). If Condition (12) is satisfied one obtains again (9) — since the integrand is nonnegative.

From Lemma 2 we derive

Lemma 3. *If in the equation*

$$\frac{dx}{dt} = Ax + Bu, \quad (16)$$

all the characteristic values of the matrix A have non-zero real parts and the function u satisfies one of the Conditions (11) or (12), then every bounded solution of the equation has the property $\lim_{t \to \infty} x(t) = 0$.

Proof. First assume that the matrix A has no characteristic value with a strictly positive real part. Then, under our assumptions, the real parts of all the characteristic values of the matrix A are strictly negative. Consider an arbitrary positive number t_1 and an arbitrary bounded solution of Eq. (16) satisfying an inequality of the form $\| x(t) \| \leqslant \delta_0$. Under the assumption adopted for A we have the property $\lim_{t \to \infty} e^{A(t-t_1)} \; x(t_1) = 0$. Hence, for every $\varepsilon > 0$ there exists $\lambda > 0$ such that

$$\text{"if } t - t_1 \geqslant \lambda \text{ and } \| x(t_1) \| \leqslant \delta_0 \text{ then } \| e^{A(t-t_1)} \; x(t_1) \| \leqslant \varepsilon\text{".} \tag{17}$$

Furthermore, by condition $\| x(t) \| \leqslant \delta_0$, the number λ which satisfies this condition can be chosen independent of $x(t_1)$ and hence of t_1. Applying Lemma 2 for the above numbers ε and λ, then writing (10) for $t = t_1 + \lambda$ and using Property (17), we obtain the condition

$$\| x\,(t_1 + \lambda) \| < 2\,\varepsilon, \text{ for every } t_1 \geqslant T_0.$$

This gives $\lim_{t \to \infty} x(t) = 0$.

Assume now that the matrix A has no characteristic value with a strictly negative real part. Then there exist strictly positive constants, α_1 and α_2, such that the inequality

$$\| e^{-AT} \| \leqslant \alpha_1 \, e^{-\alpha_2 T}$$

is satisfied. One can write the obvious equality

$$x\,(t_1) = e^{-AT}((e^{AT} x\,(t_1) - x\,(t_1 + T)) + x\,(t_1 + T)). \tag{18}$$

Since by hypothesis $x(t)$ is bounded, the inequality $\| x(t_1 + T) \| \leqslant \alpha_0$ is satisfied for every $t_1 \geqslant 0$ and $T \geqslant 0$, where α_0 is a positive constant. Let ε' be an arbitrary positive number; then there exists a number T_1 for which the inequality $\| e^{-AT_1} \| \, \alpha_0 < \varepsilon'/2$ is satisfied. By Lemma 2 applied for $\varepsilon = \alpha_0$ and $\lambda = T_1$, there exists a strictly positive number T_0 such that the inequality $\| x\,(t_1 + T_1) - e^{AT_1} \; x\,(t_1) \| \leqslant \alpha_0$ is satisfied for every $t_1 \geqslant T_0$. Using the above inequalities one obtains from (18) (for $T = T_1$) the inequality $\| x\,(t_1) \| \leqslant \varepsilon'$ for every $t_1 \geqslant T_0$. Thus one obtains again $\lim_{t \to \infty} x(t) = 0$.

Consider now the general case. By briging the matrix A to the normal Jordan form one sees that under the conditions stated in the lemma, Eq. (16) can be always partitioned into two equations of the same form

$$\left.\begin{aligned} \frac{dy}{dt} &= A_{11} y + B_1 u \\ \frac{dz}{dt} &= A_{22} z + B_2 u \end{aligned}\right\}, \quad x = R \begin{pmatrix} y \\ z \end{pmatrix}, \; (\det R \neq 0),$$

where the matrix A_{11} has no characteristic value with a strictly negative real part and A_{22} has no characteristic value with a strictly positive real part. For each of the two equations one of the reasoning used above is valid and one obtains again $\lim_{t\to\infty} x(t) = 0$; this concludes the proof of Lemma 3.

In order to extend Lemma 3 to matrices A which have some characteristic values whose real part is zero we first establish a simple result concerning polynomials:

Proposition 1. *If the polynomial with complex coefficients*

$$\pi(t) = \alpha_n + \alpha_{n-1}\, t + \ldots + \alpha_0 t^n \tag{19}$$

satisfies the inequality

$$|\pi(t)| \leqslant 1 \text{ for every } t \text{ in the interval } 0 \leqslant t \leqslant 1 \tag{20}$$

then there exists a positive number γ_1, depending only on n, such that

$$|\alpha_j| \leqslant \gamma_1,\ j = 0, 1, \ldots, n. \tag{21}$$

Proof. Introduce the numbers $\pi_j = \pi\left(\frac{j}{n}\right)$, $j=0, 1, \ldots, n$.

Then (20) implies

$$|\pi_j| \leqslant 1,\ j = 0, 1, \ldots, n. \tag{22}$$

The values π_j uniquely determine the polynomial $\pi(t)$ which must therefore be equal to the Lagrange interpolation polynomial

$$\pi(t) = \sum_{j=0}^{n} \frac{(t-t_0)(t-t_1)\ldots(t-t_{j-1})(t-t_{j+1})\ldots(t-t_n)}{(t_j-t_0)(t_j-t_1)\ldots(t_j-t_{j-1})(t_j-t_{j+1})\ldots(t_j-t_n)}\,\pi_j\,. \tag{23}$$

By identifying (19) and (23) and using Conditions (22) the existence of a number γ_1 satisfying the conditions of Proposition 1 becomes obvious.

A simple change of variable gives the following more general result:

Proposition 2. *If the polynomial (19) satisfies the inequality* $|\pi(t)| \leqslant \gamma_2$ *for every* t *in the interval*

$$0 \leqslant t \leqslant \lambda, \quad (\lambda > 0), \tag{24}$$

then there exists a positive number γ_1, depending only on n, such that the coefficients of the polynomial satisfy the inequalities

$$| \alpha_j | \leqslant \frac{\gamma_1 \gamma_2}{\lambda^{n-j}}, \quad j = 0, 1, \ldots, n. \tag{25}$$

Indeed, by introducing the new variable $\tau = t/\lambda$ and replacing $\pi(t)$ by

$$\widetilde{\pi}(\tau) = \frac{\pi(\tau\lambda)}{\gamma_2} = \sum_{j=0}^{n} \widetilde{\alpha}_j \tau^{n-j}, \text{ where } \widetilde{\alpha}_j = \frac{\alpha_j \lambda^{n-j}}{\gamma_2}, \tag{26}$$

we find that the polynomial $\widetilde{\pi}$ satisfies all the conditions required in Proposition 1 and hence we have the inequalities $|\widetilde{\alpha}_j| \leqslant \gamma_1$, which, by (26), imply (25).

We now extend Lemma 3 to systems (16) whose matrix has the form of a Jordan cell for a characteristic value whose real part is zero.

Lemma 4. *If the function u satisfies one of the conditions (11) or (12), then any bounded solution of the equation $dx/dt = Ax + Bu$, where*

$$x = \begin{pmatrix} x_1 \\ x_2 \\ \cdot \\ x_n \end{pmatrix}, \quad A = \begin{pmatrix} i\omega & 1 & 0 & . & . & 0 \\ 0 & i\omega & 1 & . & . & 0 \\ . & . & . & . & . & . \\ 0 & 0 & 0 & . & . & i\omega \end{pmatrix}, (\omega : \text{real}), \tag{27}$$

has the properties

$$1°. \lim_{t\to\infty} x_j(t) = 0 \text{ for } j = 2, 3, \ldots, n. \tag{28}$$

2°. *For every pair of numbers $\varepsilon > 0$ and $\lambda > 0$ there exists a number $T_1 > 0$ such that for every $t_1 \geqslant T_1$ and every $\tau \in [0, \lambda]$*

$$\| x_1(t_1 + \tau) - e^{i\omega\tau} x_1(t_1) \| \leqslant \varepsilon. \tag{29}$$

Proof. Since the conditions of Lemma 3 are satisfied, Condition (10) is satisfied, i.e.

$$\| x(t_1 + \tau) - e^{A\tau} x(t_1) \| < \varepsilon, \; t_1 \geqslant T_0, \; 0 \leqslant \tau \leqslant \lambda, \tag{30}$$

where T_0 depends on the arbitrary positive numbers ε and λ. From (30) it follows that for every bounded solution which satisfies an inequality of the form

$$\| x(t) \| \leqslant \delta_0, \tag{31}$$

one has

$$\| e^{A\tau} x(t_1) \| < \delta_0 + \varepsilon \, (\text{for } t_1 \geqslant T_0 \text{ and } 0 \leqslant \tau \leqslant \lambda). \tag{32}$$

Since A has the form (27), we have

$$e^{A\tau} = \begin{pmatrix} e^{i\omega\tau} & \tau\, e^{i\omega\tau} & \frac{\tau^2}{2!} e^{i\omega\tau} & \dots & \frac{\tau^{n-1}}{(n-1)!} e^{i\omega\tau} \\ 0 & e^{i\omega\tau} & \tau\, e^{i\omega\tau} & \dots & \frac{\tau^{n-2}}{(n-2)!} e^{i\omega\tau} \\ \cdot & \cdot & \cdot & \dots & \cdot \\ 0 & 0 & 0 & \dots & e^{i\omega\tau} \end{pmatrix}. \tag{33}$$

The absolute value of the first component of the vector $e^{A\tau} x(t_1)$ must be smaller than the right hand member of (32). This, together with (33) and the equality $|e^{i\omega\tau}| = 1$, imply

$$\left\| x_1(t_1) + \tau\, x_2(t_1) + \dots + \frac{\tau^{n-1}}{(n-1)!} x_n(t_1) \right\| < \delta_0 + \varepsilon \tag{34}$$

(for $t_1 \geqslant T_0$ and $0 \leqslant \tau \leqslant \lambda$).
The polynomial on the left hand member is of the form (19) (denoting $x_j(t_1) = \alpha_{n-j}\,(j-1)!$, $j = 1, 2, \dots, n$) and hence from Proposition 2 one deduces the inequalities

$$|x_j(t_1)| \leqslant \frac{\gamma_1(\delta_0 + \varepsilon)(j-1)!}{\lambda^{j-1}}, \quad j = 1, 2, \dots, \quad n, \; t_1 \geqslant T_0. \tag{35}$$

Let $\tilde{\varepsilon}$ be an arbitrary number; then one can find a sufficiently large number λ such that the right hand members of all the inequalities (35) are smaller than $\tilde{\varepsilon}$ for $j = 2, 3, \dots, n$. Then (35) shows that the left hand member is smaller than $\tilde{\varepsilon}$ if t_1 is greater than the number T_0, determined as in Lemma 3, for the numbers ε and λ (ε is arbitrary and λ is chosen as shown above). This proves (28).

We examine now the component $x_1(t)$. Notice that the absolute value of the first component of the vector of the left hand member of (30) must be smaller than ε, whence, by (33), one has

$$\left| x_1(t_1 + \tau) - e^{i\omega\tau} \left(x_1(t_1) + \tau\, x_2(t_1) + \dots + \frac{\tau^{n-1}}{(n-1)!} x_n(t_1) \right) \right| < \varepsilon$$

$$(\text{for } t_1 \geqslant T_0 \text{ and } 0 \leqslant \tau \leqslant \lambda). \tag{36}$$

By Conditions (28) (proved above) there exists a number T_1 such that if $t_1 > T_1$, then the sum of the absolute values of all the terms on the left hand member of inequality (36), excepting the terms which contain x_1, is smaller than ε. We thus obtain the inequality

$$|x_1(t_1+\tau) - e^{i\omega\tau} x_1(t_1)| < 2\varepsilon \tag{37}$$

$$(\text{for } t_1 \geqslant \max(T_0, T_1) \text{ and } 0 \leqslant \tau \leqslant \lambda).$$

Since the numbers ε and λ are arbitrary, the inequality (37) also proves Assertion 2° of Lemma 4.

If one introduces the n-dimensional vector $d'_0 = (1\ 0 \ldots 0)$, one can combine Assertions 1° and 2° of Lemma 4 into a single statement: "for every pair of positive numbers ε and ϑ there exists a positive number $T(\varepsilon, \vartheta)$ such that

$$\| x(t+\tau) - \mathrm{d}_0^* x(t)\, \mathrm{d}_0 e^{i\omega\tau} \| \leqslant \varepsilon, \tag{38}$$

for every $t \geqslant T(\varepsilon, \vartheta)$ and every $\tau \in [0, \vartheta]$.
Obviously, the vector d_0 introduced above satisfies the relations

$$A\mathrm{d}_0 = i\omega\, \mathrm{d}_0, \ \|\mathrm{d}_0\| = 1. \tag{39}$$

Generalizing the foregoing results assume that the matrix A of Eq. (16) has the form

$$A = \begin{pmatrix} A_1 & 0 & \ldots & 0 & 0 \\ 0 & A_2 & \ldots & 0 & 0 \\ . & . & . & . & . \\ 0 & 0 & \ldots & A_p & 0 \\ 0 & 0 & \ldots & 0 & A_{p+1} \end{pmatrix}, \tag{40}$$

where the matrices A_j for $j = 1, 2, \ldots, p$ are Jordan cells of the form (27) and the matrix A_{p+1} has no characteristic value whose real part is zero. Then Eq. (16) can be partitioned into $p+1$ equations of the form $dx^{(j)}/dt = A_j x^{(j)} + B_j u$, $j = 1, 2, \ldots, p+1$. Lemma 4 can be applied to the equation whose index j is equal to $1, 2, \ldots, p$, while Lemma 3 can be applied to the equation with the index $j = p+1$. The conclusions thus obtained can be expressed in a compact form by introducing the vectors d_j, $j = 1, 2, \ldots, p$ with the same number of components as the number of rows of the matrix A, (40), the component corresponding to the first line of matrix A_j

(with the same index j) being equal to unity and all the other components equal to zero. These vectors satisfy the relations

$$A\,\mathrm{d}_j = \mathrm{i}\,\omega_j\,\mathrm{d}_j,\ \|\mathrm{d}_j\| = 1,\ j = 1,\ 2, \ldots, p, \tag{41}$$

where the purely imaginary numbers $\mathrm{i}\omega_j$ are the characteristic values of the corresponding matrices A_j. Obviously these vectors satisfy also the relations

$$\mathrm{d}_j^*\,\mathrm{d}_k = \begin{cases} 1, & \text{if } j = k \\ 0, & \text{if } j \neq k. \end{cases} \tag{42}$$

Using the vectors d_j the conclusions of Lemma 3 and 4 can be expressed in the form of a single condition of the type (38):

"For every $\varepsilon > 0$ and $\vartheta > 0$ there exists $T(\varepsilon, \vartheta) > 0$ such that

$$\| x\,(t+\tau) - \sum_{k=1}^{p} \mathrm{d}_k^*\,x\,(t)\,\mathrm{d}_k\,\mathrm{e}^{\mathrm{i}\omega_k\tau} \| \leqslant \varepsilon, \tag{43}$$

for every $t \geqslant T(\varepsilon, \vartheta)$ and every $\tau \in [0, \vartheta]$".

Consider now an equation of the general form $\mathrm{d}x/\mathrm{d}t = Ax + Bu$ and assume that the matrix A admits exactly p linearly independent vectors $f_1, f_2, \ldots, f_p$ with the properties

$$Af_j = \mathrm{i}\,\omega_j f_j,\ j = 1, 2, \ldots, p, \tag{44}$$

where ω_j are real numbers. Then there exists a transformation $\tilde{x} = Rx$ such that the transformed equation $\mathrm{d}\tilde{x}/\mathrm{d}t = \tilde{A}\tilde{x} + \tilde{B}\tilde{u}$ (where $\tilde{A} = R\,AR^{-1}$ and $\tilde{B} = R\,B$) has the matrix $\tilde{A}$ of the Jordan form (40). Assuming that one of Conditions (11) and (12) is satisfied it follows, as before, that for every bounded solution the condition (43) is satisfied, i.e.

"For every $\varepsilon > 0$ and $\vartheta > 0$ there exists $\tilde{T}(0, \vartheta) > 0$ such that

$$\| \tilde{x}\,(t+\tau) - \sum_{k=1}^{p} \tilde{d}_k^*\,\tilde{x}\,(t)\,\tilde{\mathrm{d}}_k\,\mathrm{e}^{\mathrm{i}\omega_k\tau} \| \leqslant \varepsilon, \tag{45}$$

for every $t \geqslant \tilde{T}(\varepsilon, \vartheta)$ and every $\tau \in [0, \vartheta]$"

where the vectors $\tilde{\mathrm{d}}_j$ satisfy the relations (see (41) and (42)):

$$\tilde{A}\,\tilde{\mathrm{d}}_k = \mathrm{i}\omega_k\tilde{\mathrm{d}}_k,\ \tilde{\mathrm{d}}_j^*\,\tilde{\mathrm{d}}_k = \begin{cases} 1, \text{ if } j = k \\ 0, \text{ if } j \neq k. \end{cases} \tag{46}$$

Applying the transformation $\widetilde{x} = Rx$ we obtain easily the equality

$$x(t+\tau) - \sum_{k=1}^{p} \check{\mathrm{d}}_k^* x(t) \mathrm{d}_k e^{i\omega_k \tau} = R^{-1}(\widetilde{x}(t+\tau) - \sum_{k=1}^{p} \widetilde{\mathrm{d}}_k^* \widetilde{x}(t) \widetilde{\mathrm{d}}_k e^{i\omega_k \tau}), \tag{47}$$

where

$$\check{\mathrm{d}}_k^* = \widetilde{\mathrm{d}}_k^* R, \; \mathrm{d}_k = R^{-1}\widetilde{\mathrm{d}}_k. \tag{48}$$

By (45) and (47) we have the following result:

"For every $\varepsilon > 0$ and $\vartheta > 0$ there exists $T(\varepsilon, \vartheta) > 0$ such that

$$\| x(t+\tau) - \sum_{k=1}^{p} \check{\mathrm{d}}_k^* x(t) \mathrm{d}_k e^{i\omega_k \tau} \| \leqslant \varepsilon, \tag{49}$$

for every $t \geqslant T(\varepsilon, \vartheta)$ and every $t \in [0, \vartheta]$.

From (46) and (48) one obtains the relations

$$A\mathrm{d}_k = i\omega_k \mathrm{d}_k, \quad \check{\mathrm{d}}_j^* \mathrm{d}_k = \begin{cases} 1, & \text{for } j = k \\ 0, & \text{for } j \neq k. \end{cases} \tag{50}$$

The foregoing considerations prove at the same time the existence of the vectors d_j and $\check{\mathrm{d}}_j$, starting from the existence of the linearly independent vectors f_j which satisfy Relations (44).

The results obtained above are summarized in the following lemma:

Lemma 5. *If the matrix A has exactly p linearly independent vectors f_j, with the properties (44), then there exist linearly independent vectors $\mathrm{d}_1, \mathrm{d}_2, \ldots, \mathrm{d}_p$ and the vectors $\check{\mathrm{d}}_1, \ldots, \check{\mathrm{d}}_p$ such that Relations (50) are satisfied and every bounded solution of Eq. (16), for a control function u which satisfies one of the Conditions (11) or (12), has the following properties:*

1° For every pair of positive numbers ε and ϑ there exists a positive number $T(\varepsilon, \tau)$ such that:

$$\| x(t+\tau) - \sum_{j=1}^{p} \check{\mathrm{d}}_j^* x(t) \mathrm{d}_j e^{i\omega_j \tau} \| \leqslant \varepsilon, \text{ for every } t \geqslant T(\varepsilon, \vartheta)$$
$$\text{and every } \tau \in [0, \vartheta] \tag{51}$$

2° *For every constant $\vartheta_0 \geqslant 0$, one has*

$$\lim_{t \to \infty} (x(t+\vartheta_0) - \sum_{j=1}^{p} \check{\mathrm{d}}_j^* x(t) \mathrm{d}_j e^{i\omega_j \vartheta_0}) = 0. \tag{52}$$

To complete the proof of this Lemma it is sufficient to observe that Property 2° is derived from Property 1° by simply considering (51) for $\vartheta = \vartheta_0$ and $\tau = \vartheta_0$.

Using Lemma 5 one can obtain sufficient conditions of asymptotic stability for systems of the form [1])

$$\frac{\mathrm{d}x}{\mathrm{d}t} = Ax + Bh(x, y) \tag{53}$$

$$\frac{\mathrm{d}y}{\mathrm{d}t} = f(x, y), \tag{54}$$

where h and f are continuous functions. We assume that by a previous investigation we know that the solutions of the system (53), (54) are bounded. Obviously, for every solution x, y of the system (53)—(54) the function x satisfies Eq. (16) for the control function

$$u(t) = h(x(t), y(t)). \tag{55}$$

Lemma 6. *Assume that all the solutions of the system (53)—(54) are bounded and the functions h and f are continuous. Assume also that the function u, given by (55), satisfies one of the conditions (11) and (12). Furthermore, assume that the matrix A has the characteristic values $i\omega_1$, $i\omega_2, \ldots, i\omega_k$ (where $\omega_1, \ldots, \omega_k$ are real numbers) and has no other characteristic value whose real part is zero. Finally, assume, that there exists an interval $0 \leqslant t \leqslant T_0 (T_0 > 0)$ wherein the system (53)—(54) admits no solution x, y such that $x(t)$ has the expression*

$$x(t) = \sum_{j=1}^{k} v_j \, \mathrm{e}^{i\omega_j t}, \quad \max_{j=1,\ldots,k} (\| v_j \|) > 0, \tag{56}$$

where v_j are constant vectors, not all zero, and ω_j are the numbers introduced before. Under these conditions one has

$$\lim_{t\to\infty} x(t) = 0. \tag{57}$$

If all the quantities in (53)—(54) are real we may add the condition that the right hand member of (56) be real valued.

---

[1]) The general form (53), (54) includes also the case in which equation (54) and the vector y are absent (i.e. the vector is zero-dimensional) and $h(x, y)$ is replaced by $h(x)$.

The main point in the proof of Lemma 6 is to show that under the assumptions stated in the lemma, inequality (51) implies that

$$\lim_{t\to\infty} \check{d}_j^* x_j(t) = 0, \ j = 1, 2, \ldots, p. \tag{58}$$

Consider an arbitrary solution x, y of the system (53)—(54) and assume that Relation (58) is not satisfied. In other words, there exists a positive number Δ_0 such that for every $t_0 > 0$ there exists a number $t_1 \geqslant t_0$ with the property

$$\max_{j=1,2,\ldots,p} (|\check{d}_j^* x(t_1)|) > \Delta_0. \tag{59}$$

Choose a sequence of positive numbers ε_m converging monotonically to zero and apply Property 1° of Lemma 5 for $\varepsilon = \varepsilon_m$ and $\vartheta = T_0$, where T_0 is the number which occurs in the statement of Lemma 6. We find that there exists a sequence of numbers $T(\varepsilon_m, T_0)$ such that

$$\|x(t' + \tau) - \sum_{j=1}^{p} \check{d}_j^* x(t') d_j e^{i\omega_j \tau}\| \leqslant \varepsilon_m, \text{ for } 0 \leqslant \tau \leqslant T_0$$

$$\text{and for } t' \geqslant T(\varepsilon_m, T_0). \tag{60}$$

Furthermore, we can find a sequence of numbers $t_m \geqslant T(\varepsilon_m, T_0)$ such that (see (59))

$$\max_{j=1,\ldots,p} (|\check{d}_j^* x(t_m)|) > \Delta_0. \tag{61}$$

Consider the functions x_m and y_m, defined as

$$x_m(t) = x(t_m + t), \ 0 \leqslant t \leqslant T_0 \tag{62}$$

$$y_m(t) = y(t_m + t), \ 0 \leqslant t \leqslant T_0. \tag{63}$$

Since the system (53)—(54) is autonomous, these functions satisfy the equations of the system. Also, under the conditions of Lemma 6, they are uniformly bounded and equicontinuous in the interval $0 \leqslant t \leqslant T_0$. By the theorem of Arzelà-Ascoli, there exist some subsequences $x_{m_n}(t)$, $y_{m_n}(t)$ which converge uniformly in the interval $0 \leqslant t \leqslant T_0$ to certain functions whose values we denote by $x_0(t)$ and $y_0(t)$. Hence one obtains in particular the relation

$$\lim_{n\to\infty} \check{d}_j^* x_{m_n}(0) = \check{d}_j^* x_0(0), \tag{64}$$

whence (using (61)) one obtains

$$\max_{j=1,2,\ldots,p} (|\check{\mathrm{d}}_j^* x_0(0)|) > \Delta_0. \tag{65}$$

From (60) written for $m = m_n$ and $t' = t_{m_n}$ one obtains using (62) and (64), that for every $\varepsilon > 0$ there exists n such that

$$\| x_{m_n}(t) - \sum_{j=1}^{p} \check{\mathrm{d}}_j^* x_{m_n}(0)\, \mathrm{d}_j\, \mathrm{e}^{\mathrm{i}\omega_j t} \| \leqslant \varepsilon_{m_n}, \quad 0 \leqslant t \leqslant T_0, \tag{66}$$

whence it follows that the function x_0, to which the functions x_{m_n} converge, has the expression

$$\lim_{n\to\infty} x_{m_n}(t) = x_0(t) = \sum_{j=1}^{p} \check{\mathrm{d}}_j^* x_0(0)\, \mathrm{d}_j\, \mathrm{e}^{\mathrm{i}\omega_j t}, \; 0 \leqslant t \leqslant T_0. \tag{67}$$

Since the functions x_{m_n} and y_{m_n} satisfy (53) and (54), one has

$$x_{m_n}(t) = x_{m_n}(0) + \int_0^t (A x_{m_n}(\tau) + Bh(x_{m_n}(\tau), y_{m_n}(\tau)))\, \mathrm{d}\tau,$$
$$0 \leqslant t \leqslant T_0.$$

The expression under the integration sign converges uniformly in the interval $0 \leqslant t \leqslant T_0$ to $Ax_0(\tau) + Bh(x_0(\tau), y_0(\tau))$. Passing to the limit under the integration sign and taking into account the first equality of (67), one obtains

$$x_0(t) = x_0(0) + \int_0^t (A x_0(\tau) + Bh(x_0(\tau), y_0(\tau)))\, \mathrm{d}\tau, \quad 0 \leqslant t \leqslant T_0.$$

Similarly one obtains the relation

$$y_0(t) = y_0(0) + \int_0^t f(x_0(\tau), y_0(\tau))\, \mathrm{d}\tau, \qquad 0 \leqslant t \leqslant T_0.$$

Hence the functions x_0 and y_0 satisfy Eqs. (53) and (54). Furthermore, from (67) and (65) it follows that the function x_0 has the form (56) which is contrary to the conditions of Lemma 6. The contradiction show that the conditions (58) are satisfied. Hence (57) is immediately obtained by using Lemma 5, namely by writing (52) for $\vartheta_0 = 0$. Reexamining the proof we find that the last assertion of Lemma 6 is also satisfied.

Before concluding this paragraph we mention a result concerning the relations of the form (56).

Lemma 7. *Let ω_j, $j = 1, 2, \ldots, k$ be real and distinct numbers and let a_j, $j = 1, 2, \ldots, k$ be continuous and bounded vector-valued functions defined for $t \geqslant 0$, and having the following property: For every pair of numbers $\varepsilon > 0$ and $\alpha_0 > 0$ there exists a number $T(\varepsilon, \alpha_0) > 0$ such that for every $t \geqslant T(\varepsilon, \alpha_0)$ and for every $\tau \in [0, \alpha_0]$ one has*

$$\left\| \sum_{j=1}^{k} a_j(t)\, e^{i\omega_j \tau} \right\| \leqslant \varepsilon. \tag{68}$$

Then

$$\lim_{t \to \infty} a_j(t) = 0, \; j = 1, 2, \ldots, k. \tag{69}$$

The proof is based on evaluating the vectors $a_j(t)$ by the well-known method for the computation of the coefficients of a Fourier series. Consider the obvious equality

$$\int_0^{\alpha_0} \left(\sum_{j=1}^{k} a_j(t)\, e^{i\omega_j \tau} \right) e^{i\omega_m \tau}\, d\tau = a_m(t)\, \alpha_0 +$$

$$+ \sum_{j=1}^{k}{}' \, a_j(t) \frac{1}{\omega_j - \omega_m} \left(e^{i(\omega_j - \omega_m)\alpha_0} - 1\right), \tag{70}$$

where m is one of the numbers $1, 2, \ldots, k$ and Σ' means summation with respect to all the indices j excepting $j = m$. We evaluate the left hand member of (70). From the conditions of Lemma 7 it follows that for every pair of positive numbers ε and α_0 the norm of the left hand member of (70) is not greater than $\varepsilon\alpha_0$ if $t \geqslant T(\varepsilon, \alpha_0)$. The last term of (70) is also easy to evaluate. Since the functions a_j are bounded, there exists a number $\alpha_1 > 0$ such that we have, for every $t \geqslant 0$, the inequalities $\| a_j(t) \| \leqslant \alpha_1$, $j = 1, 2, \ldots, k$. Hence the norm of the last term of (70) is not greater than the number $\alpha_2 = 2k\alpha_1 \sum_{j=1}^{k} \frac{1}{\omega_j - \omega_m}$.

Using the evaluations obtained above we deduce from (70) the inequality $\| a_m(t) \| \leqslant \varepsilon - \alpha_2/\alpha_0$ which holds for every $\varepsilon > 0$ and $\alpha_0 > 0$ and every $t \geqslant T(\varepsilon, \alpha_0)$. Since α_0 is arbitrary we may choose it such that $\alpha_2/\alpha_0 = \varepsilon/2$, thus obtaining $\| a_m(t) \| \leqslant \varepsilon/2$, for every $t \geqslant T(\varepsilon, 2\alpha_2/\varepsilon)$. Since ε is arbitrary, one obtains (69).

§ 23. Conditions of asymptotic stability of single-input and multi-input systems with constant coefficients

From the results of the foregoing paragraph we can derive criteria of asymptotic stability if the control function u satisfies one of the conditions (11, § 22) and (12, § 22) However usually such conditions are not given directly. For this reason, the asymptotic stability is further studied in the present paragraph. For the systems treated in § 14 we establish

Theorem 1. *Suppose that system (1, § 14)—(2, § 14) is minimally stable and satisfies the conditions $b \neq 0$ and $\pi(\lambda, \sigma) \not\equiv 0$ as well as Condition (F) of Theorem 1, § 14* [1]*. Consider a solution of the system with the property* [2]

$$\eta(0, t) \leqslant \textit{constant (for every } t \geqslant 0). \tag{1}$$

Then in order to have also the property

$$\lim_{t \to \infty} x(t) = 0, \tag{2}$$

it is sufficient that one of the following conditions be satisfied:

1° $\pi(-i\omega, i\omega) \neq 0$ *for every real number* ω. *Furthermore, either* $\varkappa \neq 0$ *or the function* μ *is bounded.*

2° $\pi(-i\omega, i\omega)$ *vanishes for the real numbers* ω_j, $j = 1, 2, \ldots, k$ *and is different from zero for every other real number* ω. *Furthermore, the functions* μ *and* x *are given by the solution of a system of the form (see footnote 1, § 22).*

$$\frac{dx}{dt} = Ax + b\varphi(x, y) \tag{3}$$

$$\frac{dy}{dt} = f(x, y) \tag{4}$$

$$\mu(t) = \varphi(x, y), \tag{5}$$

where the functions φ *and* f *are continuous and Eqs. (3), (4) have the following property: there exists an interval* $0 \leqslant t \leqslant T_0$

---

[1]) In other words, Conditions (*i*)—(*iii*) and Property (*F*) of Theorem 1, § 14 are satisfied.

[2]) Applying Theorem 1, § 14 we find that if condition (1) is satisfied, the corresponding function x is bounded.

wherein these equations have no solution for which $x(t)$ has the form [1])

$$x(t) = \sum_{j=1}^{k} v_j \, e^{i\omega_j t}, \quad \max(\|v_j\|) \neq 0 \qquad j = 1, \ldots, k. \tag{6}$$

3° *The function $t \mapsto \mu(t)$ is bounded and there exist two constant n-vectors q and s and a function $\alpha \mapsto \rho_i(\alpha)$ of class M_i* [2]) *such that* (7)

$$\text{“the condition } \lim_{t\to\infty} s^* x(t) = 0 \text{ implies } \lim_{t\to\infty} (\mu(t) - q^* x(t)) = 0\text{”}$$

$$\int_0^{t_1} \rho_i(s^* x(t))\, dt \leqslant -\eta(0, t_1) + \textit{constant (for every } t_1 \geqslant 0) \tag{8}$$

and in addition one of the following three conditions is satisfied:

(a): *The matrix $A + bq^*$ has no characteristic value with zero real part*

(b): *The pair $(s^*, A + bq^*)$ is completely observable*

(c): *There exists a nonsingular matrix R such that the matrix $A + bq^*$ and the vector s can be expressed in the form*

$$\tilde{A} = R(A + bq^*)\, R^{-1} = \begin{pmatrix} A_{11} & 0 \\ A_{21} & A_{22} \end{pmatrix} \tag{9}$$

$$\tilde{s}^* = s^* R^{-1} = (s_1^* \; 0), \tag{10}$$

where the matrix A_{22} has no characteristic value with zero real part and where the pair (s_1^*, A_{11}) is completely observable.

The proof of the above theorem is hardly simpler than the proof of the corresponding theorem for multi-input systems.

---

[1]) The proof will show that if all the quantities of the system (3)—(4) are real, then we may add the condition that the value of the sum in the right-hand member of (6), be real.

[2]) Here α is a complex number. We remark that while in the problem of simple stability we obtain an evaluation valid for the whole family of solutions of the system, in the problem of asymptotic stability we may study the individual behaviour of an arbitrary solution independently of the other solutions of the system. Hence, the functions ρ_i may depend on the solution which is examined and may be defined only in a finite interval which contains all the values of the expression $s^* x(t)$ for the examined solution. The system (3)—(5) may also depend on the solution examined.

In order to avoid the repetition of similar arguments we shall prove only the following theorem:

Theorem 2. *Suppose that the system (1, § 16)—(2, § 16) satisfies Conditions (i)—(iii) and Property (F) of Theorem 1, § 16. Consider a solution of the system such that inequality (1) is satisfied. Then in order that the corresponding function have the Property (2), it is sufficient that one of the following conditions be satisfied:*

1° $\pi(-i\omega, i\omega) \neq 0$ *for every real number* ω. *Furthermore we either have* $\det K \neq 0$ *or else the function* $t \mapsto \|u(t)\|$ *is bounded.*

2° $\pi(-i\omega, i\omega)$ *vanishes for the real numbers* ω_j, $j = 1, 2, \ldots, k$ *and is different from zero for every other real number* ω. *At the same time the functions* u *and* x *are obtained as solutions of a system of the form (53, § 22)—(55, § 22) — where the vector-valued functions* h *and* f *are continuous — and there exists an interval wherein this system has no solution for which* x *has the form (6)* [1].

3° *The function* $t \mapsto \|u(t)\|$ *is bounded and there exist two constant* $n \times m$ *matrices* Q *and* S *and a function* $w \mapsto \rho_i(w)$ *of class* M_i *(where* w *is an* m*-vector), such that*

$$\text{“the condition } \lim_{t\to\infty} S^*x(t)=0 \text{ implies } \lim_{t\to\infty} (u(t)-Q^*x(t))=0\text{”} \tag{11}$$

$$\int_0^{t_1} \rho_i(S^*x(t))\, dt \leqslant -\eta(0, t_1) + \text{constant (for every } t_1 > 0); \tag{12}$$

and moreover, one of the following three conditions [2] *is satisfied:*

(a): The matrix $A + BQ^*$ *has no characteristic value whose real part is zero*

(b): The pair $(S^*, A + BQ^*)$ *is completely observable*

(c): There exists a nonsingular matrix R *such that the matrices* $A + BQ^*$ *and* S *can be written in the form*

$$\tilde{A} = R(A + BQ^*)R^{-1} = \begin{pmatrix} A_{11} & 0 \\ A_{21} & A_{22} \end{pmatrix} \tag{13}$$

$$\tilde{S}^* = S^*R^{-1} = (S^* \ 0), \tag{14}$$

where the matrix A_{22} *has no characteristic value whose real part is zero and the pair* (S_1^*, A_{11}) *is completely observable.*

It is easy to see that Theorem 1 is a particular case of Theorem 2. We first show that it is sufficient to prove Theorem 2

---

[1] The remark of footnote 1, page 228 remains valid.

[2] Using Property 3° of Consequence 1, § 35, it can be easily seen that Conditions (a)—(c) can be replaced by the following condition: “There exist no vector $x_0 \neq 0$ and no real number ω_0 such that the relations $S^*x_0 = 0$ and $(A + BQ^*)x_0 = e^{i\omega_0} x_0$ be simultaneously satisfied. A similar remark is valid for Conditions (a)—(c) of Theorem 1.

under the assumption that the pair (A, B) is completely controllable. By Lemma 5, § 16 the system can be always split into a completely controllable part of the form (60, § 16)—(61, § 16) and a residue of the form (62, § 16) (63, § 16). Furthermore, A_{22} of Eq. (62, § 16) is a Hurwitz matrix and hence the solution of this equation has the property $\lim_{t \to \infty} w(t) = 0$. Hence, it is sufficient to prove that if inequality (1) is satisfied, then the solution $y(t)$ of Eq. (60, § 16) has also the property $\lim_{t \to \infty} y(t) = 0$. From inequality (1) it follows that the expression $\eta_c(0, t_1)$ given by (61, § 16) satisfies a similar inequality. Indeed, by Theorem 1, § 16 the system under investigation is hyperstable and hence, by inequality (1), $x(t)$ is bounded. Hence $y(t)$, of Eq. (60, § 16), is also bounded. As in § 16, from (1) we deduce an inequality of the form (81, § 16) whence it follows that the integral $\int_0^{t_1} \| u_0 \|^2 \mathrm{d}t$ is also bounded. Hence, from (65, § 16) it follows that the function $-\eta_r(0, t_1)$ is bounded from above. Since from (64, § 16) we have $\eta_c(0, t_1) = \eta(0, t_1) - \eta_r(0, t_1)$ it immediately follows from Condition (1) that $\eta_c(0, t_1)$ satisfies a similar condition. Therefore, without loss of generality, we may assume in the sequel that the pair (A, B) is completely controllable. Then the system is of the form studied in § 9 and can be brought to special form from § 9.

Let us examine Condition 1° of Theorem 2. If $\det K \neq 0$ the system can be brought to the form (44, § 9). Since $\pi(-\mathrm{i}\omega, \mathrm{i}\omega) \neq 0$ for every real number ω, from (3, § 9) it follows that $\psi(\mathrm{i}\omega) \neq 0$ for every real number ω. Using also (45, § 9), we find that the matrix A of the system (44, § 9) has no characteristic value with zero real part. Furthermore using (44, § 9) and the fact that $\| x(t) \|$ is bounded, we find that the function u of the system (44, § 9) satisfies Condition (12, § 22). Applying now Lemma 3, § 22 one obtains (2).

Consider now the same Condition 1° of Theorem 2 but assume that $\det K = 0$ and that u is bounded. We have $\pi(-\sigma, \sigma) \not\equiv 0$ since otherwise $\pi(-\mathrm{i}\omega, \mathrm{i}\omega)$ would vanish for every real number ω, contrary to Condition 1° of Theorem 2. Hence, the system can be brought to the form (49, § 9)—(51, § 9). From Inequality (1) and Relation (51, § 9) and from the fact that the solutions of the system are bounded we derive

$$\int_0^{t_1} \| u_0(t) \|^2 \mathrm{d}t < \infty \tag{15}$$

$$\int_0^{t_1} \| y_1^{\mathrm{i}}(t) \|^2 \mathrm{d}t < \infty, \quad \mathrm{i} = 1, 2, \ldots, s. \tag{16}$$

Furthermore, since u is bounded, the solutions x of the system are uniformly continuous, a fact which will be used below, when we apply Barbălat's lemma. From (16) and from the quoted lemma (see § 21) it follows that

$$\lim_{t\to\infty} y_{1i}(t) = 0, \quad i = 1, 2, \ldots, s. \tag{17}$$

Similar relations can be established for the components $y_{kj}(t)$ with $k > 1$ by applying the Barbălat's lemma in its more general form 1′, § 21. Indeed, for an arbitrary $\vartheta > 0$ we have (see (49, § 9)):

$$\int_t^{t+\vartheta} y_{(k+1)j}(\tau)\,\mathrm{d}\tau = y_{kj}(t+\vartheta) - y_{kj}(t) - \\ -\sum_{i=1}^{j-1}\int_t^{t+\vartheta} T_{kji}\, y_{1i}(t)\,dt - \int_t^{t+\vartheta} B_{kj}\, u_0(\tau)\,\mathrm{d}\tau. \tag{18}$$

Using the inequality of Cauchy-Buniakowski-Schwarz we obtain

$$\left|\int_t^{t+\vartheta} T_{kji}\, y_{1i}(\tau)\,\mathrm{d}\tau\right| \leqslant \beta_{kji}\,\vartheta\int_t^{t+\vartheta} \|y_{1i}(\tau)\|^2\,\mathrm{d}\tau$$

$$\left|\int_t^{t+\vartheta} B_{kj}\, u_0(\tau)\,\mathrm{d}\tau\right| \leqslant \beta_{kj}\,\vartheta\int_t^{t+\vartheta} \|u_0(\tau)\|^2\,\mathrm{d}\tau,$$

where β_{kji} and β_{kj} are positive constants. Using (15) and (16) we get

$$\lim_{t\to\infty}\int_t^{t+\vartheta} T_{kji}\, y_{1i}(\tau)\,\mathrm{d}\tau = 0, \quad \lim_{t\to\infty}\int_t^{t+\vartheta} B_{kj}\, u_0(\tau)\,\mathrm{d}\tau = 0.$$

Hence, if the relation

$$\lim_{t\to\infty}(y_{kj}(t+\vartheta) - y_{kj}(t)) = 0 \tag{19}$$

is satisfied, then from (18) one obtains that $\lim_{t\to\infty}\int_t^{t+\vartheta} y_{(k+1)j}(\tau)\,\mathrm{d}\tau = 0$, and by applying Lemma 1′, § 21, one obtains the relation

$$\lim_{t\to\infty} y_{(k+1)j}(t) = 0. \tag{20}$$

From (17) it follows that Property (19) holds for $k = 1$ and hence, from (20) one obtains $\lim_{t\to\infty} y_{2j}(t) = 0$. Therefore Condition (19) is also satisfied for $k = 2$. Applying the same reasoning step by step one obtains the conditions

$$\lim_{t\to\infty} y_{kj}(t) = 0\,, \quad k = 1, 2, \ldots, p_j\,, \quad j = 1, 2, \ldots, s\,. \tag{21}$$

Eq. (50, § 9) can be treated in the same way as we proceeded above (in the case $\det K \neq 0$) with the differential equation of (44, § 9). Since for every real ω we have $\pi(-\mathrm{i}\omega, \mathrm{i}\omega) \neq 0$ we also have $\psi(\mathrm{i}\omega) \neq 0$ for every real ω (see (3, § 9)). Hence, from (52, § 9) we deduce that the matrix A_{22} has no characteristic value with vanishing real part. Eq. (50, § 9) can be written in the form (16, § 22) if we consider y_{1i}, $i = 1, 2, \ldots, s$ and u_0 as components of a new control function u. Owing to Conditions (15) and (16) the function u satisfies (12, § 22). Hence Lemma 3, § 22 gives $\lim_{t\to\infty} w(t) = 0$.

Consider now Condition 2° of Theorem 2. Applying the same reasoning as above one finds that in this case too, the system can be written in the form (49, § 9)—(51, § 9). The only difference is that now the matrix A_{22} of Eq. (50, § 9) has some characteristic values whose real parts are zero (this follows from Eqs. (52, § 9) and (3, § 9)). Hence, the Conclusion (21) remains valid. Concerning Eq. (50, § 9) it is easy to see that if Condition 2° of Theorem 2 is satisfied, then we can apply Lemma 6, § 22 and obtain the relation $\lim_{t\to\infty} w(t) = 0$. The remarks of the footnotes 1, page 228 and 1, page 229 follow from the last part of Lemma 6, § 22.

We finally examine Condition 3° of Theorem 2. Applying Theorem 1, § 16 one finds that, under the conditions of Theorem 2, the system is hyperstable and hence, if Condition (1) is satisfied, then $x(t)$ is bounded. By Condition 3°, $\|u(t)\|$ is also bounded. Since the system under investigation is hyperstable the integral $-\eta(0, t_1)$ is bounded from above (see 9, § 13). From (12) it follows that the integral of the left hand member of this relation is also bounded from above. Hence we can apply Barbălat's lemma (§ 21) and obtain (with arguments similar to those used in the proof of Lemma 2, § 21) the relation

$$\lim_{t\to\infty} S^* x(t) = 0\,. \tag{22}$$

Then from (11) one obtains that

$$\lim_{t \to \infty} (u(t) - Q^* x(t)) = 0 . \tag{23}$$

If we introduce a new control function $\widetilde{u} = u - Q^* x$, the initial equation $\mathrm{d}x/\mathrm{d}t = Ax + Bu$ can be written as

$$\frac{\mathrm{d}x}{\mathrm{d}t} = (A + BQ^*)\, x + B\widetilde{u} , \tag{24}$$

and Condition (23) becomes

$$\lim_{t \to \infty} \widetilde{u}(t) = 0 . \tag{25}$$

Equation (24) can be treated like Eq. (16, § 22) where the matrix A is to be replaced by the matrix $A + BQ^*$. If the case (a) of Condition 3° is satisfied one can apply Lemma 3, § 22 and obtain Conclusion (2). In the case (b) assume that the matrix $A + BQ^*$ has the characteristic value $\mathrm{i}\omega'_1, \mathrm{i}\omega'_2, \ldots, \mathrm{i}\omega'_k$ and has no other characteristic values with zero real part. Applying Lemma 5, § 22 we find that for every pair of numbers $\varepsilon > 0$ and $\vartheta > 0$ there exists $T(\varepsilon, \vartheta) > 0$ such that

$$\left\| x(t + \tau) - \sum_{j=1}^{p} \check{\mathrm{d}}_j^* \, x(t)\, \mathrm{d}_j \, \mathrm{e}^{\mathrm{i}\omega_j \tau} \right\| \leqslant \varepsilon,$$

$$(t \geqslant T(\varepsilon, \vartheta),\ 0 \leqslant \tau \leqslant \vartheta). \tag{26}$$

In the above relation the vectors d_j, $j = 1, 2, \ldots, p$ satisfy the relations

$$(A + BQ^*)\, \mathrm{d}_j = \mathrm{i}\omega_j \, \mathrm{d}_j , \tag{27}$$

which are obtained from (50, § 22), taking into account that now the matrix A is replaced by the matrix $A + BQ$. Furthermore, each number ω_j, $j = 1\ 2, \ldots, p$, $(p \geqslant k)$ is equal to one of the numbers ω'_j, $j = 1, 2, \ldots, k$, introduced before. From (27) it follows that Eq. (24) admits the solution $x_j(t) = \mathrm{d}_j \, \mathrm{e}^{i\omega_j t}$ for $\widetilde{u} = 0$. If the equality $S^* \mathrm{d}_j = 0$ were satisfied we would obtain the identity $S^* x_j(t) \not\equiv 0$; this contradicts the hypothesis of (b) that the pair $(S^*, A + BQ^*)$ is completely observable. Therefore the conditions

$$S^* \mathrm{d}_j \neq 0, \ j = 1, 2, \ldots, p \tag{28}$$

are satisfied. From (26) we deduce the inequality [1])

$$\left\| S^* x(t+\tau) - \sum_{j=1}^{p} \check{\mathrm{d}}_j^* x(t) S^* \mathrm{d}_j \, \mathrm{e}^{\mathrm{i}\omega_j \tau} \right\| \leqslant \| S \| \, \varepsilon , \qquad (t \geqslant T(\varepsilon, \vartheta), \; 0 \leqslant \tau \leqslant \vartheta) . \tag{29}$$

Thus from (22) it follows that for every $\varepsilon > 0$ and $\vartheta > 0$ there exists $T'(\varepsilon, \vartheta) > 0$ such that

$$\left\| \sum_{j=1}^{p} \check{\mathrm{d}}_j^* x(t) S^* \mathrm{d}_j \, \mathrm{e}^{\mathrm{i}\omega_j \tau} \right\| \leqslant \varepsilon, \; (t \geqslant T'(\varepsilon, \vartheta), \; 0 \leqslant \tau \leqslant \vartheta) . \tag{30}$$

Then applying Lemma 7, § 22, we obtain

$$\lim_{t \to \infty} \check{\mathrm{d}}_j^* x(t) S^* \mathrm{d}_j = 0 , \quad j = 1, 2, \ldots , p .$$

From this and from (28) we deduce the relations $\lim\limits_{t \to \infty} \check{\mathrm{d}}_j^* x(t) = 0$, $j = 1, 2, \ldots, p$. Using again (26) (for $\tau = 0$) we obtain (2).

Finally, in the case (c), Eq. (24) can be split into two equations

$$\frac{\mathrm{d}y}{\mathrm{d}t} = A_{11} y + B_1 \tilde{u} \tag{31}$$

$$\frac{\mathrm{d}z}{\mathrm{d}t} = A_{21} y + A_{22} z + B_2 \tilde{u} \tag{32}$$

and Condition (22) becomes $\lim\limits_{t \to \infty} S_1^* y(t) = 0$. Using the same reasoning as before (applied now to Eq. (31)) we conclude that $\lim\limits_{t \to \infty} y(t) = 0$. Using this result and Condition (25) and applying Lemma 3, § 22 to Eq. (32) we also obtain the conclusion $\lim\limits_{t \to \infty} z(t) = 0$.

This concludes the proof of Theorem 2, whence it follows, as we remarked before, that Theorem 1 is also true. We need not stress the fact that these results can be also extended to discrete systems.

---

[1]) Remark that $\check{\mathrm{d}}_j^* x(t)$ is a scalar number.

§ 24. Characterization of the hyperstability property by the stability of systems with negative feedback

It has been shown in § 13 that any system with negative feedback (see Fig. 4.1.c, for $\mu = 0$) consisting of hyperstable blocks, has a stable trivial solution (provided the existence of the solutions is ensured). From Property (h_λ), stated in Theorem 1, § 15, it follows that, conversely, the hyperstability property can be characterized by the stability of the trivial solution of a family of systems with negative feedback. The systems with negative feedback used in the property (h_λ) were not symmetric (since block B_1 of Fig. 4.1.c was equal to the block under investigation, whereas block B_2 was of a particular type). In this paragraph we establish a result similar to the property (h_λ); however, here blocks B_1 and B_2 play symmetrical parts.

Consider the blocks of the form (studied also in § 15):

$$\frac{\mathrm{d}x}{\mathrm{d}t} = Ax + b\mu \tag{1}$$

$$\nu = a^* x + \alpha\mu . \tag{2}$$

Assume that Conditions (i)—(iii) of Theorem 1, § 15 are satisfied. Assume further that all the coefficients of the system (1)—(2) (i.e. A, b, a and α) are real. Then, it is easy to see that Condition (iii) of Theorem 1, § 15 is equivalent to the condition

$$\gamma(\sigma) \not\equiv 0, \tag{3}$$

where

$$\gamma(\sigma) = \alpha + a^* (\sigma I - A)^{-1} b . \tag{4}$$

Consider now the class $\mathscr{H}'$ of the blocks of the form described above, having the following properties:

1° *The block*

$$\frac{\mathrm{d}\xi}{\mathrm{d}t} = -\xi + \mu, \qquad \nu = \xi \tag{5}$$

belongs to the class $\mathscr{H}'$.

2° *If*

$$B_j : \begin{cases} \dfrac{\mathrm{d}x_j}{\mathrm{d}t} = A_j x_j + b_j \mu_j \\ \nu_j = a_j^* x_j + \alpha_j \mu_j \end{cases}, \quad j = 1, 2,$$

are two arbitrary blocks of class $\mathcal{H}'$ and if λ is an arbitrary strictly positive number, then every solution of the system

$$\frac{dx_1}{dt} = A_1 x_1 + b_1 \mu_1 \tag{6}$$

$$\nu_1 = a_1^* x_1 + \alpha_1 \mu_1 \tag{7}$$

$$\frac{dx_2}{dt} = A_2 x_2 + b_2 \mu_2 \tag{8}$$

$$\nu_2 = \lambda (a_2^* x_2 + \alpha_2 \mu_2) \tag{9}$$

$$\mu_2 = \nu_1 , \quad \mu_1 = - \nu_2 \tag{10}$$

is bounded for $t \geqslant 0$.

We observe that the system (6)—(10) is of the type of the system with negative feedback shown in Fig. 4.1.c (for $\mu = 0$). It is easy to see that the class of the hyperstable blocks satisfying the conditions of Theorem 1, § 15, has Properties 1° and 2° stated above. We shall show that, conversely the class $\mathcal{H}'$ defined above contains only the blocks which satisfy the conditions of Theorem 1, § 15.

Observe first that from Property 2° it follows that the transfer function of an arbitrary block belonging to class $\mathcal{H}'$ must satisfy one of the following conditions:

$$\operatorname{Re} \gamma(\sigma) \geqslant 0, \text{ for every } \sigma \text{ such that } \operatorname{Re} \sigma > 0 \tag{11}$$

$$\operatorname{Re} \gamma(\sigma) \leqslant 0, \text{ for every } \sigma \text{ such that } \operatorname{Re} \sigma > 0. \tag{12}$$

Indeed, if none of the Conditions (11) and (12) is satisfied then there exist two numbers σ_1 and σ_2 such that

$$\operatorname{Re} \sigma_1 > 0, \quad \operatorname{Re} \sigma_2 > 0, \quad \operatorname{Re} \gamma(\sigma_1) < 0, \quad \operatorname{Re} \gamma(\sigma_2) > 0. \tag{13}$$

Let us join in the plane of the variable σ the points σ_1 and σ_2 by a straight segment (therefore located in the half-plane $\operatorname{Re} \sigma > 0$). Modifying, if necessary, the position of the points σ_1 and σ_2 by sufficiently small amounts such that Relations (13) be still satisfied for the modified positions, no zero and no pole of the transfer function will be situated on the mentioned segment. By continuity it follows that the segment contains a point σ_i with the properties

$$\operatorname{Re} \sigma_i > 0 , \quad \operatorname{Re} \gamma(\sigma_i) = 0 , \quad \gamma(\sigma_i) \neq 0 ,$$

and therefore there exists $\lambda > 0$ such that

$$\lambda (\gamma (\sigma_i))^2 = -1 . \tag{14}$$

Consider now the system (6)—(10) where we take $A_1 = A_2 = A$, $b_1 = b_2 = b$, $a_1 = a_2 = a$, $\alpha_1 = \alpha_2 = \alpha$ and λ equal to the value of (14). The system is satisfied by

$$\mu_1 (t) = e^{\sigma_i t} , \quad x_1 (t) = (\sigma_i I - A)^{-1} b \, e^{\sigma_i t} ,$$

$$\nu_1 (t) = \gamma (\sigma_i) \, e^{\sigma_i t} = \mu_2 (t) ,$$

$$x_2 (t) = (\sigma_i I - A)^{-1} b \gamma (\sigma_i) \, e^{\sigma_i t}, \ \nu_2 (t) = \lambda (\gamma (\sigma_i))^2 \, e^{\sigma_i t} = - \mu_1 (t)$$

which are not bounded for $t \geqslant 0$ thus contradicting Property 2°. Thus we have either (11) or else (12).

We shall show that the alternative (12) contradicts Property 1°. Since the coefficients of the system are real, $\gamma(\sigma)$ is real for every real number σ. From (4) it follows that γ is a quotient of polynomials and hence (3) implies that there exists $\sigma_0 > 0$ for which $\gamma(\sigma_0) \neq 0$. Then (12) implies $\gamma(\sigma_0) < 0$. Hence there exists $\lambda > 0$ such that

$$\frac{\lambda \gamma (\sigma_0)}{\sigma_0 + 1} = -1. \tag{15}$$

Consider again system (6)—(10) where $A_1 = A$, $b_1 = b$, $a_1 = a$ and $\alpha_1 = \alpha$ and Eqs. (8) and (9) have the form

$$\frac{d\xi}{dt} = - \xi + \mu_2 , \quad \nu_2 = \lambda \xi , \tag{16}$$

(obtained by taking instead of B_2 the block (5)). The system is satisfied by

$$\mu_1 (t) = e^{\alpha_0 t} , \ x_1 (t) = (\sigma_0 I - A)^{-1} b \, e^{\sigma_0 t}$$

$$\mu_2 (t) = \nu_1 (t) = \gamma (\sigma_0) \, e^{\sigma_0 t}, \ \xi (t) = \frac{\gamma (\sigma_0)}{\sigma_0 + 1} e^{\sigma_0 t}$$

$$\nu_2 (t) = \frac{\lambda \gamma (\sigma_0)}{\sigma_0 + 1} e^{\sigma_0 t} = - \mu_1 (t) .$$

These expressions are not bounded for $t \geqslant 0$, and this again contradicts Property 2°. Thus the only alternative left is (11). Hence Condition (f') of Theorem 1, § 15 is satisfied for every block of class $\mathscr{H}'$. Since Conditions (i)—(iii) of Theorem 1, § 15 have been adopted here by hypothesis we obtain the conclusion that every block of class $\mathscr{H}'$ belongs to the class of hyperstable blocks considered in Theorem 1, § 15.

We observe that if Eqs. (5) had been replaced by equations $\mathrm{d}\xi/\mathrm{d}t = -\xi + \mu$, $\nu = -\xi$ then instead of the alternative (11) we would have obtained alternative (12).

APPLICATIONS

This chapter illustrates different ways of using the results of the foregoing chapters to treat some well-known problems. We first consider (in § 25) the problem of absolute stability which not only remains at the origin of the present work but also supplies the simplest and most direct examples for applications. In § 26, we determine some Liapunov functions as an application of the results of Chapter 3. In § 27 one treats the stability problem for initial conditions belonging to some finite sets in the state space; we also give an example of application of the methods of the present work, to the classical problem of stability in the first approximation. Stability problems of a more special character are examined in §§ 28 and 29. We finally give an application to the optimization of control systems (§ 30).

Most of the problems examined in this chapter have been stated and studied in the last 10—20 years. As mentioned before, the problem of absolute stability can be traced to the work [1] of A. I. Lur'e and V. N. Postnikov; it was subsequently extensively treated by A. I. Lur'e [1], A. M. Letov [1], V. A. Yakubovich [1] and other authors who have used Liapunov second method. The solution presented in § 25 was given by the author in [3] using a special method which was subsequently extended in various ways by the author [4], [5], [6], [8], [9], [10], A. Halanay [1], [2], [3], C. Corduneanu [1], [2], I. Z. Tsypkin [1], [2], [3], [4], [5], [6], E. I. Jury and B. W. Lee [1], R. W. Brockett and J. L. Willems [1], C. A. Desoer [1] and others. In addition to the known results, we show that the frequency criterion implies the stability of a larger collection of systems and that the frequency criterion is equivalent to the stability of a certain collection of *linear* systems — a result of the type required in Aizerman problem [1] (see Section 2, Chapter 1 and Section 5, § 13).

The results presented in § 26 are partly derived from author's papers [4] and [8] in which it has been proved — for a general type of systems — that the most general Liapunov functions used before in the problem of absolute stability cannot give better results than the frequency criterion. A second source of these results is found in the works of V. A. Yakubovich [2] and R. E. Kalman [5] where the frequency criterion has been proved again, by ingeniously constructed Liapunov functions (see also Section 5, § 8). The fact that the frequency criterion cannot be surpassed using Liapunov functions of the standard form was proved later for a more general class of systems, by the author [13] and by V. A. Yakubovich [4]. The proof presented in § 26 is new. Needless to say the results from § 27 concerning the stability in the first approximation belong to H. Poincaré and A. M. Liapunov.

The results of § 28 have been treated before in the work of T. Welton [1] whose stability criterion for nuclear reactors — one of the precursors of the frequency criterion — has been expressed by H. Smets [1], [2], [3] in the language of frequency characteristics. The criterion presented in § 28 is more general than Welton's criterion, but less general than the one previously established by the author [15]. The proof given in § 28 is new and also permits clarification of the problem of determining the Liapunov function for the systems under investigation. Paragraph 29 contains new applications to stability problems and can be further developed by the interested reader. The treatment of optimization problems of the type shown in § 30 by the methods derived from the theory of absolute stability and using the frequency language has been effected by R. E. Kalman [6]. A general result of the type considered in § 30 has been published subsequently by the author [15].

The present chapter does not aim to exhaust the field of applications of the results of the book; the reader will find additional information of particular interest both in the works quoted and in the special monographs of S. Lefschetz [1] and M. A. Aizerman and F. R. Gantmacher [1].

§ 25. Inclusion of the problem of absolute stability in a problem of hyperstability

1. The absolute stability problem for systems with one nonlinearity

Consider the system [1])

$$\frac{dx}{dt} = Ax - b\ \varphi\ (\nu) \tag{1}$$

$$\nu = c'\ x\ , \tag{2}$$

[1]) According to our usual convention, Eqs. (1) and (2) are an abbreviation for $dx(t)/dt = Ax(t) - b\varphi(\nu(t))$, $\nu(t) = c'x(t)$ $(t \geqslant 0)$. Thus $\varphi(\nu)$ will always mean $\varphi(\nu(t))$. We introduce the symbol δ and denote by $\varphi(\delta)$ the value of the function φ corresponding to the real number δ.

where x is an n-dimensional vector-valued function, b and c are constant n-vectors, A is a constant $n \times n$-matrix, ν is a scalar-valued function and $\delta \mapsto \varphi(\delta)$ is a continuous function defined for every real δ and having the property

$$\varphi\,(0) = 0. \tag{3}$$

All the quantities in (1) and (2) as well as in the rest of this chapter, (except § 27), are real. Therefore, when applying the formulas established in the previous chapters we may replace the sign* by the sign′ (used to denote "transpose"), and we may cancel the symbols Re and the bars (used to denote complex conjugate). This will be systematically done, without further mention.

Owing to Condition (3), Eqs. (1)—(2) are satisfied by the trivial solution $x = 0$. From the general theory of ordinary differential equations it follows that under the conditions stated, Eqs. (1)—(2) have at least one solution for every initial condition $x(0)$.

We look for some conditions under which all the solutions of Eqs. (1)—(2) are bounded and the trivial solution $x = 0$ is stable in the sense of Liapunov. Furthermore (see the works quoted in the introduction of this chapter) these properties must hold for every function φ belonging to a certain family of functions which will be precisely described in each case. If the properties specified above are satisfied, the trivial solution of the system (1)—(2) is said to be absolutely stable (the family of functions for which the respective properties are satisfied will be mentioned in each case). If all the solutions of the systems have also the property $\lim\limits_{t \to \infty} x(t) = 0$ the trivial solution is said to be asymptotically absolutely stable.

The problem of absolute stability has many variants differing by the conditions imposed on the function φ. As mentioned in Section 2, Chapter 1, a condition frequently used is that the graph of the function φ be contained in a sector of the form shown in Fig. 1.3 ; this happens if one has the inequalities

$$\varphi_1\, \delta^2 \leqslant \varphi(\delta)\, \delta \leqslant \varphi_2\, \delta^2\,, \text{ for every real } \delta \tag{4}$$

where φ_1 and φ_2 are two constants, $(\varphi_2 > \varphi_1)$.

Suppose now that system (1)—(2) is absolutely stable for the set of functions φ satisfying (4). Then this implies, in particular, the stability of the trivial solution of every system of the form

$$\frac{dx}{dt} = Ax - b\ \varphi_0\ \nu \tag{5}$$

$$\nu = c'\,x\,, \tag{6}$$

where φ_0 is an arbitrary constant satisfying

$$\varphi_1 \leqslant \varphi_0 \leqslant \varphi_2, \tag{7}$$

is stable. Indeed, from (7) it follows that the function $\delta \mapsto \mapsto \varphi(\delta) = \varphi_0\,\delta$ satisfies (4).

2. Definition of an auxiliary problem of hyperstability

The problem stated in the previous section may be treated either with the help of the "hyperstable systems" of the form studied in § 14 or by means of the "hyperstable blocks" examined in § 15. The results of § 14 can be directly applied to our problem whereas the results of § 15 can be applied only after bringing the system to a form in which the constant φ_1 of (4) is zero. However the results of § 15 allow one to obtain more information concerning the behaviour of our system. In the sequel we shall first use the method of hyperstable systems (§ 14) and then obtain additional information with the help of hyperstable blocks (§ 15).

The first method results clearly from the definition of hyperstability (see Definition 8, § 13). Assume we found a hyperstable system of the form (1, § 14)—(2, § 14) with the property that for every solution x of system (1)—(2), there exist some functions $\tilde{x}$, u and η satisfying (1, § 14)—(2, § 14) and the conditions

$$\tilde{x}(t) \equiv x'(t) \tag{8}$$

$$\eta(0, t) \leqslant \rho^2(x(0)) \quad \text{for all } t \geqslant 0 \tag{9}$$

where $x \mapsto \rho(x)$ is a continuous function with $\rho(0) = 0$. Then inequality (12, § 13) is satisfied for $t_0 = 0$ and $\beta_0 = |\rho(x(0))|$. Hence, by Definition 8, § 13 inequality (13, § 13) must be also satisfied, i.e.

$$\alpha\ (x(t)) \leqslant |\rho\ (x(0))| + \beta\ (x(0))\,.$$

Hence one readily sees that the property of absolute stability is satisfied. Therefore, it is sufficient to determine a system of the form (1, § 14), (2, § 14) with the properties stated above.

In order to obtain Eq. (1, § 14) we observe that if, besides any solution x of system (1)—(2), we introduce the control function μ as

$$\mu(t) = -\varphi(c'x(t)) = -\varphi(\nu(t)), \tag{10}$$

then the functions x and μ satisfy the equation

$$\frac{dx}{dt} = Ax + b\mu \tag{11}$$

which coincides with (1, § 14). It remains to define η such that for every pair of functions x, μ corresponding to a solution of the system (1)—(2), an inequality of the form (9) be satisfied. We use the inequalities (4) which can be also written as a single inequality

$$(\varphi(\delta) - \varphi_1\delta)(\varphi_2\delta - \varphi(\delta)) \geqslant 0, \text{ for every real } \delta.$$

Using the Definition (10) of the function μ we see that the above inequality implies

$$(\mu + \varphi_1\nu)(\varphi_2\nu + \mu) \leqslant 0, \text{ for every } t \geqslant 0$$

(here we wrote μ and ν instead of $\mu(t)$ and $\nu(t)$). Hence the expression

$$\eta_1(0, t_1) = \int_0^{t_1} (\mu + \varphi_1\nu)(\varphi_2\nu + \mu)\,dt \tag{12}$$

satisfies the inequality

$$\eta_1(0, t) \leqslant 0, \text{ for every } t \geqslant 0 \tag{13}$$

for every functions x, ν and μ which correspond to a solution of system (1)—(2). Furthermore, (12) can be written in the form (2, § 14) if we replace ν by $c'x$.

Another expression with the Property (9) is obtained by using the function

$$\psi(\delta) = \int_0^{\delta} (\varphi(\rho) - \varphi_1\rho)\,d\rho\,. \tag{14}$$

From the first of the inequalities (4) it follows that the function $\delta \mapsto \psi(\delta)$ satisfies the inequality $\psi(\delta) \geqslant 0$ for every real δ.

Using this inequality, – introduced by A. I. Lur'e and V. N. Postnikov [1] simultaneously with the problem of absolute stability – , we find that the expression

$$\int_0^{t_1} (-\varphi(\nu) + \varphi_1 \nu) \frac{d\nu}{dt} dt = -\psi(\nu(t_1)) + \psi(\nu(0)) \quad (15)$$

is not greater than $\psi(\nu(0))$, for every $t_1 \geqslant 0$. Taking into account (10) we deduce that

$$\eta_2(0, t_1) = \int_0^{t_1} (\mu + \varphi_1 \nu) \frac{d\nu}{dt} dt \quad (16)$$

satisfies the inequality

$$\eta_2(0, t) \leqslant \psi(\nu(0)), \text{ for every } t \geqslant 0 \quad (17)$$

for every triple of functions x, ν and μ which correspond to a solution of Eqs. (1)–(2). Furthermore (16) takes the form (2, § 14) if we use the relations $\nu = c'x$ and $d\nu/dt = c'Ax + c'b\mu$ (see (1), (2)).

In exactly the same way and using the second of the inequalities (4) one finds that

$$\tilde{\psi}(\delta) = \int_0^{\delta} (\varphi_2 \rho - \varphi(\rho)) \, d\rho \quad (18)$$

has the same properties as (14). Hence, the expression

$$\eta_3(0, t_1) = -\int_0^{t_1} (\mu + \varphi_2 \nu) \frac{d\nu}{dt} dt \quad (19)$$

satisfies the inequality

$$\eta_3(0, t) \leqslant \tilde{\psi}(\nu(0)), \text{ for every } t \geqslant 0 \quad (20)$$

under the same conditions as in (17).

We have thus obtained three different expressions of the form (2, § 1) which satisfy (9) for every triple of functions x,

ν and μ which correspond to a solution of system (1)—(2). It is easily seen that the expression

$$\eta(0, t) = \alpha_1 \eta_1(0, t) + \alpha_2 \eta_2(0, t) + \alpha_3 \eta_3(0, t) \qquad (21)$$

possesses the same property for every nonnegative constants α_1, α_2 and α_3. Obviously, expression (21) can be also written in the form (2, § 14). Hence, if the system (11), (21) is hyperstable then the trivial solution of the system (1)—(2) is absolutely stable under the conditions (4).

3. A frequency criterion

By Theorem 1, § 14, assuming that the assumptions of the theorem are satisfied, the condition that the system (11), (21) be hyperstable reduces to a certain frequency condition (F). We shall first consider the principal condition (F).

Using expressions (12), (16) and (19) one can write (21) in the form

$$\eta(0, t_1) = \int_0^{t_1} \Bigl(\alpha_1 (\mu + \varphi_1 \nu)(\varphi_2 \nu + \mu) + ((\alpha_2 - \alpha_3)\mu + \\ + (\alpha_2 \varphi_1 - \alpha_3 \varphi_2)\nu) \frac{d\nu}{dt}\Bigr) dt, \qquad (22)$$

by integrating the term containing $\nu d\nu/dt$ and using the relations $\nu = c'x$ and $d\nu/dt = c'Ax + c'b\mu$ we can also write (21) as

$$\eta(0, t_1) = \left[\frac{1}{2}(\alpha_1 \varphi_1 - \alpha_3 \varphi_2)(c' x)^2\right]_0^{t_1} + \\ + \int_0^{t_1} (\alpha_1 \mu^2 + \alpha_1 (\varphi_1 + \varphi_2) \mu c' x + \qquad (23) \\ + (\alpha_2 - \alpha_3)(c' Ax + c' b \mu) \mu + \alpha_1 \varphi_1 \varphi_2 (c' x)^2)\, dt .$$

Comparing this expression with (2, § 14) we obtain [1])

$$J = \frac{1}{2}(\alpha_2 \varphi_1 - \alpha_3 \varphi_2) c c', \quad \varkappa = \alpha_1 + (\alpha_2 - \alpha_3) c' b, \\ l = \frac{1}{2}(\alpha_1 (\varphi_1 + \varphi_2) c + (\alpha_2 - \alpha_3) A' c), \quad M = \alpha_1 \varphi_1 \varphi_2 c c'. \qquad (24)$$

[1]) We also use the fact that all the coefficients are real.

Hence, the characteristic function of the system (11), (22) given by formula (13, § 3) can be written as

$$\chi(\lambda, \sigma) = \frac{1}{2}(\alpha_2 \varphi_1 - \alpha_3 \varphi_2)(\lambda + \sigma)\gamma(\lambda)\gamma(\sigma) + \alpha_1 \chi_1(\lambda, \sigma) +$$

$$+ (\alpha_2 - \alpha_3)\chi_2(\lambda, \sigma), \tag{25}$$

where

$$\gamma(\sigma) = c'(\sigma I - A)^{-1} b, \tag{26}$$

$$\chi_1(\lambda, \sigma) = 1 + \frac{1}{2}(\varphi_1 + \varphi_2)(\gamma(\lambda) + \gamma(\sigma)) + \varphi_1\varphi_2 \gamma(\lambda)\gamma(\sigma) \tag{27}$$

$$\chi_2(\lambda, \sigma) = c' b + \frac{1}{2} c' A(\lambda I - A)^{-1} b + \frac{1}{2} c' A(\sigma I - A)^{-1} b. \tag{28}$$

One can also express χ_2 in terms of γ using the obvious identity

$$c' b = \frac{1}{2} c'(\lambda I - A)(\lambda I - A)^{-1} b + \frac{1}{2} c'(\sigma I - A)(\sigma I - A)^{-1} b$$

which yields

$$\chi_2(\lambda, \sigma) = \frac{1}{2}(\lambda \gamma(\lambda) + \sigma \gamma(\sigma)). \tag{29}$$

Using (25), (27) and (29), the inequality from Property (F), Theorem 1, § 14, can be written in our case as

$$\alpha_1(1 + (\varphi_1 + \varphi_2)\mathrm{Re}[\gamma(i\omega)] + \varphi_1\varphi_2 |\gamma(i\omega)|^2) + \alpha_0 \mathrm{Re}[i\omega\, \gamma(i\omega)] \geqslant 0, \tag{30}$$

for every real ω which satisfies the condition $\det(i\omega I - A) \neq 0$. In relation (30) we have introduced the real number

$$\alpha_0 = \alpha_2 - \alpha_3, \tag{31}$$

(which therefore can be also negative).

Since the constants α_2 and α_3 appear in (30) only through the agency of the constant $\alpha_0 = \alpha_2 - \alpha_3$ it is sufficient to examine only the following two cases:

$$\text{case 1:} \quad \alpha_2 \geqslant 0, \quad \alpha_3 = 0, \quad (\text{hence} \quad \alpha_0 \geqslant 0) \tag{32}$$

$$\text{case 2:} \quad \alpha_2 = 0, \quad \alpha_3 > 0, \quad (\text{hence} \quad \alpha_0 < 0). \tag{33}$$

(because by taking $\alpha_2 > 0$ and $\alpha_3 > 0$ we do not obtain a more general condition (30).

We further observe that if the constant α_1 is zero, Condition (30) is independent of the numbers φ_1 and φ_2 of inequalities (4) whence it naturally follows that the statement of the problem, as given in Section 1, must be modified [1]).

Therefore, in treating the problem stated in Section 1 we shall assume that the condition $\alpha_1 > 0$ is satisfied. Then we can divide the inequality (30) by α_1 or we may assume that we have

$$\alpha_1 = 1. \tag{34}$$

4. Discussion of the condition of minimal stability

The frequency condition (30) constitutes a criterion of absolute stability if system (11), (22) satisfies Conditions (*i*)—(*iii*) of Theorem 1, § 14. Condition (*i*) of minimal stability needs a more detailed examination.

From the remark at the end of Section 1 it follows that without any loss of generality in our problem, we can assume that the trivial solution of every system of the form (5)—(7) is stable.

We now introduce the following (somewhat stronger) condition: there exists a number φ_0 in the interval (7) such that the trivial solution of the system (5), (6) is asymptotically stable [2]). We now show that if this condition is satisfied, then the condition of minimal stability given by Definition 2, § 13

[1]) Besides, in this case, $\eta(0, t)$ is independent of $\eta_1(0, t)$ (see (21)) and hence (13) is not used. Hence, any criterion of absolute stability which has the form (30) for $\alpha_1 = 0$ is valid for a more general class of functions characterized either by an inequality of the form (15), where $\psi(\nu)$ has the expression (14), or by a similar inequality satisfied by (18) (the inequalities (4) need no longer to be satisfied).

[2]) Obviously, the cases in which this condition is not satisfied but the previous condition holds are quite exceptional.

is also satisfied. Indeed, if the above condition is satisfied, then there exists a number φ_0 satisfying the strict inequalities

$$\varphi_1 < \varphi_0 < \varphi_2 \tag{35}$$

and for which the trivial solution of the system (5)—(6) is asymptotically stable. Then $A - \varphi_0 bc'$ is a Hurwitz matrix and therefore, if $t \mapsto \rho(t)$ is an arbitrary continuous and bounded function satisfying the condition

$$\lim_{t\to\infty} \rho(t) = 0 , \tag{36}$$

then all the solutions of the equation

$$\frac{\mathrm{d}x}{\mathrm{d}t} = Ax - \varphi_0 b c' x + b \rho(t) \tag{37}$$

have the property $\lim_{t\to\infty} x(t) = 0$ [1].

Let $x(0)$ be an arbitrary initial condition and ρ a function with the properties stated above. (This function will be taken of a particular form, during the proof). Then the solution x of Eq. (37) has the property $\lim_{t\to\infty} x(t) = 0$. If we also introduce the function $t \mapsto \mu(t) = - \varphi_0 \nu(t) + \rho(t)$, (where $\nu(t) = c'x(t)$), the functions defined above satisfy also Eq. (11). Hence, to obtain the property of minimal stability (see Definition 2, § 13) it is sufficient to show that the function can be chosen such that the corresponding expression $\eta(0, t)$, (22) satisfy the inequality $\eta(0, t) \leqslant 0$ for every $t \geqslant 0$.

Introducing the functions x, ν and μ in (22) one obtains for $\alpha_1 = 1$

$$\eta(0,t_1) = \int_0^{t_1}\Big(((\varphi_1 - \varphi_0)\nu + \rho)((\varphi_2 - \varphi_0)\nu + \rho) + (-(\alpha_2 - \alpha_3)\varphi_0 \nu +$$

$$+ (\alpha_2 - \alpha_3)\rho + (\alpha_2\varphi_1 - \alpha_3\varphi_2)\nu)\frac{\mathrm{d}\nu}{\mathrm{d}t}\Big)\mathrm{d}t. \tag{38}$$

[1]) We apply Lemma 3, § 22 (The condition of the boundedness of the solutions results from the boundedness of ρ, since $A - \varphi_0 bc'$ is a Hurwitz matrix).

Consider first the case (32) : $\alpha_3 = 0$, $\alpha_2 \geqslant 0$. Then (38) becomes

$$\eta(0,t_1) = \int_0^{t_1} ((\varphi_1 - \varphi_0)\nu + \rho)\left((\varphi_2 - \varphi_0)\nu + \rho + \alpha_2 \frac{d\nu}{dt}\right) dt. \quad (39)$$

If $\alpha_2 = 0$, we can take $\rho(t) \equiv 0$ and the inequality $\eta(0, t_1) \leqslant 0$ follows from (39) and (35). If $\alpha_2 > 0$ the form of expression (39) suggests the introduction of a new variable

$$\tilde{\nu} = \nu + \frac{1}{\varphi_1 - \varphi_0}\rho. \quad (40)$$

Then (39) becomes

$$\eta(0, t_1) = \int_0^{t_1} (\varphi_1 - \varphi_0)\tilde{\nu}\left((\varphi_2 - \varphi_0)\tilde{\nu} + \alpha_2 \frac{d\tilde{\nu}}{dt}\right) dt -$$
$$- \int_0^{t_1} \tilde{\nu}\left(\alpha_2 \frac{d\rho}{dt} + (\varphi_2 - \varphi_1)\rho\right) dt. \quad (41)$$

By choosing the function ρ such that

$$\alpha_2 \frac{d\rho}{dt} + (\varphi_2 - \varphi_1)\rho = 0, \quad (42)$$

the last integral of (41) vanishes. Owing to the conditions $\alpha_2 > 0$ and $\varphi_2 - \varphi_1 > 0$ all the solutions of Eq. (42) have the Property (36). (A particular initial condition for Eq. (42) will be taken below).

Under these conditions one obtains from (41)

$$\eta(0, t_1) = (\varphi_1 - \varphi_0)(\varphi_2 - \varphi_0)\int_0^{t_1} \tilde{\nu}^2\, dt + \alpha_2(\varphi_1 - \varphi_0)\left[\frac{1}{2}\tilde{\nu}^2(t)\right]_0^{t_1}$$

whence, using Condition (35) and $\alpha_2 > 0$, we deduce $\eta(0, t_1) \leqslant$ $\leqslant -\frac{1}{2}\alpha_2(\varphi_1 - \varphi_0)\tilde{\nu}^2(0)$. Using (40) we can choose the initial condition $\rho(0)$ such that $\tilde{\nu}(0) = 0$ and hence we obtain the inequality $\eta(0, t_1) \leqslant 0$ required for minimal stability.

It remains to examine the case (33) : $\alpha_2 = 0$ and $\alpha_3 > 0$. This case is treated in exactly the same way as above by introducing the function $\tilde{\nu} = \nu + \rho/(\varphi_2 - \varphi_0)$. Here the function ρ must satisfy the equation $\alpha_3 \, d\rho/dt + (\varphi_2 - \varphi_1)\rho = 0$; the rest of the proof follows the same line.

Thus the condition of minimal stability (*i*) reduces to the condition that the trivial solution of Eqs. (5), (6) be asymptotically stable for a number φ_0 satisfying the inequalities (7).

We examine now Conditions (*ii*) and (*iii*) of Theorem 1, § 14. The condition (*ii*) ($b \neq 0$) is natural since obviously the case $b = 0$ is deprived of interest. We remark, however, that it is not necessary to introduce condition $b \neq 0$ as an additional assumption since in the case $b = 0$, the matrix A of (1) is a Hurwitz matrix (as follows from the property of minimal stability) and hence the absolute stability of the trivial solution of the system (1)—(2) (where $b = 0$) is ensured for any function φ.

Condition (*iii*) is satisfied if the left hand member of the frequency Condition (30) is not identically zero. But even if that expression is identically zero, Condition (*iii*) might be satisfied; for this it is sufficient that the corresponding characteristic function given by (25), (27) and (29) should not be identically zero.

5. Sufficient conditions for absolute stability

We proved in fact the following result:

Theorem 1. *System (1), (2) is absolutely stable (under Conditions (4)) if the following conditions are satisfied.*

(A) *There exists a number φ_0 in the interval $\varphi_1 \leqslant \varphi_0 \leqslant \varphi_2$ such that the trivial solution of system (1). (2) is asymptotically stable for $\varphi(\nu) = \varphi_0 \nu$.*

(B) *There exists a real number α_0 such that*

$$1 + (\varphi_1 + \varphi_2)\,\mathrm{Re}\,[\gamma(i\omega)] + \varphi_1\varphi_2\,|\,\gamma(i\omega)\,|^2 + \alpha_0\,\mathrm{Re}[i\omega\,\gamma(i\omega)] > 0, \quad (43)$$

for every real number ω for which $\det\,(i\omega I - A) \neq 0$.

(C) *The left hand member of (43) is not identically zero or, more generally, the corresponding characteristic function (given by (25), (27) and (29) for $\alpha_1 = 1$, $\alpha_2 = \max\,(\alpha_0, 0)$ and $\alpha_3 = \max\,(-\alpha_0, 0)$) is not identically zero.*

Inequality (43) is obtained from (30) using the Condition (34), whose introduction was motivated at the end of Section 3.

6. Sufficient conditions for asymptotic stability

Using the results of § 23 we can find sufficient conditions for the asymptotic stability of the trivial solution of system (1), (2). We introduce the polynomial

$$\pi_1(\mathrm{i}\omega) = \det(-\mathrm{i}\omega I - A)\det(\mathrm{i}\omega I - A)\ \chi(-\mathrm{i}\omega, \mathrm{i}\omega), \tag{44}$$

where $\chi(-\mathrm{i}\omega, \mathrm{i}\omega)$ represents the expression of the left hand member of (43). The polynomial (44) is connected to the characteristic polynomial π of system (11), (22) (where $\alpha_1 = 1$), by the relation

$$\pi_1(\mathrm{i}\omega) = \nu_0\ \pi(-\mathrm{i}\omega, \mathrm{i}\omega), \tag{45}$$

where ν_0 is the norming factor which also occurs in relation (18, § 3). From Theorem 1, § 23 we obtain the following result:

Theorem 2. *The trivial solution of system (1)—(2) is asymptotically absolutely stable (under Conditions (4)) if besides Conditions (A), (B) and (C) of Theorem 1, one of the following conditions is also satisfied.*

1° $\pi_1(\mathrm{i}\omega) \neq 0$, *for every real number* ω [1])

2° $\pi_1(\mathrm{i}\omega)$ *vanishes for the real numbers* ω_j, $j = 1, 2, \ldots, k$ *and is different from zero for every other real number* ω. *Furthermore, there exists an interval* $0 \leqslant t \leqslant T_0$ *wherein Eqs. (1)—(2) (under Conditions (4)) have no solution of the form (6, § 23).*

2°′ $\pi_1(\mathrm{i}\omega)$ *vanishes only for the real numbers* $\omega = \omega_1 \neq 0$ *and* $\omega = -\omega_1$ [2]). *Furthermore for every number* φ_0 *in the interval* $\varphi_1 \leqslant \varphi_0 \leqslant \varphi_2$ *the equation* $\mathrm{d}x/\mathrm{d}t = Ax - \varphi_0 bc'x$ *has no solution of the form (6, § 23).*

3° *Instead of Conditions (4) the following strict inequalities*

$$\varphi_1 \delta^2 < \varphi(\delta)\,\delta < \varphi_2 \delta^2, \text{ are satisfied for every real } \delta \neq 0. \tag{46}$$

Conditions 1°, 2° and 3° correspond to the conditions of Theorem 1, § 23 bearing the same number. Condition 2°′ implies Condition 2° as we are going to prove later.

[1]) From (44) it follows that this condition holds if (43) is satisfied as a strict inequality for every real ω and if $\det(\mathrm{i}\omega I - A) \neq 0$ for every real number ω.

[2]) Since all the quantities of the system (1)—(2) are real it follows that if the polynomial π_1 vanishes for a real number $\omega = \omega_1$ it must also vanish for $\omega = -\omega_1$ (see also Relation (28, § 8)).

In order to show that Condition 1° is a sufficient condition for asymptotic stability we make use of the fact that the function x is bounded (a result obtained by applying Theorem 1). Hence, the corresponding control function given in (10) is also bounded and therefore the second additional condition required in the statement of Condition 1° of Theorem 1, § 23, is satisfied. Hence, the asymptotic stability is obtained by applying Condition 1° of Theorem 1, § 23 to the system (11), (22) (with $\alpha_1 = 1$) and using Relation (45).

If Condition 2° is satisfied, then Condition 2° of Theorem 1, § 23 is satisfied for system (11), (22). (In our case the system (3, § 23)—(5, § 23) is identical to system (1), (2)).

We now show that Condition 2°′ implies Condition 2°. Assume that Condition 2°′ is satisfied and that Condition 2° of Theorem 1, § 23 (written for the corresponding system (11), (22) and for a system (3, § 23)—(5, § 23) of the form (1), (2) under the Conditions (4)), is not satisfied. Then, for every interval of the form $0 \leqslant t \leqslant T_0$ system (1), (2) has a solution of the form (6, § 23) i.e. (under our assumptions)

$$x(t) = v_1 e^{i\omega_1 t} + v_2 e^{-i\omega_1 t}, \quad \max(\|v_1\|, \|v_2\|) \neq 0. \tag{47}$$

In order to obtain a contradiction it will be sufficient to show that for the solution (47) there exists a constant φ_0 such that

$$\varphi(\nu(t)) = \varphi_0 \, \nu(t). \tag{48}$$

(Then, from Conditions (4) it follows that φ_0 must satisfy the inequalities $\varphi_1 \leqslant \varphi_0 \leqslant \varphi_2$ required in the statement of Condition 2°′).

Since all the quantities of the system (1), (2) are real the sum of the right hand member of Relation (47) must be also real (see footnote 1, p. 228). Hence the vectors v_1 and v_2 can be written as $2v_1 = w_1 + iw_2$, $2v_2 = w_1 - iw_2$, where w_1 and w_2 are real vectors. Then, (47) becomes

$$x(t) = w_1 \cos \omega_1 t - w_2 \sin \omega_1 t, \quad (\max(\|w_1\|, \|w_2\|) \neq 0). \tag{49}$$

Hence there exist two constants ν_0 and ϑ_0 such that

$$\nu(t) = c' x(t) = \nu_0 \cos(\omega_1 t + \vartheta_0), \quad (|\nu_0| \neq 0). \tag{50}$$

Writing that the functions (49) and (50) satisfy Eq. (1) one obtains

$$(Aw_2 - \omega_1 w_1) \sin \omega_1 t = (Aw_1 + \omega_1 w_2) \cos \omega_1 t = \\ = b\varphi (\nu_0 \cos (\omega_1 t + \vartheta_0)). \tag{51}$$

If $b = 0$, Eq. (1) is satisfied by (47) for every function φ hence also for the systems which appear in the Condition 2°. If $b \neq 0$ we multiply on the left Relation (51) with $b'/b'b$ obtaining a relation of the form

$$\varphi (\nu_0 \cos (\omega_1 t + \vartheta_0)) = \nu_1 \cos (\omega_1 t + \vartheta_1). \tag{52}$$

Since the above properties can be obtained for every interval $0 \leqslant t \leqslant T_0$ the number T_0 may be taken for example greater than $2\pi/\omega_1$. Then there exists at least one point t_1 in that interval with the property $\cos (\omega_1 t_1 + \vartheta_0) = 0$. Using Condition (3) we find that for $t = t_1$ the left member of Relation (52) is zero and hence we have either $\nu_1 = 0$ or $\vartheta_1 = \vartheta_0 + k\pi$, where k is an integer. In the first case we obtain equality (48) for $\varphi_0 = 0$ and in the second case an equality of the form $\cos(\omega_1 t + \vartheta_0) = \varepsilon \cos (\omega_1 t + \vartheta_1)$, where $\varepsilon = \pm 1$. In both cases Relation (52) is of the form (48) which is, as we have seen, contradictory. The contradiction shows that Condition 2°′ implies Condition 2° and therefore it also implies the asymptotic stability.

We show now that if Condition 3° of Theorem 2 is satisfied, then Condition 3° of Theorem 1, § 23 is also satisfied (for system (11), (22) corresponding to our problem). Define the vectors q and s of Theorem 1, § 23 by the relations $q = 0$ and $s = c$. Then (7, § 23) becomes

$$\text{“If } \lim_{t\to\infty} \nu(t) = 0, \quad \text{then } \lim_{t\to\infty} \mu(t) = 0\text{”}. \tag{53}$$

From (10) and (4) one obtains $|\mu(t)| \leqslant \max (|\varphi_1|, |\varphi_2|) |\nu(t)|$ and hence Condition (53) is satisfied. Furthermore from (46) and (3) it follows that the function ρ_0, defined as :

$$\rho_0(\delta) = (\varphi(\delta) - \varphi_1 \delta)(\varphi_2 \delta - \varphi(\delta))$$

satisfies the conditions $\rho_0(\delta) > 0$ – for every real $\delta \neq 0$ – and $\rho_0(0) = 0$. Since the solution examined is bounded there exists a positive number ν_0 such that $|\nu(t)| \leqslant \nu_0$. It is easy to see that the function ρ_i defined as:

$$\rho_i(\delta) = \frac{|\delta|}{\nu_0} \sup_{0 \leqslant \tau \leqslant |\delta|} [\min(\rho_0(\tau), \rho_0(-\tau))]$$

has — for $|\delta| \leqslant \nu_0$ — the properties required to the functions of class M_i (Definition 5, § 13). Furthermore, for $|\delta| \leqslant \nu_0$ we have $\rho_0(\delta) \geqslant \rho_i(\delta)$. Therefore, from (10) and (12) and the fact that for the solution considered the inequality $|\nu(t)| \leqslant \nu_0$ is satisfied, one obtains

$$-\eta_1(0, t_1) \geqslant \int_0^{t_1} \rho_i(\nu(t))\, dt. \tag{54}$$

Since the function ν is bounded and expression (16) is equal to (15) the integral $|\eta_2(0, t_1)|$ is bounded. Similarly we find that $|\eta_3(0, t_1)|$ is bounded. Using these results and Relation (21) (for $\alpha_1 = 1$) we obtain (8, § 23) (see also footnote 2, p. 228). It remains to show that one of the conditions (a)—(c) of Theorem 1, § 23 is also satisfied. Observe that we may take $c \neq 0$. Indeed, if $c = 0$ then $\varphi(\nu) \equiv 0$ and hence Eq. (1) becomes $dx/dt = Ax$. From Condition (A) of Theorem 1 it follows that A is a Hurwitz matrix and hence the asymptotic stability of the system considered is ensured.

If $c \neq 0$ we apply Proposition 1, § 32 and bring system (1), (2) to the form (see (6, § 32)—(8, § 32))

$$\frac{dy}{dt} = A_{11}\, y - b_1\, \varphi(\nu) \tag{55}$$

$$\frac{dz}{dt} = A_{21}\, y + A_{22}\, z - b_2\, \varphi(\nu) \tag{56}$$

$$\nu = c_1'\, y, \tag{57}$$

where the pair (c_1', A_{11}) is completely observable. Furthermore, from Condition (A) of Theorem 1 (required also in Theorem 2) it follows that A_{22} is a Hurwitz matrix. Indeed, for $\varphi(\delta) = \varphi_0 \delta$, system (55)—(57) becomes:

$$\frac{dy}{dt} = (A_{11} - \varphi_0 b_1 c_1') y$$

$$\frac{dz}{dt} = (A_{21} - \varphi_0 b_2 c_1') y + A_{22} z$$

and hence in order that the trivial solution of this system be asymptotically stable it is necessary that A_{22} be a Hurwitz matrix. Thus Condition (c) of Theorem 1, § 23 is satisfied; this concludes the proof of Theorem 2.

7. Simplifying the frequency criterion

If in Eqs. (1), (2) one introduces a new function defined by

$$\tilde{\varphi}(\delta) = \beta_1 \varphi(\delta) + \beta_2 \delta, \quad (\beta_1 \neq 0), \tag{58}$$

(where β_1 and β_2 are constants), the system becomes

$$\frac{dx}{dt} = \tilde{A} x - \tilde{b} \tilde{\varphi}(\nu), \quad \nu = c' x, \tag{59}$$

(where $A = A + bc' \beta_2/\beta_1$ and $\tilde{b} = b/\beta_1$). The new function $\tilde{\varphi}$ satisfies the condition $\tilde{\varphi}(0) = 0$ and we have

$$\left.\begin{aligned} (\beta_1 \varphi_1 + \beta_2) \delta^2 \leqslant \tilde{\varphi}(\delta) \delta \leqslant (\beta_1 \varphi_2 + \beta_2) \delta^2 \text{ if } \beta_1 > 0 \\ (\beta_1 \varphi_2 + \beta_2) \delta^2 \leqslant \tilde{\varphi}(\delta) \delta \leqslant (\beta_1 \varphi_1 + \beta_2) \delta^2 \text{ if } \beta_1 < 0 \end{aligned}\right\} \tag{60}$$

(cf. (3) and (4)). We thus obtain a problem of the same form as in Section 1 and we may apply Theorems 1 and 2. Instead of the frequency Condition (43) we obtain in this case a similar inequality where $\gamma(i\omega)$ is replaced by $\tilde{\gamma}(i\omega) = c'(i\omega I - \tilde{A})^{-1}\tilde{b}$ and the numbers φ_1 and φ_2 are replaced by $\beta_1 \varphi_1 + \beta_2$ or $\beta_1 \varphi_2 + \beta_2$ (as one can see by comparing inequality (60) with (4)). Although the frequency condition thus obtained has a form which differs from (43), the two conditions are equivalent.

This conclusion can be verified by direct algebraic calculations, but it follows directly from the results of chapter 2.

Indeed, instead of (10) we obtain now the expression $\tilde{\mu}(t) = -\tilde{\varphi}(\nu(t)) = -\beta_1\varphi(\nu(t)) - \beta_2\nu(t)$. Hence, $\mu(t)$ and $\tilde{\mu}(t)$ are connected by the relation $\tilde{\mu}(t) = \beta_1\mu(t) - \beta_2 c'x(t)$. Therefore system (11), (22) is subjected to a transformation of the form presented in Section 2, § 2. Condition (43) represents in fact the inequality $\chi(-i\omega, i\omega) \geqslant 0$. Using (28, § 3) and (32, § 3) we immediately find that the transformed condition $\tilde{\chi}(-i\omega, i\omega) \geqslant 0$ is equivalent to Condition (43).

From these observations it follows that, without loss of generality, we may modify the statement of the problem in the way shown above, before applying Theorems 1 and 2. Using this conclusion we can simplify Condition (43). Clearly a considerable simplification is obtained if $\varphi_1 = 0$. Then, from condition $\varphi_2 > \varphi_1$ it follows that $\varphi_2 > 0$ and we can divide (43) by φ_2. Replacing the symbol φ_2 by λ_0 (and keeping the symbols φ_1 and φ_2 only for the general case) we obtain the inequality

$$\mathrm{Re}\,[(1 + \beta\, i\omega)\,\gamma(i\omega)] + \frac{1}{\lambda_0} \geqslant 0, \tag{61}$$

where $\beta = \alpha_2/\varphi_2$ (hence β can be chosen arbitrarily). Since $\varphi_1 = 0$, inequality (4) becomes (replacing again φ_2 by λ_0)

$$0 \leqslant \varphi(\delta)\,\delta \leqslant \lambda_0\,\delta^2. \tag{62}$$

In other words the graph of the function φ must be located in a sector of the form shown in Fig. 5.1.

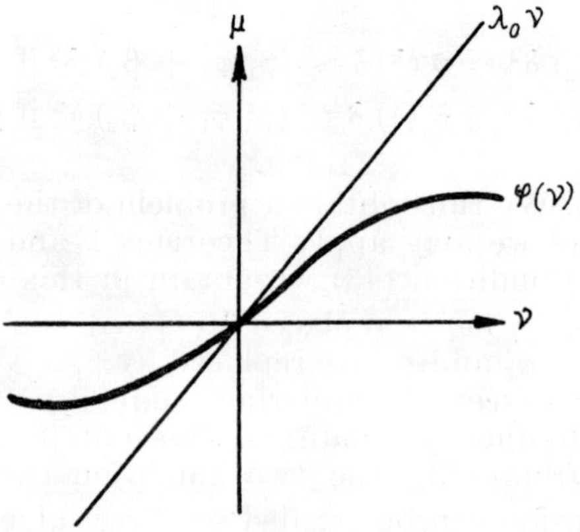

Fig. 5.1

We further observe that among the functions $\tilde{\varphi}$ of the form (58) there exists a function which satisfies (like φ) the inequalities (4). Indeed, choose

$$\tilde{\varphi}(\delta) = -\varphi(\delta) + (\varphi_1 + \varphi_2)\,\delta, \tag{63}$$

(i.e. choose, in (58), $\beta_1 = -1$ and $\beta_2 = -(\varphi_1 + \varphi_2)$) and compare the inequalities (4) and (60) (for $\beta_1 < 0$). By formula (63) we must introduce in system (11), (22) the new control function $t \mapsto \tilde{\mu}(t) = -\mu(t) - (\varphi_1 + \varphi_2)\,\nu(t)$. Substituting this in (22) we obtain an expression of the same form where μ is replaced by $\tilde{\mu}$ and the numbers α_2 and α_3 are permuted. Thus the new frequency condition we obtain differs from (43) only insofar $\gamma(i\omega)$ is replaced by $\tilde{\gamma}(i\omega)$ (for the modified problem) and the constant α_0 is replaced by $-\alpha_0$ (see (31)). As we already remarked, the new frequency condition is equivalent to the initial one. From the foregoing it also follows that if $\varphi_1 = 0$ and if, the frequency condition (61) is satisfied for a given number $\beta < 0$, then a frequency condition of the same form, but with $-\beta$ instead of β is satisfied for the modified problem obtained after introducing the function $\tilde{\varphi}(\delta) = \lambda_0\,\delta - \varphi(\delta)$.

8. Using hyperstable blocks to treat the problem of absolute stability

As shown in the previous section, if the frequency Condition (43) is satisfied for the system (1)—(2) we can always bring the problem to a form in which the inequalities (4) take the particular form (62) and the frequency Condition (43) takes the form (61) where $\beta \geqslant 0$. Hence we may examine — without loss of generality — only the frequency condition (61) for $\beta \geqslant 0$. This relation is similar to Condition (F) of Theorem 1, § 15 and, in fact, as we shall see below, reduces to the condition that a certain block of the form studied in § 15 be hyperstable.

The hyperstable blocks to be used in the treatment of our problem are suggested by inequalities (13), (17) and (20). By Conditions (62) we have

$$\int_0^{t_1} \varphi(\nu)\left(\nu - \frac{1}{\lambda_0}\,\varphi(\nu)\right) dt \geqslant 0 \tag{64}$$

$$\psi(\delta) = \int_0^{\delta} \varphi(\rho)\, d\rho \geqslant 0. \tag{65}$$

Defining $\tilde{\eta}(0, t_1)$ by

$$\tilde{\eta}(0, t_1) = \int_0^{t_1} \varphi(\nu) \left(\nu - \frac{1}{\lambda_0} \varphi(\nu) + \beta \frac{d\nu}{dt} \right) dt, \tag{66}$$

we find the inequality

$$\beta \psi(\nu(t)) \leqslant \tilde{\eta}(0, t) + \beta \psi(\nu(0)), \tag{67}$$

which is similar to the sufficient condition of hyperstability (25, § 13). Furthermore, the form of the integral (66) is similar to the expression (45, § 13) for the systems associated with the hyperstable blocks (see Section 9, § 13). Indeed, if we define the functions μ_2 and ν_2 by the relations

$$\mu_2 = \nu - \frac{1}{\lambda_0} \varphi(\nu) + \beta \frac{d\nu}{dt} \tag{68}$$

$$\nu_2 = \varphi(\nu), \tag{69}$$

the integral (66) takes the form $\int_0^{t_1} \mu_2 \nu_2 \, dt$.

If $\beta \neq 0$ Eqs. (68) and (69) can be expressed in the usual form (41, § 13), (42, § 13) by simply rearranging the terms

$$\frac{d\nu}{dt} = -\frac{1}{\beta} \nu + \frac{1}{\beta \lambda_0} \varphi(\nu) + \frac{1}{\beta} \mu_2 \tag{70}$$

$$\nu_2 = \varphi(\nu). \tag{71}$$

The system associated with this block consists of Eqs. (70), (71) and the integral (66). Hence the block (70), (71) is hyperstable whenever one can express inequality (67) in the form (25, § 13) (where x is to be replaced in our case by ν). This can be done, for example, under the following conditions [1])

$$\beta > 0, \ \lim_{\delta \to \infty} \psi(|\delta|) = \infty, \ \lim_{\delta \to \infty} \psi(-|\delta|) = \infty \tag{72}$$

$$\varphi(\delta)\, \delta > 0, \quad \text{for every real } \delta \neq 0; \tag{73}$$

[1]) These additional conditions were not needed in Theorem 1. Here also we can get rid of these conditions if we use the extended definition of a hyperstable block (see the remarks at the end of § 19).

Then from (67) one obtains the inequality (25, § 13) for the functions $\tilde{\alpha}$, $\tilde{\beta}$ and $\tilde{\gamma}$ defined by

$$\tilde{\alpha}^2(\delta) = \beta \min (\psi(\nu), \psi(-\nu))$$
$$\tilde{\beta}^2(\delta) = \beta \max (\psi(\nu), \psi(-\nu))$$
$$\tilde{\gamma}(\delta) = 0$$

and these functions satisfy the conditions required in Proposition 4, § 13. We now apply the results obtained in Section 9, § 13 and assume that the block (70), (71) (denoted in the foloŗwing by B_2) is a component of a system with negative feedback, as shown in Fig. 5.2 (which is obtained from Fig. 4.1.c for $u = \mu \equiv 0$). Therefore (see Section 9, § 13) the trivial solution of the system in Fig. 5.2 is stable if the blocks B_1 and B_2 are hyperstable and if the system has a solution for every initial condition. We determine now a block B_1 such that for every solution of the system (1), (2) we can find some additional functions to obtain a solution of the system shown in Fig. 5.2. By the definition of systems with negative feedback the functions μ_1, ν_1, μ_2 and ν_2 satisfy the relations

$$\mu_1 = -\nu_2, \quad \nu_1 = \mu_2. \tag{74}$$

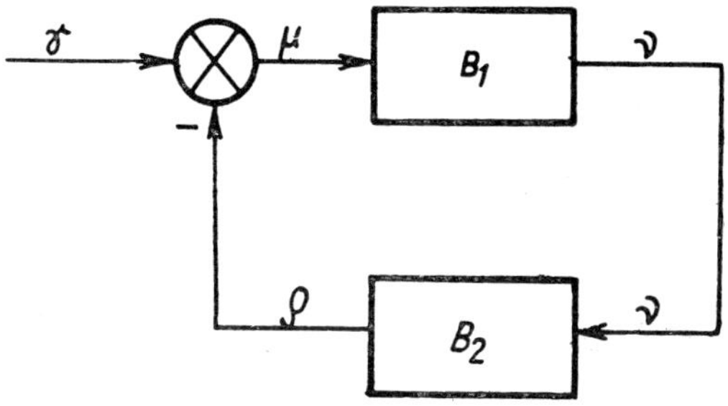

Fig. 5.2

From this and Eqs. (70), (71) of block B_2 we obtain

$$\mu_1 = -\varphi(\nu) \tag{75}$$

$$\nu_1 = \nu - \frac{1}{\lambda_0}\varphi(\nu) + \beta\frac{d\nu}{dt}. \tag{76}$$

To determine completely the block B_1 we make use of (1) and (2) by replacing $\varphi(\nu)$ by $-\mu_1$ (see (75)); this leads to equations

$$\frac{dx}{dt} = Ax + b\mu_1, \quad \nu_1 = \nu + \frac{1}{\lambda_0}\mu_1 + \beta\frac{d\nu}{dt} \tag{77}$$

i.e. using Eq. (2)

$$\frac{\mathrm{d}x}{\mathrm{d}t} = Ax + b\mu_1, \qquad \nu_1 = (c' + \beta c'A)\,x + \left(\beta\, c'\, b + \frac{1}{\lambda_0}\right)\mu_1. \qquad (78)$$

Eqs. (78) have the form (1, § 15), (2, § 15). In accordance with the foregoing remarks a sufficient condition of absolute stability is obtained by requiring that the block (78) be hyperstable. Since the transfer function of block (77) is $\sigma \mapsto (1 + \beta\sigma)\,\gamma(\sigma) + 1/\lambda_0$ (where $\gamma(\sigma)$ has the expression (26)) we immediately conclude that Condition (*F*) of Theorem 1, § 15 coincides with Condition (61). Obviously, to infer that block B_1 is hyperstable we must also check Conditions (*i*)—(*iii*) from Theorem 1, § 15. However, these conditions need not be examined here since they have been already considered in detail in the preceding sections of this paragraph. (It is sufficient to remark that the system associated to block (77) is a particular case of the systems of the form (11), (22)).

In addition to the results obtained in the previous sections, the method of the hyperstable blocks supplies some additional information about the meaning of the frequency criterion (61). This allows us to give a more complete answer to the fundamental question of finding the most general consequences of the frequency criterion (61). On the basis of the foregoing discussion this question boils down to finding the most general consequences of the assumption that block B_1, (77), is hyperstable — a problem that has been treated in §§ 13 and 15. As we know, if block B_1 is hyperstable, then all the systems of the form shown in Fig. 5.2 have a stable trivial solution (whenever the existence of the solutions is ensured) for every block B_2 which is hyperstable (in particular, for every B_2 which is the inverse of a hyperstable block, or a memoryless hyperstable block; (see Definition 9, § 13). The results of §§ 13 and 15 lead to even more general conclusions (e.g. Proposition 11, § 13).

Thus, by replacing the problem of absolute stability by a problem of hyperstability the stability problem is solved for a large class of systems which contains the systems of the form (1), (2) as a very particular case.

Reversing the question stated above, we now ask about the strongest conclusions we can draw if the frequency condition (61) is not satisfied. Whereas the answer to the first question consisted in obtaining the stability of the largest possible family *F* of systems the answer to the converse question consists in expressing the negation of Condition (61) in terms of the instability of a subfamily of *F* as restricted as possible and of a simple form.

An answer to this converse problem is obtained by using the Property (h_λ) of Theorem 1, § 15. Thus Condition (61) is not satisfied (for a given $\beta \geqslant 0$) if and only if there exists a block B_2, which is either described by the equation $\nu_2 = \frac{1}{\lambda}\, d\mu_2/dt$, $\lambda > 0$ or by the equation $d\nu_2/dt = \lambda\mu_2$, $\lambda > 0$) and for which the trivial solution of the corresponding system of Fig. 5.2 (where B_1 is described by (77)) is not stable. We recall that this property is connected with Aizerman problem, as mentioned in § 15.

The approach from the present section gave us the opportunity of introducing the non-linear hyperstable block (70), (71). This result can be also used in other stability problems by applying, for example, Proposition 9, § 13.

9. Determining the largest sector of absolute stability

As shown in Section 7, if there exists a real number β such that the frequency condition (61) is satisfied for a given number $\lambda_0 > 0$ and if the Conditions (A) and (C) of Theorem (1)[1]) are also satisfied, then the absolute stability of the trivial solution of the system (1), (2) is ensured. It is obviously desirable that the number λ_0 be large so that the sector determined by the inequalities (62) be as wide as possible. Since the maximum value of λ_0 for which inequality (61) is satisfied, depends generally on β, one must find the optimal value of β so as to obtain the highest value of λ_0 for which a condition of the form (61) is satisfied. This problem has a simple solution if one introduces a geometrical interpretation of inequality (61) (see the author's paper [6]).

Let us write the transfer function of system (1), (2) (see (26)) under the form $\gamma(i\omega) = \xi(\omega) + i\eta(\omega)$, where $\xi(\omega)$ and $\eta(\omega)$ are equal respectively to the real and imaginary part of $\gamma(i\omega)$. The set of all the points $(\xi(\omega), \omega\eta(\omega))$ for all real ω will be called "the modified transfer locus" (owing to the obvious analogy to the transfer locus usually employed in the linear theory of control systems and defined as the locus of the points $(\xi(\omega), \eta(\omega))$).

---

1) In the case $\varphi_1 = 0$ and $\varphi_2 = \lambda_0$, Condition (A) takes the following formn "There exists a number φ_0 in the interval $0 \leqslant \varphi_0 \leqslant \lambda_0$ such that the trivial solutio : of the system (1), (2) is asymptotically stable for $\varphi(\delta) = \varphi_0\delta$". In order that Condition (C) be also satisfied it is sufficient that the left hand member of (61) be not identically zero.

Using the above notations inequality (61) becomes

$$\xi(\omega) - \beta\omega\eta(\omega) + \frac{1}{\lambda_0} \geqslant 0, \qquad (\lambda_0 > 0). \tag{79}$$

If one introduces the straight line [1])

$$D : x - \beta y + \frac{1}{\lambda_0} = 0 \tag{80}$$

Condition (61) can be expressed in a geometrical language as follows: "There exists a straight line D of the form (80) such that the modified transfer locus is entirely located on the right of that line" [2]).

This supplies a graphical criterion for absolute stability similar to the Nyquist criterion [3]). It is sufficient to draw the modified transfer locus of the system (1), (2) (this can be effected even by way of a physical experiment) and to determine a straight line D with the properties stated (see Fig. 5.3). If the line D intersects the negative real semiaxis at a point $-1/\lambda_0 > 0$, the number λ_0 thus obtained determines the magnitude of the sector wherein the property of absolute stability takes place (see inequalities (62)). Clearly this graphical construction replaces only inequality (61) and hence, in order to obtain a complete criterion for absolute stability we must take into account also the other conditions specified in the statement of the Theorem 1 (see footnote 1, page 261). From the way in which the number λ_0 has been determined it follows that the sector wherein absolute stability occurs is maximum when the line D intersects the negative real semiaxis at a point which is as close to the origin as possible. The optimum position of the line D can be easily determined by graphical methods.

It is clear that if the intersection point of the line D and the negative real semiaxis (equal to $-1/\lambda_0$) belongs also to the modified transfer locus (see Fig. 5.4), then it is also a point of Nyquist's diagram and hence the number λ_0 determines the

---

[1]) This line has at least one point in common with the semiaxis $x < 0$ namely the point $x_0 = -1/\lambda_0$. If $\beta \geqslant 0$ the line is parallel to the Oy axis or inclined on the right, and if $\beta < 0$ the line D is inclined on the left and protrudes in the quadrant IV.

[2]) It can be proved that, under certain additional conditions, the case in which the transfer locus is entirely located below the realaxis is also admitted.

[3]) Another graphical method for the verification of inequalities of the form (61) has been derived by B. N. Naumov [1], using logarithmic frequency characteristics.

largest sector wherein absolute stability may occur. Hence, in the case of systems whose modified transfer locus has this property, Aizerman problem [1] is affirmatively solved (see also Section 2, Chapter 1). The analysis of particular cases

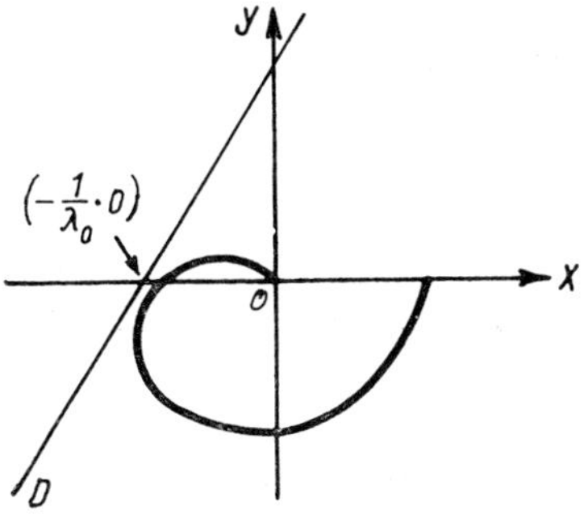

Fig. 5.3

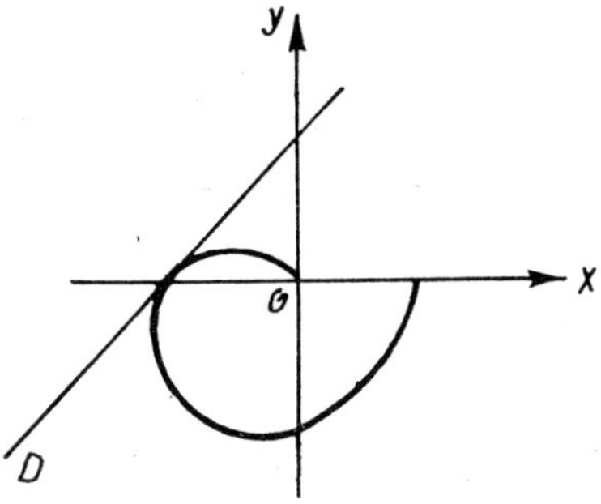

Fig. 5.4

shows that sometimes, — even if the modified transfer locus has not the property considered —, the maximum sector determined by the graphical method described above cannot be extended by any other method. The problem o establishing general necessary and sufficient conditions for absolute stability, is still open.

However, if the problem of absolute stability is treated by including it in a problem of hyperstability (as we proceeded in the previous sections of this paragraph), then, as we saw, the conditions obtained are necessary and sufficient for the corresponding problem of hyperstability.

10. Other generalizations of the problem of absolute stability

The methods used in this paragraph can be applied in the case of discrete systems or of systems with several non-linearities [1]). Even in the case of simple systems the problem can be reexamined under different hypotheses concerning the function φ. These generalizations can be done using the methods established in the foregoing chapters and applying them as we proceeded in this paragraph.

In §§ 28 and 29 we shall briefly examine a few other stability problems. However we shall confine ourselves to the method of hyperstable blocks, which will give us the opportunity of finding new nonlinear hyperstable blocks.

---

[1]) Systems with several nonlinearities have been treated in most of the author's previous papers (e.g. see [3] and [4]).

§ 26. Determination of some Liapunov functions

1. Necessary conditions for the existence of Liapunov functions of the Lur'e-Postnikov type

We consider, as in § 25, the system

$$\frac{dx}{dt} = Ax - b\varphi(\nu), \quad \nu = c'x, \quad (b \neq 0, \quad c \neq 0) \tag{1}$$

and assume that there exists a Liapunov function $z \mapsto V(z)$ of the form

$$V(z) = z'Nz - \delta_0 \int_0^{c'z} \varphi(\rho)\, d\rho, \quad (N' = N, \quad \delta_0 = \text{constant}) \tag{2}$$

whose "derivative by system (1)"

$$\begin{aligned} & z'(NA + A'N)z - \varphi(c'z)\left(\frac{1}{2}\delta_0 c'A + b'N\right)z - \\ & - z'\left(Nb + \frac{1}{2}\delta A'c\right)\varphi(c'z) + \delta_0 c'b(\varphi(c'z))^2 \end{aligned} \tag{3}$$

maintains the same sign, for every n-vector z and every function φ which satisfies the conditions $0 \leqslant \varphi(\delta)\delta < \lambda_0 \delta^2$. This property must occur, in particular, for every function φ of the form

$$\varphi(\delta) = \lambda\delta \tag{4}$$

and for every real number λ of the interval

$$0 < \lambda < \lambda_0. \tag{5}$$

We assume that the following conditions are satisfied:

(i) The derivative (3) is positive semidefinite for every function φ which satisfies Conditions (4) and (5).

(*ii*) Function V, (2), is negative definite for every function φ which satisfies Conditions (4) and (5).

(*iii*) The matrix $A - \lambda bc'$ (obtained by substituting (4) in (1)) is nonsingular for every λ in the interval (5).

(*iv*) If the derivative (3) is positive semidefinite for every λ, then the number δ_0 of (2) satisfies the inequality $\delta_0 > 0$.

(*v*) The system (1) is nondegenerate (see § 33).

Conditions (*i*) and (*ii*) are of the type usually required for Liapunov functions and amount to a choice of the sign. Conditions (*iii*) is natural since by introducing Liapunov functions with the Properties (*i*) and (*ii*) we want to establish, in particular, that all the linear systems obtained by substituting (4) in (1) have asymptotically stable trivial solutions, for every λ in the interval (5). Concerning Condition (*iv*) we see that if the derivative (3) is positive semidefinite for every λ, then from Conditions (*v*) and (*ii*) it follows that $\delta_0 \neq 0$ and then the interval (5) can be always extended (while maintaining the Properties (*i*) and (*ii*) either in the form $0 < \lambda < \infty$ or in the form $-\infty < \lambda < \lambda_0$. The latter case reduces always to the first by effecting the change of function $\tilde{\varphi}(\delta) = \lambda_0 \delta - \varphi(\delta)$. Furthermore in the case $\delta_0 \neq 0$ Condition (*ii*) cannot be satisfied in an interval $0 < \lambda < \infty$ unless $\delta_0 > 0$.

The problem treated below can be also solved without the Hypothesis (*v*) as has been shown in the author's paper [13] [1]). Using a different method, the same problem has been solved almost simultaneously by V. A. Yakubovich [4]. The method presented below is new and constitutes an application of the results obtained in Chapter 3. It may be also extended to the case of systems with several nonlinearities.

Proposition 1. *If there exists a Liapunov function of the form (2) and Properties (i)—(v) are satisfied, then there exist two real numbers* α *and* β *such that one has the inequality (61, § 25), where* $\gamma(i\omega)$ *has the expression (26, § 25). Furthermore the number* α *satisfies the inequality* $\alpha \geqslant 0$ *and in the case* $\alpha = 0$ *the number* β *is positive.*

The proof is based on the positiveness theorem of single-input systems (see Section 2, § 8) and follows the following scheme: from the condition that the quadratic form (3) be positive semidefinite we shall deduce that a certain matrix of the form (14, § 8) is positive semidefinite, thus obtaining a property of the type of Property 5° of Theorem 1, § 8. Using the equivalence of Property 5° and Property 2° of the theorem quoted, we shall express the same condition in the form of inequality (7, § 8) which will coincide with the inequality (61, § 25) we want to establish.

We shall first prove Proposition 1 under the assumption that the pair (A, b) is completely controllable (to be able to

---

[1]) We can dispense with the Hypothesis (*v*) also when using the method adopted in the present paragraph; however, this would imply somewhat ampler developments.

apply Theorem 1, §8), and that the vectors c and z and the matrix $NA + A'N$ are of the form

$$c = \begin{pmatrix} 1 \\ 0 \end{pmatrix}\}_{n-1}; \quad z = \begin{pmatrix} \zeta \\ y \end{pmatrix}; \quad NA + A'N = \begin{pmatrix} \overbrace{\rho}^{1} & \overbrace{0}^{n-1} \\ 0 & R \end{pmatrix}\begin{matrix}\}1 \\ \}n-1\end{matrix} . \tag{6}$$

We shall then show that the problem can be always stated so that the above conditions be satisfied. In (6) we introduced the $(n-1)$-vector y, the scalars ζ and ρ and the matrix R with $(n-1)$ rows and columns. Clearly $c'z = \zeta$ (see (6)). The matrix $NA + A'N$ is positive semidefinite since otherwise expression (3) would take negative values if λ is positive and sufficiently small. Therefore

$$\rho \geqslant 0 . \tag{7}$$

Writing the vector $Nb + \frac{1}{2}\delta A'c$ in the form

$$Nb + \frac{1}{2}\delta A'c = \begin{pmatrix} -\alpha_0 \\ q_0 \end{pmatrix}\begin{matrix}\}1 \\ \}n-1\end{matrix} , \tag{8}$$

and using (4), (6) and $c'z = \zeta$ the condition that the quadratic form in the right hand member of (3) be positive definite becomes

$$\begin{pmatrix} -\lambda\zeta \\ y \end{pmatrix}' \begin{pmatrix} \frac{\rho}{\lambda^2} + 2\frac{\alpha_0}{\lambda} + \delta_0 c'b & q'_0 \\ q_0 & R \end{pmatrix} \begin{pmatrix} -\lambda\zeta \\ y \end{pmatrix} \geqslant 0 . \tag{9}$$

Introducing a new variable ψ and using Condition (7) we can write the obvious inequality $\rho(\zeta - \psi)^2 \geqslant 0$, i.e.

$$\rho\zeta^2 - 2\rho\zeta\psi + \rho\psi^2 \geqslant 0 . \tag{10}$$

Adding (9) and (10) we obtain an inequality which can be written as

$$\begin{pmatrix} -\lambda\zeta \\ \psi \\ y \end{pmatrix}' \left(\begin{array}{c|cc} \frac{2\rho}{\lambda^2} + 2\frac{\alpha_0}{\lambda} + \delta_0 c'b & \frac{\rho}{\lambda} & q'_0 \\ \hline \frac{\rho}{\lambda} & \rho & 0 \\ q_0 & 0 & R \end{array}\right) \begin{pmatrix} -\lambda\zeta \\ \psi \\ y \end{pmatrix} \geqslant 0, \tag{11}$$

and the matrix of this quadratic form is clearly positive semidefinite. Using (8) one proves that

$$\begin{pmatrix} \rho/\lambda \\ q_0 \end{pmatrix} = Nb + \frac{1}{2}\,\delta_0 A'\,c + \left(\alpha_0 + \frac{\rho}{\lambda}\right)c\,, \qquad (12)$$

by introducing in the last term of the right hand member the expression of c from (6). If the matrix of the quadratic form (11) is partitioned into blocks (as indicated by the dotted lines) and if we introduce again the matrix $NA + A'N$ (by means of the last relation of (6)) and the vector $Nb + \frac{1}{2}\,\delta_0 A'c$ (using (12)), we obtain the matrix :

$$\begin{bmatrix} \dfrac{2\rho}{\lambda^2} + \dfrac{2\alpha_0}{\lambda} + \delta_0 c'\,b & \left(Nb + \dfrac{1}{2}\,\delta_0\,A'\,c + \left(\alpha_0 + \dfrac{\rho}{\lambda}\right)c\right)' \\ Nb + \dfrac{1}{2}\,\delta_0\,A'\,c + \left(\alpha_0 + \dfrac{\rho}{\lambda}\right)c & N\,A + A'\,N \end{bmatrix}. \qquad (13)$$

This matrix coincides with the matrix (14, § 8) if we define

$$\varkappa = \frac{2\,\rho}{\lambda^2} + \frac{2\,\alpha_0}{\lambda} + \delta_0 c'b,\;\; l = \frac{1}{2}\,\delta\,A'\,c + \left(\alpha_0 + \frac{\rho}{\lambda}\right)c,\; M = 0. \qquad (14)$$

We have seen that (13) is a positive semidefinite matrix. From the equivalence of Properties 5° and 2° of Theorem 1, § 8 we obtain

$$\varkappa + 2\,\mathrm{Re}\,l'(\mathrm{i}\omega\,I - A)^{-1}\,b \geqslant 0, \quad (\det\;(\mathrm{i}\omega\,I - A) \neq 0). \qquad (15)$$

(see (7, § 8) and (13, § 3)). Using (14) and the identity

$$c'A\,(\mathrm{i}\omega\,I - A)^{-1}\;b = -\,c'\,b + \mathrm{i}\omega\,\gamma\,(\mathrm{i}\omega) \qquad (16)$$

(see (26, § 13) and (26, § 25)) the above inequality can be written as

$$\frac{\alpha\,(\lambda)}{\lambda} + \mathrm{Re}\,(\alpha(\lambda) + \mathrm{i}\omega\,\beta)\,\gamma(\mathrm{i}\omega) \geqslant 0, \;\; (\det\,(\mathrm{i}\omega\,I - A) \neq 0), \qquad (17)$$

where

$$\alpha(\lambda) = \frac{2\rho}{\lambda} + 2\,\alpha_0 \tag{18}$$

$$\beta = \delta_0. \tag{19}$$

Inequality (17) must be satisfied for every number λ of the interval (5). Since this inequality, written for $\lambda = \lambda_0$, has the form (61, § 25) it only remains to show that the numbers $\alpha(\lambda_0)$ and β satisfy the sign conditions required in Proposition 1.

If A is a singular matrix there exists a vector $z_0 \neq 0$ such that $Az_0 = 0$ and hence $z_0'(NA + A'N)z_0 = 0$, whence writing the vector z_0 in the form $z_0' = (\zeta_0\, y_0')$ and then using the last relation of (6)) we obtain $\rho\zeta_0^2 + y_0' R y_0 = 0$. But we have seen that the matrix $NA + A'N$ is positive semidefinite and therefore (see (6)) the obtained equality is satisfied only if $y_0' R y_0 = 0$ and $\rho\zeta_0^2 = 0$. Furthermore the component ζ_0 of the vector $z_0' = (\zeta_0 y_0')$ satisfies the equality $c'z_0 = \zeta_0$ (see (6)). If we had $\zeta_0 = 0$ we would obtain, using $Az_0 = 0$, the equalities $c'A^k z_0 = 0$ for every integer $k \geqslant 0$; these relations — which contradict Condition 4° of Consequence 1, § 32 — would show that the pair (c', A) is not completely observable, contrary to our Hypothesis (*v*). Hence $\zeta_0 \neq 0$ and the equality $\rho\zeta_0^2 = 0$ obtained above entails $\rho = 0$. Then (18) given $\alpha(\lambda) = 2\alpha_0$, a constant independent of λ. If α_0 were negative the left hand member of (17) would be also negative for positive and sufficiently small values of λ, a contradiction. Thus $\alpha(\lambda_0) \geqslant 0$. If $\alpha(\lambda_0) = 0$ the inequality $\beta > 0$ results from (19) and Hypothesis (*iv*).

If A is not singular then one can write (17) for $\omega = 0$, thus obtaining $(1 - \lambda c'A^{-1}b)c(\lambda)/\lambda \geqslant 0$ for every λ of the interval (5). Writing the identity (29, § 3) for $\sigma = 0$ and $q = -\lambda_c$ we obtain

$$1 - \lambda\, c'\, A^{-1}\, b = \frac{\det\,(A - \lambda\, b\, c')}{\det A}\,.$$

Since the left hand member is positive for $\lambda = 0$ it remains positive for every λ of the interval (5) (otherwise we would obtain the equality $\det\,(A - \lambda bc') = 0$ for a certain λ of the interval (5), contrary to Hypothesis (*iii*)). Therefore, the inequality $(1 - \lambda c'A^{-1}b)\alpha(\lambda)/\lambda \geqslant 0$ obtained above implies $\alpha(\lambda) \geqslant 0$ for every $0 < \lambda < \lambda_0$. Since $\alpha(\lambda)$ has the form (18) we obtain by passing to the limit the inequality $\alpha(\lambda_0) \geqslant 0$. If $\alpha(\lambda_0) = 0$

the conclusion $\beta > 0$ is obtained as above, by using (19) and Hypothesis (*iv*).

We now show that we can always satisfy Conditions (6). Since $c \neq 0$ (see (1)) we can find some real vectors $c_2, c_3 \ldots, c_n$ such that the $n \times n$-matrix $S = (c\, c_2\, c_3 \ldots c_n)$ be nonsingular. Applying the transformation $\tilde{z} = S'z$ we obtain

$$\tilde{z} = \begin{pmatrix} \zeta \\ \tilde{y} \end{pmatrix}\}n-1 \quad , \quad \zeta = c'z = \tilde{c}'\tilde{z}, \; \tilde{c} = \begin{pmatrix} 1 \\ 0 \end{pmatrix}\}n-1 \quad . \tag{20}$$

Hence the vectors $\tilde{z}$ and $\tilde{c}$ have the form of (6). Instead of the matrix $NA + A'N$ we have the transformed matrix $S^{-1}(NA + + A'N)(S')^{-1}$ which can be always written in the form

$$\tilde{N}\tilde{A} + \tilde{A}'\tilde{N} = S^{-1}\,(NA + A'N)\,(S')^{-1} = \begin{pmatrix} \tilde{\rho} & r' \\ r & R \end{pmatrix}\begin{matrix}\}1 \\ \}n-1\end{matrix} \quad . \tag{21}$$

From (21) it follows in particular that the real $(n-1) \times \times (n-1)$ — matrix R is symmetric and hence has $(n-1)$ linearly independent eigenvectors $y_1, y_2, \ldots, y_{n-1}$ satisfying the relations

$$R\, y_j = \lambda_j y_j, \; j = 1, 2, \ldots, n-1 \; ; \; y_j' y_k = \begin{cases} 0 \text{ if } j \neq k \\ 1 \text{ if } j = k, \end{cases} \tag{22}$$

where the eigenvalues λ_j are real (e.g. see F. R. Gantmacher [1] Chapter IX, § 13). The n-1-vector r can be uniquely written as

$$r = \alpha_1\, y_1 + \alpha_2 y_2 + \ldots + \alpha_{n-1}\, y_{n-1}. \tag{23}$$

We remark that if the equality $\lambda_j = 0$ is satisfied for a certain index j the corresponding coefficient of (23) is zero. Otherwise we would obtain, using (22), the equalities $Ry_j = 0$ and $y_j'\, r = \alpha_j \neq 0$. Hence we would have

$$\begin{pmatrix} -\varepsilon\,\alpha_j \\ y_j \end{pmatrix}' \begin{pmatrix} \tilde{\rho} & r' \\ r & R \end{pmatrix} \begin{pmatrix} -\varepsilon\,\alpha_j \\ y_j \end{pmatrix} = (\,\varepsilon^2 \tilde{\rho} - 2\,\varepsilon)\; |\alpha_j|^2,$$

an expression which is negative for a sufficiently small ε. We would thus obtain the contradictory conclusion that the derivative (3) could also take negative values if λ is positive and sufficiently small. The obtained contradiction shows that the expression (23) of the vector r takes the form

$$r = \alpha_{i1}\, y_{i1} + \alpha_{i2}\, y_{i2} + \ldots + \alpha_{i_p}\, y_{i_p}, \tag{24}$$

where p is a positive integer smaller or equal to $n - 1$ and the integers i_j, $j = 1, 2, \ldots, p$ take values in the interval $[1, n - 1]$ and satisfy the conditions

$$\lambda_{i_j} \neq 0, \quad j = 1, 2, \ldots, p. \tag{25}$$

Hence, we can define the vector

$$q = \sum_{j=1}^{p} \frac{\alpha_{i_j}}{\lambda_{i_j}} y_{i_j}, \tag{26}$$

which, by (22) and (24), satisfies the equality

$$Rq = r. \tag{27}$$

We now replace the vector $\tilde{y}$ of (20) by

$$\tilde{\tilde{y}} = \tilde{y} + \zeta q, \tag{28}$$

the variable ζ being unchanged. Then the vector $\tilde{z}$ of (20) is replaced by

$$\tilde{\tilde{z}} = \begin{pmatrix} \zeta \\ \tilde{\tilde{y}} \end{pmatrix}, \text{ where } \zeta = \tilde{\tilde{c}}' \, \tilde{\tilde{z}} \,, \tilde{\tilde{c}} = \begin{pmatrix} 1 \\ 0 \end{pmatrix}. \tag{29}$$

These relations agree with the first two relations of (6). The new form of the matrix $\tilde{N}\tilde{A} + \tilde{A}'\tilde{N}$ is obtained from the equalities

$$\tilde{z}'(\tilde{N}\,\tilde{A} + \tilde{A}'\,\tilde{N})\,\tilde{z} = \begin{pmatrix} \zeta \\ \tilde{y} \end{pmatrix}' \begin{pmatrix} \tilde{\rho} & r' \\ r & R \end{pmatrix} \begin{pmatrix} \zeta \\ \tilde{y} \end{pmatrix} = \begin{pmatrix} \zeta \\ \tilde{\tilde{y}} \end{pmatrix}' \begin{pmatrix} \tilde{\rho} - r'q & 0 \\ 0 & R \end{pmatrix} \begin{pmatrix} \zeta \\ \tilde{\tilde{y}} \end{pmatrix}, \tag{30}$$

where the first equality is obtained by using (20) and (21) and the second equality follows from (27) and (28). Thus, the matrix $\tilde{N}\tilde{A} + \tilde{A}'\tilde{N}$ is replaced by the last expression of (6). Thus Conditions (6), under which we have solved our problem, do not restrain the generality of the results.

2. Functions of the Liapunov type for systems with a single non-linearity

We now show that if there exist two numbers α and β with the properties of Proposition 1 and if the pair (A, b) is completely controllable, then there exists a function V of the form (2) whose derivative (3) is positive semidefinite for $0 \leqslant \varphi(\delta)\delta \leqslant \leqslant \lambda_0 \delta^2$.

Defining

$$J = 0, \ \varkappa = \frac{\alpha}{\lambda_0} + \beta c' b, \ l = \frac{1}{2} \alpha c + \frac{1}{2} \beta A' c, \ M = 0, \qquad (31)$$

Condition (61, § 25) can be written in the form (7, § 8) where $\chi(-i\omega, i\omega)$ has the expression (13, § 3). In other words (see Property 2° of Theorem 1, § 8) the system

$$\frac{dx}{dt} = Ax + b\mu, \ \eta(0, t_1) = \int_0^{t_1} \left(\left(\frac{\alpha}{\lambda_0} + \beta c' b \right) \mu^2 + \mu(\alpha c' + \beta c' A) x \right) dt \qquad (32)$$

is positive. By Property 3° of the same theorem, the integral $\eta(0, t_1)$ can be written in the form (12, § 8), i.e. (since $J = 0$)

$$\eta(0, t_1) = -\Big[x' N x \Big]_0^{t_1} + \int_0^{t_1} |\gamma \mu + w' x|^2 \, dt. \qquad (33)$$

(Recall that all the quantities are now real). But the integral of (32) can be also written as

$$\eta(0, t_1) = -\alpha \int_0^{t_1} \varphi(\nu) \left(\nu - \frac{1}{\lambda_0} \varphi(\nu) \right) dt - \beta \int_0^{t_1} \varphi(\nu) \frac{d\nu}{dt} \, dt, \qquad (34)$$

where we have used the relations (see (1)) :

$$\mu = -\varphi(\nu), \quad \nu = c' x. \qquad (35)$$

Eliminating $\eta(0, t_1)$ between (33) and (34) we find for every solution of system (32),

$$\left[x'(t) N x(t) - \beta \int_0^{\nu(t)} \varphi(\rho) \, d\rho \right]_0^{t_1} =$$

$$= \int_0^{t_1} \left((\gamma\mu + w'x)^2 + \alpha \varphi(\nu) \left(\nu - \frac{1}{\lambda_0} \varphi(\nu) \right) \right) dt,$$

and further, after differentiating

$$\frac{dV}{dt} = (\gamma \mu + w' x)^2 + \alpha \varphi(\nu) \left(\nu - \frac{1}{\lambda_0} \varphi(\nu) \right) \qquad (36)$$

where V stands for the function $t \to V(x(t))$, obtained from (2) (with δ_0 replaced by β). The right hand member of (36), is non-negative for every function φ with the property $0 \leqslant \varphi(\delta)\delta \leqslant \lambda_0 \delta^2$; this proves the assertion made at the beginning of this section.

The Liapunov function obtained above is not yet sufficient for obtaining conclusions on the asymptotic stability of the trivial solution; for this we need one of the various extensions of Liapunov theory contributed by E. A. Barbashin and N. N. Krasovskiy [1], J. P. LaSalle [1], V. A. Yakubovich [3—7], etc. However, these developments are outside the scope of the present monograph and we refer to the works quoted for further treatment of the problem.

§ 27. Stability in finite domains of the state space

1. An auxiliary lemma

We have seen in § 25 that the main step in applying the hyperstability theory to the treatment of a concrete problem of stability consists in obtaining an inequality of the form (12, § 13). Sometimes we can obtain such an inequality only if $x(t)$ (representing the state of the system under investigation at time t), does not leave a certain finite domain (containing the origin) of the phase space. Then we can obtain the stability of the trivial solution of the system and the boundedness of all the solutions whose initial condition is contained in a certain neighbourhood of the origin in the state space. A simple case is shown in the following lemma:

Lemma 1. *Consider system (10, § 13), (11, § 13) and let $\mathcal{S}$ be a certain family of solutions of this system (determined by the condition that an additional relation between x, u and t be satisfied). Assume that system (10, § 13), (11, § 13) is hyperstable (see Definition 8, § 13) and that the following property is satisfied: "There exists a constant $\alpha_0 > 0$ such that for every solution from the family $\mathcal{S}$ for which one has*

$$\sup_{t_0 \leqslant t \leqslant t_1} \| x(t) \| \leqslant \alpha_0, \tag{1}$$

one also has

$$\eta(t_0, t_1) \leqslant 0, \tag{2}$$

where t_1 is the same as in (1)."

Then there exist a constant $\alpha_1 > 0$ and a continuous function $z \mapsto \rho(z)$ of the n-vector z such that $\rho(0) = 0$ and any solution *from $\mathcal{S}$ whose initial condition satisfies the inequality* $\| x(t_0) \| \leqslant \alpha_1$, *satisfies also the inequality* $\| x(t) \| \leqslant \rho(x(t_0))$ *for every* $t \geqslant t_0$ *for which* $x(t)$ *is defined.*

Proof. Using the functions α and β which occur in Definition 8, § 13 we can always find a function ρ with the properties required in Lemma 1 such that the inequality $\| x(t) \| \leqslant \rho(x(t_0))$ be implied by the inequality $\alpha(x(t)) \leqslant \beta(x(t_0))$ (which is obtained from (13, § 13) for $\beta_0 = 0$). Then we can find a number α_1 such that for every $x(t_0)$ for which the condition $\| x(t_0) \| \leqslant \alpha_1$ is satisfied, the inequality $| \rho(x(t_0)) | < \alpha_0$ be also satisfied. Applying Property H_s of Definition 8, § 13 it is easy to see that the assumption that a solution satisfying the conditions of the Lemma leaves the set $\| x \| \leqslant \rho(x(t_0))$, is contradictory.

2. Stability in the first approximation

Lemma 1 can be applied to obtain a hyperstability approach to the classical problem of stability in the first approximation. Consider the equation

$$\frac{dx}{dt} = A\,x + f(x, t), \tag{3}$$

assuming that the function $(z, t) \mapsto f(z, t)$ is defined and continuous for every complex n-vector z and every real $t \geqslant 0$ and has the properties:

$$1^\circ : f(0, t) = 0 \text{ for every } t \geqslant 0 \tag{4}$$

2° : "for every $\varepsilon > 0$ there exists $\alpha > 0$ such that the inequality $\|z\| \leqslant \alpha$ implies that $\| f(z, t) \| \leqslant \varepsilon \|z\|$, for every

$$t \geqslant 0\text{''}. \tag{5}$$

We also assume that A is a Hurwitz matrix. Owing to Condition (4), Eq. (3) admits the trivial solution $x = 0$ whose stability is under investigation. (The existence of the solutions of Eq. (3) results from the continuity of the function f.

Together with Eq. (3) we consider the system

$$\frac{dx}{dt} = Ax + u, \quad \eta(0, t_1) = \int_0^{t_1} (u^* u - \varepsilon^2 x^* x)\, dt. \tag{6}$$

where the positive number ε is, for the moment, undetermined. This system is of the form (1, § 16), (2, § 16) for the following values

$$B = I, \quad J = 0, \quad K = I, \quad L = 0\,, \quad M = -\,\varepsilon^2 I. \qquad (7)$$

This system satisfies Conditions (*i*)—(*iii*) of Theorem 1, §16. In order to obtain Condition (*i*) of minimal stability we consider an arbitrary initial condition $x(0)$ and the corresponding solution x of the differential equation of (6) for $u(t) \equiv 0$. Then, since A is a Hurwitz matrix, $\lim_{t\to\infty} x(t) = 0$. Furthermore from the expression (6) we obtain the inequality $\eta(0, t) \leqslant 0$ (since we have chosen $u(t) \equiv 0$) for every $t \geqslant 0$; thus the property of minimal stability is satisfied. Condition (*ii*) ($B \neq 0$) follows from $B = I$. For Condition (*iii*) consider the matrix $H(-\,\mathrm{i}\omega, \mathrm{i}\omega)$ of system (6)

$$H(-\,\mathrm{i}\omega, \;\; \mathrm{i}\omega) = I - \varepsilon^2(-\,\mathrm{i}\omega I - A^*)^{-1}\,(\mathrm{i}\omega I - A)^{-1} \qquad (8)$$

(which results from (37, § 5) and from (7)). Since A is a Hurwitz matrix, all the entries $(\mathrm{i}\omega I - A)^{-1}$ are continuous functions of ω defined for every real ω, and approaching zero when $|\,\omega\,|$ tends to infinity. Therefore $\lim H(-\,\mathrm{i}\omega, \mathrm{i}\omega) = I$ and $\lim_{\omega\to\infty} \det H\,(-\mathrm{i}\omega, \mathrm{i}\omega) = 1$. By (56, § 5), the characteristic polynomial π is not identically zero and hence Condition (*iii*) of Theorem 1, § 16 is also satisfied.

We now show that the system is hyperstable if ε is small. By Theorem 1, § 16 it is sufficient to obtain Property (F) of the theorem quoted. From (8) it follows that the equality

$$x^* H(-\,\mathrm{i}\omega, \mathrm{i}\omega)\, x = 1 - \varepsilon^2 \|\,(\mathrm{i}\omega I - A)^{-1} x\,\|^2, \quad (x^* x = 1) \qquad (9)$$

is satisfied for every vector x with the property $\|x\| = 1$. Using again the properties of the matrix $(\mathrm{i}\omega I - A)^{-1}$ specified above, it follows that there exists a number $\varepsilon_0 > 0$, such that the inequality

$$\|\,(\mathrm{i}\omega\, I - A)^{-1}\, z\,\| \leqslant \frac{1}{2\,\varepsilon_0}, \qquad (10)$$

is satisfied for every real ω and every n-vector z with the property $\|z\| = 1$. Therefore, for $\varepsilon = \varepsilon_0$ the left hand member of (9) is positive and hence the matrix $H(-\,\mathrm{i}\omega, \mathrm{i}\omega)$ is positive definite for every real number ω. We have thus obtained Property (F) of Theorem 1, § 16 and the conclusion that system (6) is hyperstable.

Using the solutions of Eq. (3) we define now a family $\mathcal{S}$ of solutions of system (6) with the properties required in Lemma 1. Consider an arbitrary solution x of Eq. (3). If we define the function u by the relation $u(t) = f(x(t), t)$ the differential equation of (6) is satisfied and we have, for $\varepsilon = \varepsilon_0$:

$$\eta(0, t_1) = \int_0^{t_1} (\|f(x(t), t)\|^2 - \varepsilon_0^2 \|x\|^2) \, dt. \tag{11}$$

The defined above set of solutions of system (6) forms the family $\mathcal{S}$. Applying Property (5) for $\varepsilon = \varepsilon_0$ we find there exists a number α_0 such that if $\|z\| \leqslant \alpha_0$, then $\|f(z, t)\| \leqslant \varepsilon_0 \|z\|$. Therefore from (11) one obtains

$$\eta(0, t_1) \leqslant 0, \quad \text{if} \sup_{0 \leqslant t \leqslant t_1} \|x(t)\| \leqslant \alpha_0. \tag{12}$$

Thus the conditions of Lemma 1 are satisfied and hence we find that the trivial solution of Eq. (3) is stable. Furthermore, using Theorem 2, § 23 and remarking that Condition 1° of this theorem is satisfied we find that the trivial solution of Eq. (3) is even asymptotically stable.

§ 28. Stability of systems containing nuclear reactors

The case of the control systems of nuclear reactors presents certain particular features worth examining. Without going into a detailed exposition of the subject which would entail too ample a development (see T. Welton [1], H. Smets [1, 2], etc.) we shall confine ourselves to give the simplest stability criteria for the systems considered; incidentally we shall obtain a new type of non-linear hyperstable block. There is no need to go into details concerning the physical meaning of the equations and symbols we use. We shall write the equations in a simplified form (cf. (5, § 17)—(5, § 19) in the author's paper [15]):

$$\frac{dx}{dt} = Ax + b\mu_1 \tag{1}$$

$$\nu_1 = c'x \tag{2}$$

$$\frac{d\rho}{dt} = -\sum_{j=1}^{p} \delta_j (\rho - \eta_j) + (1 + \rho)\mu_2 \tag{3}$$

$$\frac{d\eta_j}{dt} = \lambda_j (\rho - \eta_j), \; j = 1, 2, \ldots, p \tag{4}$$

$$\nu_2 = \rho \tag{5}$$

$$\mu_1 = -\nu_2 = -\rho, \quad \mu_2 = \nu_1 = c'x. \tag{6}$$

System (1)—(6) may be considered as a system with negative feedback made up of the block B_2 : (3)—(5) (which describes the operation of the nuclear reactor) and the block B_1 : (1), (2) (which represents the elements of the control system outside the reactor and takes also into account some secondary physical effects); the connection between the two blocks is given by (6) which are typical of systems with negative feedback. In Eqs. (3) and (4) δ_j and λ_j are positive constants and p may be an arbitrary positive integer.

Block (3)—(5) has a property which is important in the sequel : any solution of this block corresponding to initial conditions which satisfy the inequalities $1 + \rho(0) > 0$ and $1 + \eta_j(0) > 0$, $j = 1, 2, \ldots, p$, satisfies also the inequalities

$$1 + \rho(t) > 0, \quad 1 + \eta_j(t) > 0, \quad j + 1, 2, \ldots, p \tag{7}$$

for every $t > 0$. Indeed, Eq. (4) can be written in the form $d(1 + \eta_j)/dt = -\lambda_j(1 + \eta_j) + \lambda_j(1 + \rho)$, whence it follows that $1 + \eta_j(t) \geqslant e^{-\lambda_j t}(1 + \eta_j(0))$ over any interval in which the inequality $1 + \rho(t) > 0$ holds. Hence, if the property we want to prove does not occur, there exists a positive number t_0 for which the equality $1 + \rho(t_0) = 0$ and the inequalities $1 + \eta_j(t_0) > 0$, $j = 1, 2, \ldots, p$, are satisfied and at the same time we have

$$1 + \rho(t_0 + \varepsilon) < 0, \quad (1 + \rho(t_0) = 0), \tag{8}$$

for every positive and sufficiently small number ε. But from (3) it follows that at the point t_0 the derivative of ρ is positive thus contradicting (8) and proving (7).

Block (3)—(5) has a peculiar form. We show that this block is hyperstable. From (3) we obtain

$$\int_0^{t_1} \mu_2 \nu_2 \, dt = \int_0^{t_1} \frac{\rho}{1 + \rho} \frac{d\rho}{dt} dt + \sum_{j=1}^{p} \delta_j \int_0^{t_1} \frac{\rho}{1 + \rho} (\rho - \eta_j) \, dt \tag{9}$$

and from each of the Eqs. (4) we obtain

$$0 = \frac{\delta_j}{\lambda_j} \int_0^{t_1} \frac{\eta_j}{1 + \eta_j} \frac{d\eta_j}{dt} dt - \delta_j \int_0^{t_1} \frac{(\rho - \eta_j)}{1 + \eta_j} \eta_j \, dt, \; j = 1, 2, \ldots, p. \tag{10}$$

Introduce the function

$$\psi(\zeta) = \int_0^{\zeta} \frac{\tau}{1 + \tau} d\tau = \zeta - \ln(1 + \zeta), \tag{11}$$

which clearly satisfies the conditions

$$\Psi(0) = 0,\ \psi(\zeta) > 0 \text{ for } -1 < \zeta < 0 \text{ and } 0 < \zeta < \infty. \tag{12}$$

Adding (9) and (10) and using (11) we obtain

$$\int_0^{t_1} \mu_2 \nu_2 \, dt = [\psi(\rho(t))]_0^{t_1} + \sum_{j=1}^{p} \frac{\delta_j}{\lambda_j} [\psi(\eta_j(t))]_0^{t_1} + \\ + \sum_{j=1}^{p} \delta_j \int_0^{t_1} \frac{(\rho - \eta_j)^2}{(1+\rho)(1+\eta_j)} \, dt. \tag{13}$$

Using Conditions (12) and the fact that the expression under the integral of the right hand member of (13) is nonnegative, we obtain from (13) the inequality

$$\eta_2(0, t_1) = \int_0^{t_1} \mu_2 \nu_2 \, dt \geqslant -\psi(\rho(0)) - \sum_{j=1}^{p} \frac{\delta_j}{\lambda_j} \psi(\eta_j(0)) \tag{14}$$

which shows that block (3)—(5) is hyperstable. Using the function ψ (11), we replace the variables ρ and η_j by the new variables

$$\tilde{\rho} = \psi(\rho) \operatorname{sign} \rho,\ \tilde{\eta}_j = \psi(\eta_j) \operatorname{sign}, \eta_j,\ j = 1,2, \ldots, p.$$

Obviously, these variables are defined if (7) holds and we have seen that the examined solutions of the system (1)—(6) have this property. After introducing $\tilde{\rho}$ and $\tilde{\eta}_j$ the right hand member of (14) becomes equal to $-|\tilde{\rho}(0)| - \sum_{j=1}^{p} \frac{\delta_j}{\lambda_j} |\tilde{\eta}_j(0)|$. Hence, by Proposition 4, § 13 block B_2 (3)—(5) is hyperstable. Therefore system (1)—(6) can be represented as in Fig. 5.2 where B_2 is hyperstable. By Proposition 9, § 13 if the block B_1, (1)—(2) is hyperstable then all the solutions of the system (1)—(6) are bounded and the trivial solution of the system is stable.

The obtained criterion can be stated explicitly using Theorem 1, § 15. We thus find the frequency criterion

$$\operatorname{Re} \gamma(i\omega) \geqslant 0, \quad (\text{where } \gamma(i\omega) = c'(i\omega I - A)^{-1} b), \tag{15}$$

which must be satisfied for every real number ω for which $\det(i\omega I - A) \neq 0$. The other conditions required in Theorem 1, § 15 are satisfied if 1° : the transfer function γ is not identically zero ; 2° : $b \neq 0$, and 3° : there exists a number $\lambda_0 \geqslant 0$ such that the trivial solution of equation $dx/dt = (A - \lambda_0 bc')x$ is asymptotically stable (whence the condition of minimal stability required in Theorem 1, § 15 follows immediately).

Using Theorem 1, § 23 we find that in order to have also asymptotic stability it is sufficient, for example, that A be a Hurwitz matrix and the inequality (15) be satisfied for every real ω as a strict inequality (in which case $\pi(-i\omega, i\omega)$ is different from zero for every real ω). The obtained frequency criterion, (15), is in fact Welton's criterion [1] in the form given by H. Smets [1].

A more general criterion is easily obtained by introducing a new input function

$$\tilde{\mu} = (1 + \rho)\, c'x + \beta\rho, \quad \beta > 0 \tag{16}$$

and by writing the system (1)—(6) in the form

$$\frac{d\rho}{dt} = -\beta\rho - \sum_{j=1}^{p} \delta_j (\rho - \eta_j) + \tilde{\mu} \tag{17}$$

$$\frac{d\eta_j}{dt} = \lambda_j (\rho - \eta_j), \qquad j = 1, 2, \ldots, p \tag{18}$$

$$\frac{dx}{dt} = Ax - b\rho \tag{19}$$

$$\tilde{\nu} = -c'x. \tag{20}$$

The Eqs. (17)—(20) define a new block with the input $\tilde{\mu}$ and the output $\tilde{\nu}$. Substituting (16) and (17) it is easy to see that Eqs. (17)—(19) coincide with Eqs. (1)—(6). From (16) and (20) we obtain

$$\tilde{\eta}(0, t) = \int_0^{t_1} \tilde{\mu}\, \tilde{\nu}\, dt = -\int_0^{t_1} (c'x)^2 (1 + \rho)\, dt - \beta \int_0^{t_1} c'x\, \rho\, dt. \tag{21}$$

Owing to Condition (7) the first integral of the right hand member of (21) is $\leqslant 0$ for every $t_1 > 0$, the second integral is equal to the product of $-\beta$ and the integral (9) and hence satisfies an inequality which is obtained from (14) by taking into account this remark. Therefore the integral (21) satisfies the inequality $\tilde{\eta}(0, t_1) \leqslant \gamma^2$, where γ^2 is equal to the product of $-\beta$ and the right hand member of inequality (14). Therefore Condition (8, § 13) is satisfied and hence, if block (17)—(20) is hyperstable then, by applying Property h_s (Definition 3, § 13) we obtain the same conclusion as above concerning the stability of the trivial solution of the system. Instead of Condition (15) we thus obtain a similar but more general frequency condition

$$\operatorname{Re} \frac{\gamma(i\omega)}{\dfrac{1}{\gamma_R(i\omega)} + \beta} \geqslant 0, \quad (\beta > 0) \tag{22}$$

where

$$\gamma_R(\mathrm{i}\omega) = \frac{1}{\mathrm{i}\omega + \sum_{j=1}^{p} \delta_j \dfrac{\mathrm{i}\omega\,\lambda_j}{\mathrm{i}\omega + \lambda_j}} . \tag{23}$$

The inequality (22) can be written as

$$\mathrm{Re}\,(\beta|\gamma_R(\mathrm{i}\omega)|^2\,\gamma(\mathrm{i}\omega) + \gamma(\mathrm{i}\omega)\,\gamma_R(\mathrm{i}\omega)) \geqslant 0,\ (\beta > 0). \tag{24}$$

To the above form one can also bring the criterion (11.1) from the author's paper [15], where this result was obtained as a limit case of a more general criterion securing the stability in a limited domain of the state space.

All these results can be further developed as in § 25.

§ 29. Stability of some systems with non-linearities of a particular form

1. Systems with monotone non-linear characteristics

Let us consider again system (1, § 25), (2, § 25) assuming that φ is continuous, satisfies the Condition (62, § 25) and in addition is also monotone, i.e., it has the property

"for every pair of numbers δ_1 and $\delta_2 > \delta_1$ the inequality

$$\varphi(\delta_2) - \varphi(\delta_1) > 0 \tag{1}$$

is satisfied." (see V. A. Yakubovich [6—8], R. W. Brockett and J. L. Willems [1]). Then the block

$$\frac{d\xi}{dt} = -\rho_2\,\xi + \mu_2,\ (\rho_2 \geqslant 0) \tag{2}$$

$$\nu = \rho_1\,\xi + \mu_2,\ (\rho_1 > 0) \tag{3}$$

$$\nu_2 = \varphi\,(\nu) \tag{4}$$

is hyperstable. Indeed, the expression

$$\eta_2(0, t_1) = \int_0^{t_1} \mu_2\,\nu_2\,dt = \int_0^{t_1} (\nu - \rho_1\,\xi)\,\varphi(\nu)\,dt \tag{5}$$

(see (3)) can be written as

$$\eta_2(0,\ t_1) = \eta_{2,1} + \eta_{2,2}\,, \tag{6}$$

where

$$\eta_{2,1} = \int_0^{t_1} (\nu - \rho_1\,\xi)\,(\varphi\,(\nu) - \varphi(\rho_1\,\xi))\,dt \tag{7}$$

$$\eta_{2,2} = \int_0^{t_1} (\nu - \rho_1\,\xi)\,\varphi(\rho_1\,\xi)\,dt. \tag{8}$$

Using the relation

$$\nu - \rho_1\,\xi = \frac{d\xi}{dt} + \rho_2\,\xi \tag{9}$$

(see (2) and (3)), expression (8) becomes

$$\eta_{2,2} = \int_0^{t_1} \frac{d\xi}{dt}\varphi(\rho_1\,\xi)\ dt + \rho_2 \int_0^{t_1} \xi\varphi(\rho_1\,\xi)\,dt =$$

$$= \frac{1}{\rho_1}\,[\psi\,(\rho_1\,\xi)]_0^{t_1} + \rho_2\int_0^{t_1} \xi\varphi\,(\rho_1\,\xi)\,dt\,, \tag{10}$$

where the function ψ has the expression (65, § 25). By (7) and Property (1) we have $\eta_{2.1} \geqslant 0$ and from (10) and the inequalities $\rho_1 > 0$, $\rho_2 \geqslant 0$ and $\varphi(\delta)\delta \geqslant 0$ (see (3, § 25) and (7, § 25)) we obtain $\eta_{2,2} \geqslant (\psi(\rho_1\xi(t_1)) - \psi(\rho_1\xi(0)))/\rho_1$. From these inequalities and (6), we deduce

$$\eta_2\,(0,\, t_1) \geqslant \frac{1}{\rho_1}\,\psi(\rho_1\xi\,(t_1)) - \frac{1}{\rho_1}\,\psi(\rho,\ \xi(0)). \tag{11}$$

Owing to the inequality $\varphi(\delta)\delta \geqslant 0$ and to (1), Conditions (72, § 25) are satisfied. Therefore inequality (11) shows that block (2)—(4) is hyperstable (see Proposition 4, § 13).

The conclusion extends immediately to more general blocks of the form [1]:

$$\sum_{j=1}^{p} \delta_j (\nu - \rho_{1,j} \xi_j) = \mu_2, \quad (\delta_j > 0, \rho_{1,j} > 0) \tag{12}$$

$$\frac{d\xi_j}{dt} = -(\rho_{1,j} + \rho_{2,j}) \xi_j + \nu, \quad (\rho_{2,j} \geqslant 0), \quad j = 1, 2, \ldots, p \tag{13}$$

$$\nu_2 = \varphi(\nu) \tag{14}$$

Then the integral $\eta_2(0, t)$ can be written as a sum of integrals of the form

$$\eta_2(0, t) = \sum_{j=1}^{p} \delta_j \int_0^{t_1} (\nu - \rho_{1,j} \xi_j) \varphi(\nu) \, dt$$

and each term can be treated as in (5)—(10). (Instead of (9) we use the similar relations $\nu - \rho_{1,j}\xi_j = d\xi_j/dt + \rho_{2,j}\xi_j$, deduced from (13)).

Taking into account (64, § 25) and (65, § 25) we find that if Eq. (12) is replaced by

$$\beta \frac{d\nu}{dt} + \alpha \left(\nu - \frac{1}{\lambda_0} \varphi(\nu) \right) + \sum_{j=1}^{p} \delta_j (\nu - \rho_{1,j} \xi_j) = \mu_2,$$
$$(\alpha \geqslant 0, \beta \geqslant 0, \delta_j > 0, \rho_{1,j} > 0), \tag{15}$$

the block described by Eqs. (15), (13) and (14) is hyperstable. Thus the results of § 25, can be generalized to the case of monotone characteristics by simply replacing the non-linear hyperstable block B_2 considered in Section 8, § 25 by the hyperstable block introduced above. Proceeding as we did in § 25 we define the linear block B_1 with the input μ_1 and the output ν_1 connected to μ_2 and ν_2 by the relations $\mu_1 = -\nu_2$, $\mu_2 = \nu_1$ (typical for systems with negative feedback). From these relations and Eqs. (14) and (15) we have

$$\mu_1 = -\varphi(\nu) \tag{16}$$

$$\nu_1 = \alpha \left(\nu + \frac{\mu_1}{\lambda_0} \right) + \beta \frac{d\nu}{dt} + \sum_{j=1}^{p} \delta_j (\nu - \rho_{1,j} \xi_j). \tag{17}$$

[1] By introducing in (13) the expression of ν, derived from (12), we obtain an ordinary system of differential equations whose solutions, introduced in (12), determine $\nu(t)$ and hence, — by equation (14) — also $\nu_2(t)$.

Block B_1 is completely defined if one considers also the Relations (1, § 25) and (2, § 25)

$$\frac{\mathrm{d}x}{\mathrm{d}t} = Ax + b\,\mu_1, \quad \nu = c'\,x \tag{18}$$

(see also (16)), and Eqs. (13). (Then, if the function μ_1 is given, one obtains x and ν from Eqs. (18) and ξ_j from Eqs. (13), and then ν_1 can be determined by (17)).

Thus we have defined a system with negative feedback consisting of the block B_1 — described by Eqs. (18), (13) and (17)—, and the non-linear hyperstable block B_2, — described by Eqs. (15), (13) and (14). By Proposition 9, § 13 the trivial solution of this system is stable if the block B_1 : (18), (13), (17) is hyperstable. It is easy to see that this block can be brought to the form studied in § 13. The transfer function of this block has the following expression (derived from (18), (13) and (17)) :

$$\gamma_1(\sigma) = \left(\alpha + \beta\sigma + \sum_{j=1}^{p} \delta_j \frac{\sigma + \rho_{2,j}}{\sigma + \rho_{1,j} + \rho_{2,j}}\right)\gamma(\sigma) + \frac{\alpha}{\lambda_0}, \tag{19}$$

where $\gamma(\sigma) = c'(\sigma I - A)^{-1}b$.

Using Theorem 1, § 15 to express the hyperstability of the block B_1 we obtain the frequency condition [1]) $\operatorname{Re}\gamma_1(\mathrm{i}\omega) \geqslant 0$, for every real ω for which the left hand member is defined. The obtained condition generalizes the criterion given in § 25, provided that the function φ is monotone. The other conditions of Theorem 1, § 15 are satisfied if, 1° : the expression $\operatorname{Re}\gamma_1(\mathrm{i}\omega)$ is not identically zero, 2° : $b \neq 0$ and 3° : A is a Hurwitz matrix (then we easily obtain the property of minimal stability). In order to have also asymptotic stability it is sufficient that (for instance) besides the conditions specified above, the expression $\operatorname{Re}\gamma_1(\mathrm{i}\omega)$ be strictly positive for every real number ω (then the polynomial $\pi(-\mathrm{i}\omega, \mathrm{i}\omega)$ is different from zero for every real number ω and Condition 1° of Theorem 1, § 23 is satisfied).

Without discussing the criterion obtained (which may contain any number of arbitrary parameters) we observe only that it can be considerably improved by making use of Theorems 1, § 14 and 1, § 15 (see also V. A. Yakubovich [6—8]. R. W. Brockett and J. L. Willems [1]).

---

[1]) More exactly, it is sufficient that there exists constants $\alpha \geqslant 0$, $\beta \geqslant 0$, an integer $p \geqslant 0$ and some constants $\delta_j > 0$, $\rho_{1,j} > 0$, $\rho_{2,j} \geqslant 0$ such that the inequality $\operatorname{Re}\gamma_1(\mathrm{i}\omega) \geqslant 0$ be satisfied.

2. Stability of a system with a non-linearity depending on two variables

Sometimes the stability of a system containing a non-linear function of several variables can be deduced from the hyperstability of a single-input system. This is the case of the following example [1])

$$\frac{dx}{dt} = Ax - b\varphi_1(\nu - \varphi_2(\xi)) \tag{20}$$

$$\frac{d\xi}{dt} = \varphi_1(\nu - \varphi_2(\xi)) \tag{21}$$

$$\nu = c'x. \tag{22}$$

Assume that in these equations φ_1 is a continuous function which satisfies the inequality

$$\varphi_1(\rho)\left(\rho - \frac{1}{\lambda_0}\varphi_1(\rho)\right) \geqslant \varepsilon\rho^2, \quad (\lambda_0 > 0, \varepsilon > 0), \text{ for every real } \rho. \tag{23}$$

(see also (8, § 25)) and φ_2 is a differentiable function which satisfies the conditions

$$\varphi_2(0) = 0, \qquad \frac{d\varphi_2(\delta)}{d\delta} > 0, \text{ for every real } \delta. \tag{24}$$

We shall show that under these conditions system (20)—(22) can be treated in the same way as the system studied in § 25. We introduce the integral

$$\eta_1'(0, t_1) = \int_0^{t_1} \varphi_1(\nu - \varphi_2(\xi))\left(\nu - \frac{1}{\lambda_0}\varphi_1(\nu - \varphi_2(\xi))\right)dt, \tag{25}$$

which can be written as

$$\eta_1'(0, t_1) = \eta_{1,1} + \eta_{1,2}, \tag{26}$$

---

[1]) This system generalizes the equations of some time-optimal systems.

where

$$\eta_{1,1} = \int_0^{t_1} \varphi_1(\rho)\left(\rho - \frac{1}{\lambda_0}\varphi_1(\rho)\right) dt, \quad (\rho = \nu - \varphi_2(\xi)) \tag{27}$$

$$\eta_{1,2} = \int_0^{t_1} \varphi_1(\nu - \varphi_2(\xi))\, \varphi_2(\xi)\, dt. \tag{28}$$

From (27) and (23) one obtains

$$\eta_{1,1} \geqslant 0, \tag{29}$$

and from (28) and (21),

$$\eta_{1,2} = \int_0^{t_1} \frac{d\xi}{dt}\, \varphi_2(\xi)\, dt = \psi_2(\xi(t_1)) - \psi_2(\xi(0)), \tag{30}$$

where $\psi_2(\tau) = \int_0^{\tau} \varphi_2(\delta)\, d\delta$ has the same properties as the function (65, § 25) and hence it cannot take negative values. From this remark and from (26), (29) and (30) one obtains the inequality

$$\eta_1'(0, t_1) \geqslant \psi_2(\xi(t_1)) - \psi_2(\xi(0)). \tag{31}$$

Another expression with similar properties is the integral

$$\eta_2'(0, t_1) = \int_0^{t_1} \varphi_1(\nu - \varphi_2(\xi)) \frac{d\nu}{dt} dt, \tag{32}$$

which can be written as:

$$\eta_2'(0, t_1) = \eta_{2,1} + \eta_{2,2}, \tag{33}$$

where

$$\eta_{2,1} = \int_0^{t_1} \varphi_1(\nu - \varphi_2(\xi)) \frac{d(\nu - \varphi_2(\xi))}{dt} dt \tag{34}$$

$$\eta_{2,2} = \int_0^{t_1} \varphi_1(\nu - \varphi_2(\xi)) \frac{d\,\varphi_2(\xi)}{dt} dt. \tag{35}$$

The integral (34) can be written in the form

$$\eta_{2,1} = \psi_1(\nu(t_1) - \varphi_2(\xi(t_1))) - \psi_1(\nu(0) - \varphi_2(\xi(0))), \qquad (36)$$

where $\psi_1(\tau) = \int_0^\tau \varphi_1(\rho)d\rho$ has the same properties as the function (65, § 25). From (35) and from (21) and (24) we have

$$\eta_{2,2} = \int_0^{t_1} \frac{d\varphi_2(\xi)}{d\xi}\left(\frac{d\xi}{dt}\right)^2 dt \geqslant 0. \qquad (37)$$

From (33), (36) and (37) one obtains the inequality

$$\eta_2'(0, t_1) \geqslant \psi_1(\nu(t_1) - \varphi_2(\xi(t_1))) - \psi_1(\nu(0) - \varphi_2(\xi(0))). \qquad (38)$$

The above results show that the block

$$\beta\frac{d\nu}{dt} + \alpha\left(\nu - \frac{1}{\lambda_0}\varphi(\nu - \varphi_2(\xi))\right) = \mu_2, \quad (\beta > 0\,, \quad \alpha > 0) \qquad (39)$$

$$\frac{d\xi}{dt} = \varphi_1(\nu - \varphi_2(\xi)) \qquad (40)$$

$$\nu_2 = \varphi_1(\nu - \varphi_2(\xi)) \qquad (41)$$

is hyperstable. Indeed, using (25) and (32) we find that the integral associated to the block (39)—(41) has the expression

$$\eta_2(0, t_1) = \int_0^{t_1} \mu_2 \nu_2 \, dt = \alpha\,\eta_1'(0, t_1) + \beta\eta_2'(0, t_1), \qquad (42)$$

and hence, by (31) and (38), we have

$$\eta_2(0, t_1) \geqslant \alpha\,\psi_2(\xi(t_1)) + \beta\,\psi_1(\nu(t_1) - \varphi_2(\xi(t_1))) -$$
$$- \alpha\,\psi_2(\xi(0)) - \beta\,\psi_1(\nu(0) - \varphi_2(\xi(0))). \qquad (43)$$

Using the conditions required above for the functions φ_1 and φ_2, we apply Proposition 4, § 13.

Proceeding as we repeatedly did before, we define a linear block B_1 with the input μ_1 and the output ν_1 satisfying the relations

$$\mu_1 = -\nu_2, \qquad \mu_2 = \nu_1, \qquad (44)$$

which imply that

$$\mu_1 = -\varphi_1(\nu - \varphi_2(\xi)) \tag{45}$$

$$\nu_1 = \alpha\left(\nu + \frac{1}{\lambda_0}\mu_1\right) + \beta\frac{d\nu}{dt} \cdot \tag{46}$$

Using (45), (20) and (22) we obtain

$$\frac{dx}{dt} = Ax + b\mu_1, \quad \nu = c'x. \tag{47}$$

Block B_1 is completely determined by (47) and (46). As in § 25, if this block is hyperstable then, by applying Proposition 9, § 13, one finds that all the solutions of the system are bounded and that the trivial solution is stable. Applying Theorem 1, § 15 we obtain the frequency condition

$$\operatorname{Re}[(\alpha + \beta i\omega)\,\gamma(i\omega)] + \frac{\alpha}{\lambda_0} \geqslant 0, \; (\gamma(i\omega) = c'(i\omega\, I - A)^{-1}\, b),$$

which must be satisfied for every real number ω which satisfies the condition det $(i\omega I - A) \neq 0$. If the left hand member of this frequency condition is not identically zero, $b \neq 0$, A is a Hurwitz matrix, then the other conditions of Theorem 1, § 15 are also satisfied. If, in addition, the left hand member of the frequency condition is different from zero for every real number ω, then Condition 1° of Theorem 1, § 23 is satisfied and hence the asymptotic stability is ensured.

More general conclusions can be also obtained from Theorems 1, § 14 and 1, § 15; however, we consider that the simplified treatment presented above is sufficient.

§ 30. Optimization of control systems for integral performance indices

The application presented now differs completely from the previous ones and makes use of the property of positiveness treated in Chapter 3. Consider a system described by the equation

$$\frac{dx}{dt} = Ax + b\,\mu \tag{1}$$

and a given initial condition $x(0) = x_0$; we are looking for function μ (belonging to a certain class of functions to be described later), such that an integral performance index of the form

$$\eta\,(0,\infty) = \int_0^\infty (\chi\mu^2 + 2\,\mu l' x + x' Mx)\,\mathrm{d}t, \tag{2}$$

be minimal (see A. M. Letov [2], R. E. Kalman [1], [6], I. Kurzweil [1]). If in (2) we replace the upper integration limit, α, by an arbitrary positive number t_1, Relations (1) and (2) constitute a system of the type considered in §§ 2—4 and 8 (for $J = 0$) [1]).

We assume that

$$\varkappa \neq 0 \tag{3}$$

$$\pi(-\,\mathrm{i}\omega,\,\mathrm{i}\omega) \neq 0, \text{ for every real } \omega. \tag{4}$$

Here $\pi(-\,\mathrm{i}\omega,\,\mathrm{i}\omega)$ results from relation (13, § 3) and (18, § 3) (the positive norming constant ν of (16, § 3) does not influence the solution of the problem). We shall determine the function μ which minimizes the integral (2), in the class of the piecewise continuous functions for which the corresponding solution of Eq. (1) satisfies the condition

$$\lim_{t\to\infty} x(t) = 0, \tag{5}$$

and the integral (2) is convergent. Finally, we assume that the defined above system is positive (Definition 1, § 8). [2])

Since Condition (3) is satisfied the system can be brought to the form (51, § 8)

$$\frac{\mathrm{d}x_c}{\mathrm{d}t} = A_c\,x_c + b_c\,\mu_c \tag{6}$$

$$\eta_c\,(0,\,t_1) = x_c'\,(t_1)\,U\,x_c\,(t_1) - x_c'\,(0)\,U\,x_c\,(0) + \int_0^{t_1} |\,\mu_c\,|^2\,\mathrm{d}t. \tag{7}$$

---

[1]) Throughout this paragraph we use the simplifications arising from the fact that the coefficients of the system considered here are real.

[2]) Usually this condition is automatically satisfied since the quadratic form under the integral of (2) is positive semidefinite.

The polynomial ψ, (47, § 8), — for which this system is constructed and which satisfies the relation of factorization (8, § 8) — , will be chosen such as to have only roots with a negative real part; this is possible owing to Condition (4). Then from (49, § 8) it follows that the matrix A_c of (6) is a Hurwitz matrix. Since system (6)—(7) and the initial system belong to the same class (Definition 1, § 2) the two systems are related by a transformation of the form (23, § 2)—(28, § 2) for some parameters N, ρ, q and R. (Throughout this paragraph whenever a formula of § 2 is quoted the superscript $\sim$ is replaced by the subscript c).

Consider the initial condition $x_c(0) = Rx_0$ and solve for system (6), (7) the following problem, which is similar to the initial problem : to determine a function μ_c such that : 1° : μ_c is piecewise continuous, 2° : the corresponding solution of equation (6), for $x_c(0) = Rx_0$, satisfies the condition

$$\lim_{t\to\infty} x_c(t) = 0 \tag{8}$$

and 3° : the integral $\eta_c(0, \infty)$ exists and is minimum. Owing to Condition (8) this means that the expression

$$\eta_c(0, \infty) = - x_c'(0)\, U\, x_c(0) + \int_0^\infty (\mu_c(t))^2 \, dt \tag{9}$$

should exist and be minimum. Formula (9) shows that the solution of the problem is given by $\mu_c(t) \equiv 0$, since then the integral (9) takes the smallest possible value and at the same time Condition (8) is satisfied by the corresponding solution of Eq. (6) since A_c is a Hurwitz matrix.

Going back to the initial system (1), (2) and using Proposition 1, § 2 we easily find that the integral (2) is minimum for the function μ obtained from (31, § 2) i.e. for

$$0 \equiv \rho\mu(t) - \rho q' x(t) = \rho\mu(t) - \rho q' R^{-1} x_c(t),$$

or

$$\mu(t) = q'x(t) = q' R^{-1} x_c(t). \tag{10}$$

Using once again Proposition 1, §2 we see that any other piecewise continuous function μ leads, by (31, § 2) to a piecewise continuous function μ_c which is not identically zero and hence the integral (9) takes a higher value than the minimum one. It immediately follows that, under our assumptions the initial integral (2) takes also a higher value than the minimal value and hence the solution determined above is unique [1]).

It also appears that the function μ which minimizes the integral (2) satisfies (10), irrespective of the initial condition $x(0)$ for which the problem has been stated. Substituting relation (10) in (1) we obtain the optimal system in the form

$$\frac{\mathrm{d}x}{\mathrm{d}t} = Ax + bq'x. \tag{11}$$

The same approach can be used even if we have to determine simultaneously the optimal expressions of several distinct functions. Then instead of the system (1), (2) we have to consider the multi-input system (1, § 5), (2, § 5), the treatment of the problem and the results obtained being in every way similar (see the author's paper [16]).

---

[1]) It also follows that, under the hypotheses concerning $\varkappa$, $\pi(-i\omega, i\omega)$ and $\psi(\sigma)$, system (6), (7), constructed in Section 6, § 8, is unique.

CONTROLLABILITY; OBSERVABILITY; NONDEGENERATION

This appendix contains results which are very frequently used throughout the work. Of the three concepts which are treated in the appendix, we choose "controllability" as the fundamental one and consider "observability" and "nondegeneration" as concepts derived from it[1]). Theorem 1, § 31 which contains different equivalent forms of the property of controllability of single-input systems, gives also the main consequences of this property. The results of § 32 concerning the complete observability of single-input systems are immediate consequences of the Theorem 1, § 31. The property of nondegeneration resulting from the association of the properties of controllability and observability possesses new characteristics which are described in the case of simple systems, in § 33. Most of the results obtained for single-input systems can be extended to multi-input systems as shown (partially) in §§ 34—35. A special form, to which the multi-input blocks can be brought, is presented in § 36.

The references used in the appendix include recent papers by different authors. The concept of complete controllability, introduced at first as an algebraic conditions for simplifying some problems — and found as such, for instance, in the early theory of optimal control systems of L. S. Pontryaghin and co-workers — has become an independent concept owing particularly to R. E. Kalman [2—4][2]), who has also stated the property of complete observability as a dual property relative to complete controllability. Other formulations and consequences of the property of complete controllability are derived from the works of M. A. Aizerman and F. R. Gantmacher [1], the author's papers [11] and [16] and the monograph of S. Lefschetz [3]. The term of "nondegeneration" is used in the sense introduced by the author in [11][3]) and the refer-

---

[1]) Other interpretations are possible as well.

[2]) See also R. E. Kalman, Y. C. Ho., K. S. Narendra [1].

[3]) In literature the term "nondegeneration" is also used with other meanings.

ences used in the treatment of this problem consist of the paper quoted and the work [4] by R. E. Kalman.

The following exposition is limited to the requirements of the present work and gives a fairly complete survey of the examined properties only in the case of relatively simple systems.

§ 31. Controllability of single-input systems

1. Definition of the complete controllability of single-input systems

Let us consider a system of the form

$$\frac{\mathrm{d}x}{\mathrm{d}t} = Ax + b\mu(t). \tag{1}$$

Here, as throughout the rest of the work, x is an n-dimensional vector, μ a piecewise continuous scalar function, A a constant $n \times n$-matrix and b a constant n-vector. All the quantities in [1] (except t) may take complex values.

Definition 1. *System (1) (or the pair (A, b)) is said to be completely controllable if for every n-vector x_0 and every interval $t_1 < t < t_2$ there exists a continuous function μ_{x_0} defined on the interval $t_1 \leqslant t \leqslant t_2$, such that the solution of the system (1), for $\mu = \mu_{x_0}$ and $x(t_1) = x_0$, satisfies the condition $x(t_2) = 0$.*

In other words and more intuitively, the system (1) is completely controllable if we can — within a finite time interval — bring it from any state $x = x_0$ to the "stationary state" $x = 0$ through a suitable chosen control function.

The foregoing definition, formulated essentially after R. E. Kalman is by no means the only possible definition. Any of the equivalent statements of the Theorem 1 below could be used as a definition of complete controllability. One important equivalent formulation is related to the following concept of "complete reachability".

Definition 2. *The system (1) is "completely reachable" if for any vector x_0 and any interval $t_1 < t < t_2$ there exists a continuous function $\tilde{\mu}_{x_0}$ defined on the interval $t_1 \leqslant t \leqslant t_2$, such that the solution of the system (1), for $\mu = \tilde{\mu}_{x_0}$ and $x(t_1) = 0$, satisfies the condition $x(t_2) = x_0$.*

Definitions 1 and 2 may be extended to more complex systems. Thus extended, the two definitions are distinct. However, in the case of the system (1) the two definitions are equivalent as stated in the following proposition :

Proposition 1. *System (1) is completely controllable if and only if it is completely reachable.*

Proof. We introduce the symbol $x_1 \xrightarrow{\mu_0} x_2$ the meaning of which is : "the solution of the system (1), for $\mu = \mu_0$ and $x(t_1) = x_1$" satisfies the equality $x(t_2) = x_2$. By virtue of the linearity of the system (1), if we have $x_1 \xrightarrow{\mu_0} x_2$ and $\tilde{x}_1 \xrightarrow{\tilde{\mu}_0} \tilde{x}_2$, then we have also $(x_1 + \tilde{x}_1) \xrightarrow{(\mu_0+\tilde{\mu}_0)} (x_2 + \tilde{x}_2)$. In the following we shall assume that the interval $t_1 < t < t_2$ is fixed. We shall also assume that the system (1) is completely reachable. Let us consider an arbitrary vector x_0 and arbitrary function μ_0 (e.g. identically zero) ; then there exists a vector x_2 such that we have $x_0 \xrightarrow{\mu_0} x_2$; owing to complete reachability there exists a $\tilde{\mu}_0$ such that $0 \xrightarrow{\tilde{\mu}_0} (-x_2)$; by virtue of linearity the relations obtained imply $x_0 \xrightarrow{(\mu_0+\tilde{\mu}_0)} 0$ which shows that system (1) is completely controllable. Conversely, let us assume, that the system (1) is completely controllable. For an arbitrary vector x_0 and an arbitrary control function μ_0 (e.g. identically zero) there exists a vector x_1 such that we have $x_1 \xrightarrow{\mu_0} x_0$; owing to the complete controllability there exists a $\tilde{\mu}_0$ such that $(-x_1) \xrightarrow{\tilde{\mu}_0} 0$; by virtue of linearity the relations obtained imply $0 \xrightarrow{(\mu_0+\tilde{\mu}_0)} x_0$, which shows that the system (1) is completely reachable.

2. Theorem of complete controllability of single-input systems

The theorem which follows contains 16 equivalent formulations of the property of complete controllability. In fact, it combines in a single statement several separate theorems. In this way the proof becomes considerably simpler. Moreover it emphasizes the deep unity of the various properties involved.

Theorem 1. *The property: "the system (1) is completely controllable" (or "the pair (A, b) is completely controllable") may be stated in any of the following equivalent forms:*

1°. *There does not exist any non-singular matrix R such that the matrix $\tilde{A} = R\,A\,R^{-1}$ and the vector $\tilde{b} = Rb$ be of the form*

$$\tilde{A} = \begin{pmatrix} A_{11} & A_{12} \\ 0 & A_{22} \end{pmatrix}; \qquad \tilde{b} = \begin{pmatrix} b_1 \\ 0 \end{pmatrix}, \tag{2}$$

where A_{11} *is a* $k \times k$*-matrix* ($k < n$, *possibly* $k = 0$) *and* b_1 *is a* k*-vector* [1]).

2°. *The vector* b *does not belong to a subspace invariant under matrix* A *of a dimension smaller than* n.

3°. *There exists no eigenvector of the matrix* A *orthogonal to the vector* b [2]).

4°. *The matrix with* n *rows and* n *column*

$$C = (b \;\; Ab \;\; A^2 b \; \ldots \; A^{n-1}\, b) \tag{3}$$

is non-singular ($\det C \neq 0$).

5°. *The equations*

$$RA - AR = 0, \tag{4}$$

$$Rb = d \tag{5}$$

admit a unique matrix solution R, *for every vector* d. (*Consequence: if the matrix*

$$(d \;\; Ad \;\; A^2 d \; \ldots \; A^{n-1}\, d)$$

is non-singular [3]), *then the matrix* R *which satisfies equations* (4) *and* (5) *is also non-singular*).

6°. *For any other pair* $(\widetilde{A}, \widetilde{b})$ *which satisfies the conditions*

$$\det (\widetilde{b} \;\; \widetilde{A}\widetilde{b} \; \ldots \; \widetilde{A}^{n-1}\, \widetilde{b}) \neq 0 \tag{6}$$

and

$$\det (\sigma I - \widetilde{A}) = \det (\sigma I - A) \tag{7}$$

there exists a non-singular matrix R *such that*

$$\widetilde{A} = RAR^{-1}; \;\; \widetilde{b} = Rb. \tag{8}$$

(*Consequence: the matrix* R *from* (8) *is uniquely determined.*)

---

[1]) It follows that the matrix A_{12} has k rows and $n - k$ columns, the matrix A_{22} has $n - k$ rows and $n - k$ columns, the symbol 0 in the expression of the matrix A represents a zero matrix with $n - k$ rows and k columns, and the symbol 0 in the expression of the vector b represents a vector with $n - k$ components all of which are zero.

[2]) This property may be also expressed in the following form: the rank of the matrix (with n rows and $n + 1$ columns) $(\sigma I - A\; b)$ is equal to n for every σ.

[3]) This condition is of the type stated in Property 5° and can be therefore replaced (by virtue of the proof of Theorem 1) by the condition that the pair (A, d) be completely controllable.

7°. *There exists a non-singular matrix R such that the matrix $\tilde{A} = RAR^{-1}$ and the vector $\tilde{b} = Rb$ have the form*

$$\tilde{A} = \begin{pmatrix} 0 & 1 & 0 & \dots & 0 \\ 0 & 0 & 1 & \dots & 0 \\ . & . & . & \dots & . \\ 0 & 0 & 0 & \dots & 1 \\ \gamma_1 & \gamma_2 & \gamma_3 & \dots & \gamma_n \end{pmatrix}; \quad \tilde{b} = \begin{pmatrix} 0 \\ 0 \\ . \\ 0 \\ 1 \end{pmatrix}, \tag{9}$$

where γ_j are the coefficients of the characteristic equation

$$\sigma^n = \sum_{j=1}^{n} \gamma_j \, \sigma^{j-1}$$

of the matrix A. (Consequence: the matrix R is unique.)

8°. *There exists a non-singular matrix R such that the matrix $\tilde{A} = RAR^{-1}$ and the vector $\tilde{b} = Rb$ have the Jordan-Lur'e-Lefschetz form* [1])

$$\tilde{A} = \begin{pmatrix} A_1 & 0 & . & . & 0 \\ 0 & A_2 & . & . & 0 \\ . & . & . & . & . \\ 0 & 0 & . & . & A_k \end{pmatrix}; \quad \tilde{b} = \begin{pmatrix} b_1 \\ b_2 \\ . \\ b_k \end{pmatrix}, \tag{10}$$

where A_j, $j = 1, 2, \dots, k$ are Jordan cells with n_j rows and columns of the form

$$A_j = \begin{pmatrix} \lambda_j & 1 & 0 & . & . & 0 \\ 0 & \lambda_j & 1 & . & . & 0 \\ . & . & . & . & . & . \\ 0 & 0 & 0 & . & . & \lambda_j \end{pmatrix} \qquad j = 1, 2, \dots, k, \tag{11}$$

---

[1]) Property 8° is stated following S. Lefschetz [3]. In the case where the Jordan form of the matrix A is diagonal and the characteristic values are distinct, the hypothesis that the system can be brought to the form (10)—(12) has been introduced and used by A. I. Lur'e [1].

the vectors b_j, $j = 1, 2, \ldots, k$ with n_j components have the expression

$$b_j^* = (0 \quad 0 \quad 0 \quad \ldots \quad 0 \quad 1), \tag{12}$$

and the numbers λ_j and n_j are obtained from the expression of the characteristic polynomial

$$\det(\sigma I - A) = \prod_{j=1}^{k} (\sigma - \lambda_j)^{n_j}, \; (\lambda_p \neq \lambda_q \text{ whenever } p \neq q) \tag{13}$$

of the matrix A. (Consequence: the matrix R is unique.)

9°. *For every polynomial of the form*

$$\pi_1(\sigma) = \sigma^n - \sum_{j=1}^{n} \beta_j \sigma^{j-1}, \tag{14}$$

where β_j are constants there exists an n-vector q_0 such that [1])

$$\det(\sigma I - A - b q_0^*) = \pi_1(\sigma). \tag{15}$$

(Consequence: the vector q_0 is unique).

10°. *For every polynomial of the form*

$$\pi_2(\sigma) = \alpha_n \sigma^{n-1} + \alpha_{n-1} \sigma^{n-2} + \ldots + \alpha_1, \tag{16}$$

there exists an n-vector q_0 such that [2])

$$q_0^* (\sigma I - A)^{-1} b = \frac{\pi_2(\sigma)}{\det(\sigma I - A)}. \tag{17}$$

(Consequence: the vector q_0 is unique.)

11°. *There exists an n-vector q_0 for which the expression $q_0^*(\sigma I - A)^{-1} b$ is irreducible* [3]).

---

[1]) In other words the characteristic polynomial of the matrix $A + b q_0^*$ is equal to the given polynomial.

[2]) A property used by the author in [11] and by R. E. Kalman in [5].

[3]) In other words the expression $q^*(\sigma I - A)^{-1} b$ is a quotient of two polynomials which have no common root; moreover the degree of the polynomial which constitutes the denominator is equal to the order of the system.

12°. *There does not exist any non-zero vector* q_0 *such that the expression* $q_0^*(\sigma I - A)^{-1}b$ *be identically zero* [1]).

13°. *For every set of distinct complex numbers* σ_j, $j = 1, 2, \ldots, n$, *different from the eigenvalues of the matrix* A, *the vectors* $(\sigma I - A)^{-1}b$, $j = 1, 2, \ldots, n$ *are linearly independent.*

14°. *There exist* n *complex numbers* σ_j, $j = 1, 2, \ldots, n$ *such that the vectors* $(\sigma_j I - A)^{-1}b$, $j = 1, 2, \ldots, n$ *are linearly independent.*

15°. *The identity* $q_0^* e^{At} b \equiv 0$ *in an interval* $t_1 < t < t_2$ *is possible only for* $q_0 = 0$ [2]).

16°. *For every pair of numbers:* t_1 *and* $t_2 > t_1$ *the matrix*

$$W(t_1, t_2) = \int_{t_1}^{t_2} e^{At}\, bb^*\, e^{A^* t}\, dt \tag{18}$$

is positive definite [3]).

3. Discussion

Before proving Theorem 1, we note that Properties 1°—16° can be divided into three distinct classes. Some of the properties are expressed in the "parameter form", i.e. they involve only the parameters A and b and possibly a transformation matrix R. Other properties have the "frequency-domain form", that is, they involve in addition an arbitrary complex variable σ, in some expressions which are generally obtained when using the frequency domain techniques. Finally, other properties, as Definition 1, itself, have the "time-domain form", i.e. they contain explicitly the time variable t. One can "translate" some properties from one language into another thus obtaining other equivalent properties which are useful in some applications. Thus, Property 1° can be expressed in the time-domain form as follows:

1°. *There does not exist a non-singular transformation* $\tilde{x} = Rx$, *such that the transformed system takes the form*

$$\tilde{x} = \begin{pmatrix} y \\ z \end{pmatrix}; \qquad \left.\begin{aligned} \frac{dy}{dt} &= A_{11} y + A_{12} z + b_1 \mu(t) \\ \frac{dz}{dt} &= A_{22} z \end{aligned}\right\}, \tag{19}$$

---

1) This condition of complete controllability was stated by M. A. Aizerman and F. R. Gantmacher [1].

2) One of the formulations used by S. Lefschetz in [3].

3) This property is formulated after R. E. Kalman.

where the matrices A_{11}, A_{12}, A_{22} and the vector b_1 have the dimensions shown in the statement of Property 1° of Theorem 1, and the vectors y and z have k and $n-k$ components respectively.

Similarly, Property 7° can be stated in the following time domain-form :

7°'. *There exists a uniquely determined, non-singular transformation $\tilde{x} = Rx$, such that the transformed system can be written in the form*

$$\tilde{x} = \begin{pmatrix} x_1 \\ x_2 \\ \cdot \\ \cdot \\ x_n \end{pmatrix}; \quad \left.\begin{aligned} \frac{dx_1}{dt} &= x_2 \\ \frac{dx_2}{dt} &= x_3 \\ &\cdots\cdots\cdots\cdots \\ \frac{dx_n}{dt} &= \gamma_1 x_1 + \gamma_2 x_2 + \ldots + \gamma_n x_n + \mu(t) \end{aligned}\right\} . \qquad (20)$$

Obviously, Property 8° admits a similar formulation. Remark also that Property 12° is the translation of Property 15° in the frequency-domain form.

Some of the properties stated in Theorem 1 are obviously particular cases of other properties (e.g. Property 14° follows immediately from Property 13°). It may seem curious at first sight, that Property 14°, which states less than Property 13°, is nevertheless equivalent to the latter[1]). This is due to the fact that the part of Property 13°, which is not included in Property 14° is a consequence of Property 14°. Both types of properties are useful in their own ways. Property 14° may serve to establish the fact that a given system is completely controllable by using a minimum amount of information relative to the system ; Property 13° may serve to deduce additional information concerning a system which is already known to be completely controllable. The parts played by Properties 13° and 14° are reversed if we want to negate the property of complete controllability. Then Property 13° is useful in establishing that a system is not completely controllable and Property 14° will serve to obtain additional information concerning a system which is not completely controllable. By combining Properties 13° and 14° we may state the following criterion of complete controllability :

---

[1]) A simple illustration of this situation is given by the equivalence of the statement "the polynomial $ax + b$ vanishes at two distinct points" and the statement (which is equivalent to the first, although it seemingly expresses more) : "the polynomial $ax + b$ is identically zero".

"One chooses n arbitrary distinct complex numbers σ_j, $j = 1, 2, \ldots, n$ *different from the eigenvalues of the matrix* A. *If the vectors* $(\sigma_j I - A)^{-1}b$, $j = 1, 2, \ldots, n$ *are linearly independent, then the pair* (A, b) *is completely controllable; if the vectors are not linearly independent, the pair* (A, b) *is not completely controllable.*

We shall now state a simple consequence of Theorem 1 concerning incompletely controllable systems.

Proposition 2. *Assume that the pair* (A, b) *of Eq.(1) is not completely controllable and that* $b \neq 0$. *Then there exists a non-singular transformation* $\tilde{x} = Rx$ *such that Eq.(1) can be written in the form (19) where the pair* (A_{11}, b) *is completely controllable.*

Proof. Since the pair (A, b) is not completely controllable, it follows from Property 1° of Theorem 1, that there exists a non-singular matrix R, such that the transformed pair $(\tilde{A}, \tilde{b})$ can be written in the form (2). Furthermore, condition $b \neq 0$ yields $b_1 \neq 0$. Let n_1 be the number of rows of the matrix A_{11}. Since $b_1 \neq 0$ we have $n_1 \geqslant 1$. If the pair (A_{11}, b_1) obtained above is not completely controllable, we can bring it by the same procedure to the form (2). By repeating the operation whenever it is possible, we obtain a sequence of pairs $(\tilde{A}, \tilde{b})$ of the form (2), where the vector b_1 is always non-zero (hence we have always $n_1 \geqslant 1$) and at the same time the number of rows of the matrix A_{11} decreases with at least one unit at each step. Hence, after a finite number of steps, either the pair (A_{11}, b) becomes completely controllable (and Proposition 2 is proved), or the number n_1 of rows of the matrix A_{11} takes the minimum value (which is, as we have seen, equal to 1). In the latter case the matrix A_{11} reduces to a scalar and the vector b_1 becomes a non-zero scalar. Then, by Property 4° of Theorem 1 the pair (A_{11}, b_1) is also completely controllable and this concludes the proof of Proposition 2.

Finally we shall state a result which in a certain sense extends the property from Definition 1 to systems which are not completely controllable.

Proposition 3. *If* b *is a non-zero vector* [1]*, then for every scalar* α *there exists a continuous function* μ*, such that the corresponding solution of Eq. (1), for* $x(0) = 0$*, satisfies the condition* $x(1) = \alpha(\sigma I - A)^{-1}b$.

Proof. If the pair (A, b) is completely controllable, the proposition results immediately from Definition 2 and Proposition 1. If not, we use Proposition 2 and apply the transforma-

[1]) If $b = 0$ the proposition is trivial.

tion $\tilde{x} = Rx$ of that proposition. Consequently Eq. (1) is written in the form (19) where the pair (A_{11}, b_1) is completely controllable. We have to prove that there exists a function μ, such that the corresponding solution of Eq. (19), for $x = 0$ (i.e. $y(0) = 0$ and $z(0) = 0$), satisfies the condition $\tilde{x}(1) = Rx(1) = \alpha R(\sigma I - A)^{-1}b = \alpha(\sigma I - \tilde{A})^{-1}\tilde{b}$ (the last relation is easily obtained by using the relations $\tilde{A} = RAR^{-1}$ and $\tilde{b} = Rb$). From the initial condition $z(0) = 0$ and from (19) we obtain $z(t) \equiv 0$. Since $\tilde{A}$ and $\tilde{b}$ have the form (2), the final value $\tilde{x}(1)$ is equal to

$$\tilde{x}(1) = \alpha(\sigma I - \tilde{A})^{-1}\tilde{b} = \begin{pmatrix} \alpha(\sigma I_1 - A_{11})^{-1}\; b_1 \\ 0 \end{pmatrix} \tag{21}$$

where I_1 is the identity matrix with the same dimensions as A_{11}. In other words we must have $z(1) = 0$ (which is in agreement with the identity $z(t) \equiv 0$ obtained above), and also $y(1) = \alpha(\sigma I_1 - A_{11})^{-1}b_1$. The last condition can always be satisfied since y satisfies the equation

$$\frac{dy}{dt} = A_{11}y + b_1\mu(t)$$

which results from (19) taking into account $z(t) \equiv 0$, and which is completely controllable.

4. Proof of the theorem

The equivalence of the properties mentioned in Theorem 1, will be proved using the following scheme

$$\begin{aligned}&(\text{Def}) \to 1° \to 2° \to 3° \to 4° \to 5° \to 6° \to 7° \to 8° \to 9° \to 10° \to \\ &\to 11° \to 12° \to 13° \to 14° \to 15° \to 16° \to (\text{Def})\end{aligned} \tag{22}$$

where the numbers indicate the properties stated in Theorem 1, the symbol "Def" means the property of Definition 1 and the arrow has the meaning of "implies" [1]). Some of the implications included in the scheme will be indirectly proved with the help of some intermediate implications.

---

[1]) Theorem 1 could be also proved after the scheme (Def) $\to k° \to$ (Def), where $k = 1, 2, \ldots, 16$. However then one has to prove more implications than in (22).

A special remark is necessary concerning Properties 5°, 6°,..., 10° which, besides their main parts, contain some consequences (mentioned within brackets in Theorem 1). When we shall prove that $k° \to (k + 1)°$, for $k = 4, 5, \ldots, 10$, in fact, we shall prove that the main part of $k°$ (excluding the part within brackets) implies the whole property $(k + 1)°$ (including the part within brackets). In this way it is clearly proved that the properties from (22) are equivalent even if one eliminates the parts within brackets. Moreover the part within brackets of each Property $k°$ is proved to be a consequence of the remaining part of Property $k°$ (since, as remarked above, the main part of $k°$ is equivalent to the main part of $(k - 1)°$ which in turn implies the part within brackets of $k°$).

We prove, step by step, the implications of the scheme (22).

(Def) → 1°. By expressing Property 1° in the time-domain form (Property 1°′, Section 3) one sees that if Property 1° is not satisfied then the system can be written in the form (19). From the last equation of (19) it follows that the equality $z(t_2) = 0$ implies the equality $z(t_1) = 0$. Hence the property from the definition 1 cannot be satisfied.

1° → 2°. If Property 2° is not satisfied, then the vector b belongs to a subspace X of dimension $k < n$, invariant under the matrix A. One can find a basis of the n-dimensional space consisting of the vectors $e_1, e_2, \ldots, e_k, e_{k+1}, \ldots, e_n$ and such that the vectors $e_1, e_2, \ldots, e_k$ constitute a basis of the subspace X. Replacing the old basis by the new one, one finds that the matrix $\widetilde{A}$ and the vector $\widetilde{b}$ have in the new basis, the form (2) and hence Property 1° is not satisfied. The matrix R, mentioned in the statement of Property 1° has the expression $R^{-1} = (e_1\ e_2 \ldots e_n)$.

2° → 3°. If Property 3° is not true, then there exists a vector x_0 with the properties

$$x_0^* A = \alpha x_0^*, \quad x_0^* b = 0, \quad x_0 \neq 0. \tag{23}$$

The subspace X of all the vectors which are orthogonal to x_0 is invariant under A, since from $x_0^* x = 0$ and from the first of the Relations (23), it follows that $x_0^* A x = \alpha x_0^* x = 0$ Since $x_0 \neq 0$, the dimension of the subspace X is smaller than n. The second relation of (23) shows that b belongs to the subspace X thus contradicting Property 2°.

3° → 4°. If Property 4° is not satisfied, then there exists at least one vector $x_0 \neq 0$ such that $x_0^* C = 0$, or $C^* x_0 = 0$, or (using expression (3) of C)

$$b^*(A^*)^j x_0 = 0, \qquad j = 0, 1, \ldots, n - 1. \tag{24}$$

Since matrix A satisfies its own characteristic equation, Eq. (24) is valid for every integer j. It follows that Relations (24) will still hold after replacing the vector x_0 by A^*x_0. Hence, the set of all the vectors satisfying (24) constitute a subspace which is invariant under the matrix A^*. This subspace contains at least an eigenvector of the matrix A^*. From (24) for $j = 0$, one sees that this eigenvector is orthogonal to b, thus contradicting Property 3°.

4° → 5°. The matrix equation $RA - AR = 0$ is obviously satisfied by any expression of the form $R = \sum_{j=1}^{n} \gamma_j A^{j-1}$, where γ_j are scalar constants. Introducing this expression in (5) yields $Rb = \sum A^{j-1}b\ \gamma_j = d$, which can be always satisfied by a proper choice of the constants γ_j. Indeed, the obtained equation has the form $Ca = d$ where C has the expression (3) and the vector a has the components $\gamma_1, \gamma_2, \ldots, \gamma_n$. It follows — using Property 4° — that Eq. (5) is satisfied for $a = C^{-1}d$ and hence the system (4), (5) has a solution of the form shown above. The solution is unique, if the homogeneous system (for $d = 0$) has only the trivial solution $R = 0$. But this property does take place since from $Rb = 0$ and $RA - AR = 0$, one obtains successively : $RAb = 0$, $RA^2b = 0, \ldots, RA^{n-1}b = 0$ or (see (3)) $RC = 0$ and finally, $R = 0$ (since, by Property 4°, the matrix C is non-singular).

One proves now the part specified within brackets in the statement of Property 5°. One shows namely that from the relations $RA - AR = 0$ and $Rb = d$ it follows that

$$RA^j b = A^j d, \qquad j = 0, 1, \ldots, n - 1. \tag{25}$$

For $j = 0$ Relation (25) coincides with (5). If Relation (25) is true for $j = k - 1$, then it is also true for $j = k$; this results from the following sequence of equalities

$$RA^k b = (RA)(A^{k-1}b) = A(A^{k-1}d) = A^k d,$$

where the second equality is obtained by using the relation $RA = AR$ (4) and the assumption that (25) is satisfied for $j = k - 1$. Relation (25) can be also written as

$$R\,(b\ Ab \ldots A^{n-1}b) = (d\ Ad \ldots A^{n-1}d). \tag{26}$$

If the matrix from the right-hand member is non-singular, then clearly R is also non-singular.

5° → 6°. One shows first that Property 3° follows from Property 5°; Indeed, if Property 3° is not satisfied, then there exists a vector $x \neq 0$, such that $x^*A = x$ and $x^*b = 0$. Hence $x^*(A - \alpha I) = 0$ and therefore $\det(A - \alpha I) = 0$, whence one concludes that there exists a vector $y \neq 0$, such that $Ay = \alpha y$. These relations show that the matrix $R_0 = yx^* \neq 0$ satisfies the equations $R_0 A - AR_0 = 0$, and $R_0 b = 0$, whence it follows that if R is a solution of the system (4), (5), then the matrix $R + R_0$ is also a solution; this contradicts the condition of uniqueness stated in Property 5° and concludes the proof of the implication 5° → 3° [1].

Using also the implication 3° → 4° proved before, one concludes that Property 5° implies $\det C \neq 0$; i.e. (see (3))

$$\det(b\ Ab\ A^2b \ldots A^{n-1} b) \neq 0. \tag{27}$$

One proves now the implication 5° → 6°. Since the matrices A and $\tilde{A}$ satisfy their own characteristic equations which, by the assumptions of Property 6°, are identical, one can write the relations

$$A^n = \sum_{j=1}^{n} \gamma_j A^{j-1}, \quad \tilde{A}^n = \sum_{j=1}^{n} \gamma_j \tilde{A}^{j-1}, \tag{28}$$

the coefficients γ_j being the same in both equations. One introduces the matrix

$$N = \begin{pmatrix} 0 & 0 & .. & 0 & \gamma_1 \\ 1 & 0 & .. & 0 & \gamma_2 \\ 0 & 1 & .. & 0 & \gamma_3 \\ . & . & .. & . & . \\ 0 & 0 & .. & 1 & \gamma_n \end{pmatrix}, \tag{29}$$

which satisfies the equalities

$$CN = AC, \quad \text{where} \quad C = (b\ Ab \ldots A^{n-1}\ b), \tag{30}$$

$$\tilde{C}N = \tilde{A}\tilde{C}, \quad \text{where} \quad \tilde{C} = (\tilde{b}\ \ \tilde{A}\tilde{b} \ldots \tilde{A}^{n-1}\ \tilde{b}). \tag{31}$$

---

[1]) We have proved so far the implications 1° → 2° → 3° → 4° → 5° → 3°. Hence we can state that the Properties 3°—5° of Theorem 1 are equivalent.

By assumptions (see Property 6°), the matrix $\widetilde{C}$ is non-singular. As shown before, the matrix C is also non-singular (see (27)). Under these conditions one obtains from (30) and (31) the equality

$$\widetilde{A} = R_0 A R_0^{-1}, \text{ where } R_0 = \widetilde{C} C^{-1}. \tag{32}$$

The matrix R_0 is also nonsingular. Furthermore one has $R_0 C = \widetilde{C}$ and in particular $R_0 b = \widetilde{b}$, as stated. One shows now the uniqueness of the matrix R of (8), as a consequence of the main part of Property 6°. Indeed, if besides (8) one has also the relations $\widetilde{A} = \widetilde{R} A \widetilde{R}^{-1}$ and $\widetilde{b} = \widetilde{R} b$, then the matrix $R_2 = R\widetilde{R}^{-1}$ satisfies the relations $R_2 \widetilde{A} = \widetilde{A} R_2$ and $R_2 \widetilde{b} = \widetilde{b}$, whence one obtains $R_2 \widetilde{A}^k \widetilde{b} = \widetilde{A}^k \widetilde{b}$ for $k = 0, 1, \ldots, n-1$ or $R_2 \widetilde{C} = \widetilde{C}$, where $\widetilde{C}$ has the form (31). Since $\widetilde{C}$ is nonsingular (see (6)) one obtains $R = \widetilde{R}$.

Using Property 6° one can easily bring a completely controllable pair to certain special forms, as we do below in the proof of implications 6° → 7° → 8°.

6° → 7°. Clearly, the characteristic equation of the matrix A (9) can be written in the form $\sigma^n = \sum\limits_{j=1} \gamma_j \sigma^{j-1}$ and hence coincides with the characteristic equation of matrix A (see Property 7°). The matrix $\widetilde{C} = (\widetilde{b}, \widetilde{A}\widetilde{b} \ldots \widetilde{A}^{n-1}\widetilde{b})$ is triangular and has all the terms on the secondary diagonal equal to 1 and the elements above the diagonal equal to zero. Hence, the matrix $\widetilde{C}$ is nonsingular. Thus the pair (A, b), defined by Relations (9), satisfies all the conditions of Property 6°, whence Property 7° follows. The uniqueness of R is proved as at the end of the proof of the implication 5° → 6°.

7° → 8°. This implication will be indirectly proved with the help of the chain of implications 7° → 4° → 6° → 8°. The implication 4° → 6° results from the chain of implications 4° → 5° → 6° proved previously. The implication 7° → 4° has also been proved (see implications 6° → 7°) when it was shown that the matrix C is nonsingular. It remains to prove the implication 6° → 8°. Observe that the characteristic equation of matrix $\widetilde{A}$ (see (10), (11)) is equal to the right-hand member of Relation (13) and hence it coincides with the characteristic equation of matrix A. From Property 6°, it follows that in order to obtain Property 8° it is sufficient to show the matrix $\widetilde{C} = (\widetilde{b}\ \widetilde{A}\widetilde{b} \ldots \widetilde{A}^{n-1}\widetilde{b})$, in which $\widetilde{A}$ and $\widetilde{b}$ are given by Relations (10)—(12), is non-singular. With that end in view one shows first that the pair $(\widetilde{A}, \widetilde{b})$ satisfies Property 3°. From (10) and (11) it follows that the matrix A^* has only k eigenvectors [1] of the form

$$x_j^* = (y_{j1}^* \, y_{j2}^* \ldots y_{jk}^*)\,, \quad j = 1, 2, \ldots, k\,,$$

where y_{jl} are vectors with n_1 components given by the relations

$$y_{jl}^* = \begin{cases} (0 \quad 0 \, .. \, 0\,) & \text{if} \quad j \neq l\,, \\ (0 \quad 0 \, .. \, 0 \quad \alpha)\,, \quad \alpha \neq 0, & \text{for } j = l. \end{cases}$$

Hence, using (10) and (12) which define $\widetilde{b}$, one sees that all the eigenvectors x_j of the matrix $\widetilde{A}^*$ satisfy the inequalities $x_j^* \widetilde{b} \neq 0$; $j = 1, 2, \ldots, k$. Hence the pair $(\widetilde{A}, \widetilde{b})$ satisfies Property 3°. From the implication 3° → 4°, proved above, one deduces that the matrix $\widetilde{C}$, written for the pair $(\widetilde{A}, \widetilde{b})$ (10)—(13), is non-singular; thus we have obtained all the conditions required for deriving Property 8° from Property 6°. The uniqueness of R is proved as at the end of the proof of the implication 5° → 6°.

8° → 9°. This implication will be indirectly proved with the help of the chain of implications 8° → 4° → 7° → 9°; The implication 4° → 7° results from the chain of implications 4° → 5° → 6° → 7° proved previously. The implication 8° → 4° has been already proved (see the proof of the implication 7° → 8°) when it was shown that for the pair $(\widetilde{A}, \widetilde{b})$ given by (10)—(12) the matrix $\widetilde{C}$ is nonsingular. It remains to prove the implication 7° → 9°. Assuming that the pair (A, b) has been brought to the form $(\widetilde{A}, \widetilde{b})$ from (9), consider an arbitrary vector $\widetilde{q}$ of the form

$$\widetilde{q}^* = (q_1 \, q_2 \ldots q_n)\,. \tag{33}$$

Using Relations (9) one obtains

$$\widetilde{A} + \widetilde{b}\widetilde{q}^* = \begin{pmatrix} 0 & 1 & 0 & . & . & 0 \\ 0 & 0 & 1 & . & . & 0 \\ . & . & . & . & . & . \\ 0 & 0 & 0 & . & . & 1 \\ \widetilde{\gamma}_1 & \widetilde{\gamma}_2 & \widetilde{\gamma}_3 & \cdot & \cdot & \widetilde{\gamma}_n \end{pmatrix}, \tag{34}$$

---

[1]) Two eigenvectors which are obtained one from the other by multiplication with a scalar are not considered to be distinct.

where

$$\widetilde{\gamma}_j = \gamma_j + q_j , \qquad j = 1, 2, \ldots, n . \tag{35}$$

The characteristic equation of matrix (39) can be written as

$$\sigma^n = \sum_{j=1}^{n} \widetilde{\gamma}_j \, \sigma^{j-1} = \sum_{j=1}^{n} (\gamma_j + q_j) \, \sigma^{j-1}. \tag{36}$$

Given the polynomial $\pi_1(\sigma)$, (14) one can always determine uniquely the components q_j by the relations

$$q_j = \beta_j - \gamma_j , \qquad j = 1, 2, \ldots, n . \tag{37}$$

Then the characteristic polynomial of the matrix $\widetilde{A} + \widetilde{b}\widetilde{q}^*$ equals $\pi_1(\sigma)$; i.e. one has the relation

$$\det (\sigma I - \widetilde{A} - \widetilde{b}\, \widetilde{q}^*) = \pi_1 (\sigma) . \tag{38}$$

Using the relations $\widetilde{A} = RAR^{-1}$ and $\widetilde{b} = Rb$, one obtains from (38) that

$$\pi_1 (\sigma) = \det (\sigma I - RAR^{-1} - Rb\widetilde{q}^*) = \det R (\sigma I - A - bq_0^*) \, R^{-1} = $$
$$= \det (\sigma I - A - bq_0^*) , \tag{39}$$

where we have introduced the new vector $q_0^* = \widetilde{q}^* R$ and taken into account the relation $\det R . \det R^{-1} = 1$. The first and the last members of (39) yield (15) which had to be proved. The proof also shows that the vector q_0 is unique.

9° → 10°. This implication too will be indirectly proved using the scheme of implications 9° → 3° → 7° → 10°; of these, only the implications 9° → 3° and 7° → 10° are not yet proved. To prove the implication 9° → 3° observe that if Property 3° is not satisfied then there exists a vector x_0 with the Properties (23); hence, for any vector q we have the equalities $x_0^*(A + b^*q) = x_0^* A = \alpha x_0^*$. Consequently α is an eigenvalue of the matrix $A + b^*q$ for any vector q. Therefore if one chooses a polynomial π_1 (as mentioned in the statement of Property 9°), which does not vanish for $\sigma = \alpha$, then Relation (15) is violated for every vector q. This contradicts Property 9° and proves the implication 9° → 3°. It remains to prove the implication 7° → 10°. Using Relations (9) one obtains by direct calculation

$$(\sigma I - \widetilde{A})^{-1} \widetilde{b} = \frac{v(\sigma)}{\det (\sigma I - \widetilde{A})} ,$$

where the vector $v(\sigma)$ has the expression

$$v(\sigma) = \begin{pmatrix} 1 \\ \sigma \\ \vdots \\ \sigma^{n-1} \end{pmatrix}.$$

Using also the relations $\tilde{A} = RAR^{-1}$ and $\tilde{b} = Rb$, one obtains

$$(\sigma I - A)^{-1} b = \frac{R^{-1} v(\sigma)}{\det(\sigma I - A)} \cdot \tag{40}$$

Property 10° is a simple consequence of the above relation [1]). Indeed, using the vector $v(\sigma)$ one can write (16) in the form $\pi_2(\sigma) = a_0^* v(\sigma)$, where $a_0^* = (\alpha_1 \alpha_2 \ldots \alpha_n)$. Using this condition and Relation (40) we find that Relation (17) becomes $q_0^* R^{-1} v(\sigma) = a_0^* v(\sigma)$ and hence the vector q_0 whose existence is stated in Property 10°, is given by the relation $q_0^* = a_0^* R$. It also follows that this vector is unique.

10° → 11°. It is sufficient to choose the polynomial π_2 such that the right hand member of (17) is irreducible ; then, applying Property 10°, one can find a vector q_0 which satisfies Property 11°.

11° → 12°. This implication is proved using the chain of implications 11° → 1° → 10° → 12° of which only the implications 11° → 1° and 10° → 12° are not yet proved. To prove the implication 11° → 1° we observe that if Property 1° is not satisfied, then the pair (A, b) can be written in the form (2). By writing the arbitrary vector q in the form $\tilde{q}^* = (q_1^* q_2^*)$, where the vector q_1 has the same number of components as b_1, one obtains the equality

$$\tilde{q}^* (\sigma I - \tilde{A})^{-1} \tilde{b} = q_1^* (\sigma I_{11} - A_{11})^{-1} b_1 , \tag{41}$$

where I_{11} is the identity matrix with the same number of dimensions as A_{11}. The right-hand member of Relation (41) can be written in the form of a quotient of two polynomials, where the denominator is equal to $\det(\sigma I_{11} - A_{11})$ and thus its degree is smaller than the order n of the system. Hence any expression of the form of the left-hand member of (41) is reducible. From

[1]) Therefore the properties of Theorem 1 are also equivalent to the following property : "There exists a non-singular matrix R such that one has Relation (40)".

the relation $\tilde{q}^*(\sigma I - \tilde{A})^{-1}\tilde{b} = \tilde{q}^*(\sigma I - RAR^{-1})^{-1}Rb = \tilde{q}^*R(\sigma I - - A)^{-1}b$ it follows that any expression of the form $q^*(\sigma I - A)^{-1}b$ is reducible — and this contradicts Property 11°. The above proof shows that the *whole* Property 10° follows from 11°. It remains to prove the implication 10° → 12° which however is immediate since the identity

$$q^*(\sigma I - A)^{-1}b \equiv 0 \tag{42}$$

is satisfied for $q = 0$. From the uniqueness of the vector q_0 which satisfies the conditions of Property 10° it follows that the vector $q = 0$ is the only vector which satisfies (42); we have thus obtained Property 12°.

12° → 13°. If Property 13° is not satisfied then there exists a *non-zero* vector q, such that

$$q^*(\sigma_j I - A)^{-1}b = 0, \; j = 1, 2, \ldots, n, \; \det(\sigma_j I - A) \neq 0. \tag{43}$$

The left-hand member of (43) is of the form

$$q^*(\sigma I - A)^{-1}b = \frac{\pi(\sigma)}{\det(\sigma I - A)}, \tag{44}$$

where π is a polynomial of a degree not higher than $n - 1$. From (43) and (44) it follows that

$$\pi(\sigma_j) = 0, \quad j = 1, 2, \ldots, n.$$

The polynomial π has at the most the degree $n - 1$ and vanishes at n distinct points σ_j, hence is identically zero. Therefore the left-hand member of (44) is identically zero. However the vector q is non-zero, — contrary to Property 12°.

13° → 14°. Property 14° is obviously included in Property 13°.

14° → 15°. We consider the chain of implications 14° → → 1° → 4° → 15° of which only the implications 14° → 1° and 4° → 15° have not been proved. To prove the implication 14° → → 1° observe that if Property 1° is not satisfied, the pair (A, b) can be written in the form (2), whence it follows that the last $n - k$ components of any vector $(\sigma I - A)^{-1}\tilde{b}$ are equal to zero. Hence all the vectors of this form belong to a subspace of a dimension smaller than n and therefore one cannot find n linearly independent vectors of this form. Since one has the relation $(\sigma I - \tilde{A})^{-1}\tilde{b} = R(\sigma I - A)^{-1}b$, one concludes directly that there do not exist n linearly independent vectors of the form $(\sigma I - A)^{-1}b$ — contrary to Property 14°. It remains to prove the implication 4° → 15°. If Property 15° is not satisfied, then there exists an interval $t_1 < t < t_2$ wherein the identity $x_0^* e^{At}b \equiv 0$ is satisfied (for $x_0 \neq 0$), whence we deduce by successive differentiations at an arbitrary point t_i of the interval:

$$x_0^* \, e^{At_i} A^j b = 0, \; j = 0, 1, 2, \ldots, n-1$$

i.e.

$$x_0^* \, e^{At_i} \, C = 0 \qquad (x_0 \neq 0) \tag{45}$$

where C has the expression (3). Since the matrix e^{At_i} is non-singular, the vector $x_0^* e^{At_i}$ is non-zero and hence (45) can only occur if the matrix C is singular, contrary to Property 4°.

15° → 16°. From (18) of the matrix $W(t_1, t_2)$ it follows that

$$x_0^* \, W(t_1, t_2) \, x_0 = \int_{t_1}^{t_2} | b^* \, e^{A^* t_i} \, x_0 |^2 \, dt \, ; \tag{46}$$

and hence the matrix $W(t_1, t_2)$ is at least positive semidefinite. It is even positive definite because if there existed a non-zero vector x_0 such that $x_0^* \, W(t_1, t_2) x_0 = 0$, then we would obtain from (46) the conclusion

$$b^* \, e^{A^* t} \, x_0 \equiv 0 \qquad (x_0 \neq 0), \; t_1 < t < t_2 \, ; \tag{47}$$

an identity which contradicts Property 15°.

16° → (Def)[1]. The solution of the system $\dfrac{dx}{dt} = Ax + b\mu(t)$ which at $t = t_1$ satisfies the condition $x(t_1) = x_0$, satisfies the relation

$$x(t_2) = e^{-At_1} \left(e^{At_2} \, x_0 + \int_{t_1}^{t_2} e^{A\tau} \, b\mu(t_1 + t_2 - \tau) \, d\tau \right). \tag{48}$$

Hence, it is sufficient to show that if Property 16° is satisfied, then for every vector x_0 there exists a continuous function ρ such that

$$e^{At_2} \, x_0 = - \int_{t_1}^{t_2} e^{A\tau} \, b \, \rho(\tau) \, d\tau \, . \tag{49}$$

---

[1]) The proof is due to R. E. Kalman.

Since the matrix $W(t_1, t_2)$ is positive definite, it is invertible. Therefore (49) is satisfied if the function ρ is defined by

$$\rho(\tau) = - b^* e^{A^*\tau} (W(t_1, t_2))^{-1} e^{At_2} x_0 , \tag{50}$$

as one can easily see by substituting (50) in (49) and then using Definition (18) of matrix $W(t_1, t_2)$.

5. Relations between single-input completely controllable systems

We consider the pair (A, b) where A is an $n \times n$-matrix and b is an n-vector. We shall say that n is the "order" of the pair (A, b).

In the set of all the pair (A, b) of the same order n, one introduces the following relation :

"*Relation* R_c": *A pair* (A_1, b_1) *is said to be in relation* R_c *with the pair* (A_0, b_0), *if there exists an* $n \times n$-*nonsingular-matrix* R_0 *and an* n-*vector* q_0, *such that*

$$A_1 = R_0 (A_0 + b_0 q_0^*) R_0^{-1}; \; b_1 = R_0 b_0 . \tag{51}$$

From (51) it follows that (A_1, b_1) coincides with (A_0, b_0) if $R_0 = I$ and $q_0 = 0$. The inverse relations $A_0 = R_1(A_1 + b_1 q_i^*) R_i^{-1}$, $b_0 = R_i b_i$, where $R_i = R_0^{-1}$ and $q_i = -(R_0^{-1})^* q_0$ also follow. Considering a new pair (A_2, b_2) related to (A_1, b_1) by relations of the form (51)

$$A_2 = R_i (A_1 + b_1 q_i^*) R_1^{-1}; \; b_2 = R_1 b_1 ,$$

and replacing in these relations the formulas (51), one finds that the pair (A_2, b_2) is related to the pair (A_0, b_0) by the relation $A_2 = R_2 (A_0 + b_0 q_2^*) R_2^{-1}$; $b_2 = R_2 b_0$, where $R_2 = R_1 R_0$ and $q_2 = q_0 + R_0^* q_1$.

These remarks show that the relation R_c is reflexive, symmetric and transitive. Therefore R_c is a relation of equivalence and allows the partition of the set of all the pairs (A, b) into classes of equivalent pairs (see, for instance, Gr. C. Moisil [1]). The introduction of these classes points out a strong link between completely controllable pairs as shown by the following theorem:

Theorem 2. *All completely controllable pairs* (A, b) *of the same order* n *belong to a class of equivalent pairs with respect to the equivalence relation* R_c. *Therefore, if* (A_0, b_0) *is a completely controllable pair, then all the pairs* (A_1, b_1), *obtained by formula (51) where* $\det R_0 \neq 0$, *are completely controllable; conversely, if* (A_0, b_0) *and* (A_1, b_1) *are two completely controllable* *pairs of the same order, then there exists a non-singular matrix R_0 and a vector q_0 such that the two pairs are related by Relations (51).*

Proof. From (51) one obtains, for every q, the equality $A_0 + b_0 q^* = R_0^{-1}(A_1 + b_1(q^* - q_0^*) R_0^{-1}) R_0$, whence one deduces

$$\det(\sigma I - A_0 - b_0 q^*) = \det(\sigma I - A_1 - b_1(q^* - q_0^*) R_0^{-1}) \tag{52}$$

If the pair (A_0, b_0) is completely controllable, then by Property 90 of Theorem 1, the left-hand member of (52) can be made equal to any polynomial of the form (14) by an adequate choice of the vector q. In other words, by an adequate choice of the vector $q_1^* = (q^* - q_0^*) R_0^{-1}$ the right-hand member of Eq. (52) can be made equal to any polynomial. Applying again Property 9^0 of Theorem 1 one sees the pair (A_1, b_1) is also completely controllable.

Conversely, if the pairs (A_0, b_0) and (A_1, b_1) are of the same order and completely controllable, then by virtue of Property 7° of Theorem 1, there exists non-singular matrices R_0 and R_1 such that the pair $(\widetilde{A}_0, \widetilde{b}_0)$ and $(\widetilde{A}_1, \widetilde{b}_1)$ defined by the relations $\widetilde{A}_0 = \widetilde{R}_0 A_0 \widetilde{R}_0^{-1}$, $\widetilde{b}_0 = \widetilde{R}_0 b_0$, $\widetilde{A}_1 = \widetilde{R}_1 A_1 \widetilde{R}_1^{-1}$ and $\widetilde{b}_1 = \widetilde{R}_1 b_1$, have the form (9). The matrices $\widetilde{A}_0$ and $\widetilde{A}_1$ differ only by the last row. Let $(\gamma_1\ \gamma_2 \ldots \gamma_n)$ be the last row of the matrix $\widetilde{A}_0$ and $(\widetilde{\gamma}_1\ \widetilde{\gamma}_2 \ldots \widetilde{\gamma}_n)$ the last row of the matrix $\widetilde{A}_1$. Then one has the equality

$$\widetilde{A}_1 = \widetilde{A}_0 + \widetilde{b}_0\, \widetilde{q}^*, \tag{53}$$

where the components of the vector $q^* = (q_1\ q_2 \ldots q_n)$ are given by the equalities $q_j = \widetilde{\gamma}_j - \gamma_j$, $j = 1, 2, \ldots, n$. Replacing in (53) $\widetilde{A}_1$, $\widetilde{A}_0$ and $\widetilde{b}_0$ by their expressions in terms of A_1, A_0 and b_0, one obtains the equality

$$\widetilde{R}_1 A_1 \widetilde{R}_1^{-1} = \widetilde{R}_0 A_0 \widetilde{R}_0^{-1} + \widetilde{R}_0 b_0 q^*,$$

whence

$$A_1 = \widetilde{R}_1^{-1} \widetilde{R}_0 (A_0 + b_0\, q^*\, \widetilde{R}_0)\, \widetilde{R}_0^{-1} \widetilde{R}_1. \tag{54}$$

From Relations (9) one also obtains $\widetilde{b}_1 = \widetilde{b}_0$. By expressing $\widetilde{b}_1$ and $\widetilde{b}_0$ in terms of b_1 and b_0, one has

$$b_1 = \widetilde{R}_1^{-1} \widetilde{R}_0\, b_0. \tag{55}$$

Relations (54) and (55) have the form (51) where $R_0 = \widetilde{R}_1^{-1} \widetilde{R}_0$ and $q_0^* = q^* \widetilde{R}_0$.

Remark. Theorem 2 can be also used for stating the property of complete controllability in other forms equivalent to the properties of Theorem 1. To this end it is sufficient to replace the pair (A_0, b_0) by the pair (A_1, b_1) of the form (51) in any of the properties of Theorem 1. Starting from an arbitrary Property k° of Theorem 1 we can introduce the equivalent properties :

"k^0_M : For any non-singular matrix R and vector q, the pair (A_1, b_1) given by (51) satisfies the Property k°", and

"k^0_m : There exists a non-singular matrix and a vector q, such that the pair (A_1, b_1) given by (51) satisfies Property k°".

§ 32. Single-output completely observable systems

The property of complete observability can be characterized by the following definition

Definition 1. *The system* [1])

$$\mathrm{d}x/\mathrm{d}t = Ax + b\,\mu(t), \qquad \nu = q^*x \tag{1}$$

is said to be completely observable (or the pair (q^*, A) is said to be completely observable) if the identity $q^*e^{At}x_0 \equiv 0$ is possible only for $x_0 = 0$.

The definition can be restated in more intuitive terms if we consider all the "free" evolutions of the system (for an identically zero "input" μ and an arbitrary initial state x_0). A completely observable system is characterized by the property that the "output" $\nu_l(t)$ corresponding to a free evolution is identically zero only if the state of the system x coincides with stationary state $x = 0$ throughout the free evolution considered.

The property stated in Definition 1 is obtained from Property 15°, Theorem 1, § 31 if one replaces A by A^* and b by q. Hence, the complete observability of the pair (q^*, A) is equivalent to the complete controllability of the pair (A^*, q) a fact which expresses a relation of duality between controllability and observability, as pointed out by R. E. Kalman. It is sufficient to replace A by A^* and b by q in any of the properties stated in Theorem 1, § 32 in order to obtain new formulations of the property of complete observability. Such a simple rule makes it unnecessary to formulate explicitly all the various equivalent forms of the property of complete observability.

---

[1]) The notations are the same as in (1, § 31). The added symbol q represents an n-vector.

In this book, only some of these properties are used. They are stated in "Consequence 1" below and bear the same number as the corresponding properties of Theorem 1, § 31.

Consequence 1. *The property "the pair* (q^*, A) *is completely observable" is equivalent to any of the properties obtained by replacing* A *by* A^* *and* b *by* q *in the properties of Theorem 1, § 31. In particular this property is equivalent to the following properties :*

1°. *There does not exist a non-singular matrix* R *for which the matrix* $\widetilde{A} = RAR^{-1}$ *and the vector* $\widetilde{q} = (R^{-1})^* q$ *have the form*

$$\widetilde{A} = \begin{pmatrix} A_{11} & 0 \\ A_{21} & A_{22} \end{pmatrix}, \tag{2}$$

$$\widetilde{q}^* = (q_1^* \quad 0), \tag{3}$$

where A_{11} *is a* $k \times k$*-matrix* ($k < n$, *possibly* $k = 0$) *and* q *is a* k*-vector.*

4°. *The* $n \times n$*-matrix*

$$\mathcal{O} = (q \quad A^*q \quad (A^*)^2 \quad q \ldots (A^*)^{n-1} q)$$

is non-singular.

11°. *There exists a vector* b *for which the expression* $q^*(\sigma I - A)^{-1}b$ *is irreducible.*

12°. *There does not exist a non-zero vector* b *such that the expression* $q^*(\sigma I - A)^{-1}b$ *is identically zero.*

16°. *The matrix*

$$W(t_1, t_2) = \int_{t_2}^{t_1} e^{A^*t}\, q\, q^*\, e^{At}\, dt \tag{4}$$

is positive definite for any numbers t_1 *and* $t_2 > t_1$.

As in § 31, Section 3, we derive from Consequence 1 the following proposition :

Proposition 1. *Assume that the pair* (q^*, A) *is not completely observable and that* q *is a non-zero vector. Then there exists a non-singular matrix* R *such that the matrix* $\widetilde{A} = RAR^{-1}$ *and the vector* $q = (R^{-1})^* q$ *have the form (2), (3) and the pair* $(q_1, A_{11})^*$ *is completely observable. At the same time, if* *one applies the transformation* $\widetilde{x} = Rx$ *and one writes the vector* $\widetilde{x}$, *— in agreement with the formulas (2) and (3) —, in the form*

$$\widetilde{x} = \begin{pmatrix} y \\ z \end{pmatrix}, \tag{5}$$

then Eq. (1) can be written as

$$\frac{dy}{dt} = A_{11}\, y + b_1\, \mu(t) \tag{6}$$

$$\frac{dz}{dt} = A_{21}\, y + A_{22}\, z + b_2\, \mu(t) \tag{7}$$

$$\nu = q_1^*\, y, \ (q_1^*, A_{11}) = \textit{completely observable pair.} \tag{8}$$

§ 33. Nondegenerate systems

1. Definition of the property of nondegeneration and statement of the theorem of nondegeneration

Definition 1. *The system*

$$dx/dt = Ax + b\mu(t), \quad \nu = q^*x \tag{1}$$

is said to be non degenerate if it is completely controllable and completely observable (i.e., if both the pair (A, b) *and the pair* (A^*, q) *are completely controllable.)*

The usefulness of the concept of nondegeneration is evinced by the following theorem :

Theorem 1. *The property "the system (1) is nondegenerate" is equivalent to any of the following properties :*

(a) *The transfer function of the system (1) (equal to* $q^*(\sigma I - A)^{-1}b$*) is irreducible and the degree of the denominator is equal to n (the order of the system).*

(b) *If an arbitrary system of the same form (1)*

$$d\widetilde{x}/dt = \widetilde{A}\widetilde{x} + \widetilde{b}\,\widetilde{\mu}(t), \quad \widetilde{\nu} = \widetilde{q}^*\,\widetilde{x} \tag{2}$$

has the same transfer function and the same order [1]) *as (1), then there exists a non-singular matrix R such that the parameters of the two systems are connected by the relations*

$$\tilde{A} = RAR^{-1}, \quad \tilde{b} = Rb, \quad \tilde{q} = (R^{-1})^* q. \tag{3}$$

2. Remarks on the theorem of nondegeneration

Before proving Theorem 1 we shall discuss the significance of the condition (b).

In the set of all the systems of the form (1) we introduce the relation R_n defined as follows :

"*Relation* R_n": *The system (2) is said to be in relation* R_n *the system (1) if there exists a non-singular matrix R such that we have the relations (3).*

It is easy to see that R_n is reflexive, symmetric and transitive. Hence it is an equivalence relation and allows the partition of the set of all the systems of the form (1) in classes of equivalent systems. These classes are endowed with the following property : "All the systems belonging to the same class have the same transfer function ". Indeed, by using Relations (3) one obtains successively

$$\tilde{q}^* (\sigma I - \tilde{A})^{-1} \tilde{b} = q^* R^{-1} (\sigma I - RAR^{-1})^{-1} Rb =$$

$$= q^* R^{-1} (R (\sigma I - A) R^{-1})^{-1} Rb = q^* (\sigma I - A)^{-1} b. \tag{4}$$

Property (b) of Theorem (1) shows that in the case of the nondegenerated systems the inverse property is also valid.

Property (a) of Theorem 1 supplies an effective criterion for establishing when a system is nondegenerate : it is sufficient to write that the resultant of the polynomials constituting the numerator and denominator of the transfer function is different from zero [2]).

3. Proof of the theorem of nondegeneration

We shall show that Property (a) of Theorem 1 is equivalent to the property from Definition 1. Since $q^*(\sigma I - A)^{-1}b$ is irreducible, one deduces from Property (a) with the help of

---

[1]) I. e. if the vectors x and $\tilde{x}$ have the same number of components.

[2]) For a simplified calculation of the resultant the Bezout method will be found convenient (see the author's paper [1]).

Property 11° of Theorem 1, § 31 that the pair (A, b) is completely controllable. Similarly, using Property 11° from Consequence 1, § 32 one deduces that the pair (q^*, A) is completely observable. Thus the property from Definition 1 results from Property (a). In order to prove the converse statement let us assume that a system of the form (1) has simultaneously the following properties: α) it is completely controllable; β) it is completely observable and γ) its transfer function is reducible. We shall show that these assumptions lead to a contradiction. The system being completely controllable, we can find [1]) a non-singular matrix R such that (40, § 31) holds or

$$(\sigma I - A) R v (\sigma) - b \det (\sigma I - A) = 0. \tag{5}$$

Since both terms of (5) are polynomials, the relation is valid for any σ (although (40, § 31) from which we have deduced Relation (5) makes no sense for the characteristic values σ_i which satisfy the equation $\det(\sigma_i I - A) = 0$). Since, by hypothesis, the transfer function $q^*(\sigma I - A)^{-1}b$ is reducible it follows — using (40, § 31) — that the expression $q^* R v(\sigma)/\det(\sigma I - A)$ is reducible, in other words there exists a number σ_0 such that

$$\det (\sigma_0 I - A) = 0 \tag{6}$$

$$q^* R v (\sigma_0) = 0. \tag{7}$$

From (5) and (6) it follows that $(\sigma_0 I - A) R v(\sigma_0) = 0$ and hence $Rv(\sigma_0)$ is an eigenvector of the matrix A; Relation (7) shows that this eigenvector and the vector q are orthogonal; hence the pair (A^*, q) is not completely controllable (see Property 3° of Theorem 1, § 31), in other words the pair (q^*, A) is not completely observable. This contradicts the assumption(β) above and concludes the proof that Property (a) is equivalent to the property from Definition 1.

We shall show that Property (b) of Theorem 1 also is equivalent to the property from Definition 1. It is easy to see that Property (b) implies that the system is nondegenerate in the sense of Definition 1. Indeed, by Property 1° of Theorem 1, § 31, any pair (A, b) which is not completely controllable can be brought to the form (2, § 31) and then the corresponding

[1]) See Relation (40, § 31) and the corresponding footnote.

system takes the form (19, § 31). We complete the system by the relation $\nu = q^*_1 y + q^*_2 z$ and we easily see that the transfer function of the system thus obtained does not depend on the matrix A_{22}. Let us consider two systems of such a form differing only by the matrix A_{22} and having therefore the same transfer function. We may choose the matrices A_{22} of the two systems so that they have no common eigenvalues. Since the eigenvalues of the matrix A_{22} are at the same time eigenvalues of the matrix $\tilde{A}$ (see (2, § 31)) the matrices $\tilde{A}$ of the two systems have different eigenvalues and hence cannot be connected through any relation of the form (3). Hence, Property (b) cannot occur if the system is not completely controllable. Similarly, by bringing the pair (A, q) to the form (2, § 32), (3, § 32) of Consequence 1, § 32, we can show that Property (b) cannot occur if the system is not completely observable.

Conversely, if an arbitrary system of the form (1) is nondegenerate, then Property (b) of Theorem 1 is valid. Indeed, assume that two systems of the form (1) and (2) are nondegenerate and have the same transfer function , i.e.

$$q^* (\sigma I - A)^{-1} b \equiv \tilde{q}^* (\sigma I - \tilde{A})^{-1} \tilde{b} \tag{8}$$

Then by the result proved above, Property (a) is satisfied for the two systems and hence both members of the identity (8) are irreducible. Using this and the proportionality of the denominators of the two members of the identity (10), one obtains the equality

$$\det (\sigma I - A) = \det (\sigma I - \tilde{A}). \tag{9}$$

Since the systems (1) and (2) are nondegenerate, the pairs (A, b) and $(\tilde{A}, \tilde{b})$ are completely controllable. Taking this and Eq. (9) into account and applying Property 6° of Theorem 1, § 31, one obtains the first two relations of (3). With the help of these two relations one may write the right-hand member of (8) as

$$\tilde{q}^* (\sigma I - RAR^{-1})^{-1} R = \tilde{q}^* R (\sigma I - A)^{-1} b, \tag{10}$$

Introducing this relation in (8) one obtains the identity

$$(q^* - \tilde{q}^* R) (\sigma I - A)^{-1} b \equiv 0. \tag{11}$$

Since the pair (A, b) is completely controllable, from (11) and from Property 12° of Theorem 1, § 31 one derives the equality $q^* - \tilde{q}^* R = 0$, from which the last of Relations (3) follows.

4. Bringing nondegenerate systems into the Jordan-Lur'e-Lefschetz form

Property (b) of Theorem 1 can be used in order to determine rapidly whether two distinct systems of the form (1), (2) can be derived one from the other by means of some relations of the form (3); to this end it is sufficient to check whether the transfer functions of the two systems are irreducible and equal. This property may be also used as an effective means of bringing a system of the form (1) into certain particular forms which are useful for simplifying the treated problems. As an illustration we shall bring the system (1), assumed to be nondegenerate, to the Jordan-Lur'e-Lefschetz form, i.e. to a form in which the pair (A, b) has the form from Property 8°, Theorem 1, § 31.

Consider a system of the form (2) where the pair $(\tilde{A}, \tilde{b})$ can be written as in Property 8° of Theorem 1, § 31. Putting the vectors $\tilde{x}$ and $\tilde{q}$ in the form

$$\tilde{x}' = (x_1' \ \ x_2' \ \ldots \ x_k'); \quad \tilde{q}_k^* = (q_1^* \ \ q_2^* \ \ldots \ q_k^*), \tag{12}$$

where x_j and q_j are vectors with n_j components, one can write systems (2) in the form

$$\frac{\mathrm{d}x_j}{\mathrm{d}t} = A_j x_j + b_j \tilde{\mu}(t), \tag{13}$$

$$\nu = \sum_{j=1}^{k} q_j^* x_j, \tag{14}$$

where the pairs (A_j, b_j) are given by the formulas (11, § 31), (12, § 31). Taking into account these formulas and introducing the components of the vectors q_j^* by the equalities

$$q_j^* = (q_{j1} \ \ q_{j2} \ \ldots \ q_{jn_j}),$$

one obtains the transfer function of the system (13), (14)

$$\tilde{q}^* (\sigma I - \tilde{A})^{-1} \tilde{b} = \sum_{j=1}^{k} \left(\frac{q_{j1}}{(\sigma - \lambda_j)^{n_j}} + \frac{q_{j2}}{(\sigma - \lambda_j)^{n_j - 1}} + \dots + \frac{q_{jn_j}}{\sigma - \lambda_j} \right)$$

$$\lambda_j \neq \lambda_l \text{ si } j \neq l. \tag{15}$$

From the foregoing it follows that between systems of the Jordan-Lur'e-Lefschetz form and the transfer functions of the form (15) there exists a one-to-one correspondence. A transfer function of the form (15) being given, one can immediately determine the coefficients of the corresponding systems of the Jordan-Lur'e-Lefschetz form.

Let us consider an arbitrary nondegenerate system of the form (1) and write its transfer function as a sum of simple fractions, thus obtaining an expression similar to (15). In accordance with the preceding remarks one can directly write a system of the Jordan-Lur'e-Lefschetz form, whose transfer function is equal to the function (15) corresponding to the system examined. The order of the system will be equal to the order of the initial system. By Property (b) of Theorem (1) there exists a matrix R such that the two systems are connected by relations of the form (3).

The same procedure may be applied in order to bring the nondegenerate system considered to any desired particular form. To this end it is sufficient to write the special form to which we want to bring the system, and to leave the coefficients of the form undetermined; one then writes the corresponding transfer function. By identifying the transfer function thus obtained with the transfer function of the given system, one obtains a set of equations; if the latter can be solved and if the two systems are of the same order, one deduces, by using Property (b) of Theorem 1, that the system can be expressed in the desired form. Simultaneously one finds the values of the coefficients of the transformed system, which initially were undetermined.

§ 34. Controllability of multi-input systems

1. Definition and theorem of the complete controllability of multi-input systems

Most of the results obtained in the study of single-input systems can be extended to multi-input systems of the form

$$\mathrm{d}x/\mathrm{d}t = Ax + Bu(t) \tag{1}$$

where A is an $n \times n$-matrix, B is an $n \times m$-matrix, x is an n-vector and the control u is a vector-valued piecewise continuous function. The quantities $x(t)$ and $u(t)$ as well as the entries of A and B may take complex values.

The property of complete controllability of this system is defined by a direct extension of Definition 1, § 31 :

Definition 1. *The system (1) is said to be completely controllable (or the pair (A, B) is said to be completely controllable), if for any vector x_0 and any interval $t_1 < t < t_2$ there exists a control function u_{x_0} continuous and defined in the interval $t_1 \leqslant \leqslant t \leqslant t_2$ such that the solution of the system (1) for $u = u_{x_0}$ and $x(t_1) = 0$ satisfies the condition $x(t_2) = x_0$.*

The property of complete reachability is defined as in the case of single-input systems and it is shown that the system (1) is completely controllable if and only if it is completely reachable.

The property of controllability may be expressed in various other equivalent ways similar to these of Theorem 1, § 31; however, it should be noted that not all the properties of that theorem can be directly extended to the case of multi-input systems. Theorem 1, § 31 is replaced by the theorem which follows and where the different properties bear the same number as the corresponding properties of Theorem 1, § 31. We replace by dots the properties whose extension to be multi-input case is not given. More complete results are available but not needed in this book.

Theorem 1. *The property "the system (1) is completely controllable" (or "the pair (A, B) is completely controllable") is equivalent to any of the following properties:*

1°. *There does not exist a non-singular matrix R such that the transformed pair $\tilde{A} = RAR^{-1}$, $\tilde{B} = RB$ can be written in the form*

$$\tilde{A} = \begin{pmatrix} A_{11} & A_{12} \\ 0 & A_{22} \end{pmatrix}; \quad B = \begin{pmatrix} B_1 \\ 0 \end{pmatrix}, \tag{2}$$

where A_{11} is a $k \times k$-matrix ($k<n$, possibly $k = 0$), and B is a $k \times m$-matrix (it follows that A_{12} is a $k \times (n - k)$ matrix and A_{22} is a $(n - k) \times (n - k)$-matrix).

2°. *There exists no subspace invariant under matrix A, of dimension smaller than n, containing simultaneously all the vectors formed with the columns of matrix B.*

3°. *There exists no eigenvector x of the matrix A^* which satisfies the relation $x^*B = 0$.*

4°. The matrix with n rows and nm columns

$$C = (B \quad AB \quad A^2B \ldots A^{n-1} \; B) \tag{3}$$

is of rank n.

. . . .

7°. If the permutation of the components [1]) $u_1, u_2 \ldots u_m$ *of the control function u is admitted, then there exists a non-singular matrix* R_1 *for which the transformed matrices* $\tilde{A} = RAR^{-1}$ *and* $\tilde{B} = RB$ *have the form* [2])

$$\tilde{A} = \begin{pmatrix} A_{11}^0 & A_{12} & \ldots & A_{1l} \\ 0 & A_{22}^0 & \ldots & A_{2l} \\ \cdot & \cdot & \cdots & \cdot \\ 0 & 0 & \ldots & A_{ll}^0 \end{pmatrix}; \; \tilde{B} = \begin{pmatrix} b_1^0 & 0 & \ldots & 0 & B_1 \\ 0 & b_2^0 & \ldots & 0 & B_2 \\ \cdot & \cdot & \cdots & \cdot & \cdot \\ 0 & 0 & \ldots & b_l^0 & B_l \end{pmatrix}. \tag{4}$$

where every pair A_{jj}^0, b_j^0 *has the structure given by the formulas (9, § 31), and in the matrices* A_{jk}, $j < k$ *only the entries on the first column may be non-zero, all the other entries being zero. (In the expressions (4), l is a positive integer not exceeding m,* A_{jj}^0 *are* $n_j \times n_j$*-matrices* $(n_1 + n_2 + \ldots + n_l = n)$, A_{jk} *are* $n_j \times n_k$*-matrices,* b_j^0 *are* n_j*-vectors and* B_j *are* $n_j \times (m - k)$*-matrices).*

8°. If R is a nonsingular matrix such that the transformed matrix $\tilde{A} = RAR^{-1}$ *has the Jordan normal form, then, by writing the transformed matrices* $\tilde{A} = RAR^{-1}$ *and* $\tilde{B} = RB$ *in the form*

$$\tilde{A} = \begin{pmatrix} A_1 & 0 & \ldots & 0 \\ 0 & A_2 & \ldots & 0 \\ \cdot & \cdot & \cdots & \cdot \\ 0 & 0 & \ldots & A_l \end{pmatrix}, \quad \tilde{B} = \begin{pmatrix} B_1 \\ B_2 \\ \cdot \\ B_l \end{pmatrix}, \tag{5}$$

where A_j *are Jordan cells of the form (11, § 31) the matrices* B_j *thus obtained have the following property: if p cells* $A_{j_1}, A_{j_2} \ldots A_{j_p}$ *have the same eigenvalue* $(\lambda_{j_1} = \lambda_{j_2} = \ldots = \lambda_{j_p})$ *then the rank of the matrix* $(f_{j_1} f_{j_2} \ldots f_{j_p})$ *(where* $f_{j_k}^*$ *represents the last row of the matrix* B_{j_k}, $k = 1, 2, \ldots, p$*), is equal to p.*

---

[1]) The result may be immediately extended to the case where such a permutation is not admitted, but the form of the matrix B becomes less regular than in (4).

[2]) Another particularly useful form to which one can bring the matrices A and B has been given by P. Brunovski (1).

9°. *Let* π_1 *be a polynomial of degree* n, *of the form (14, § 31) (with the leading coefficient equal to the unity). Then there exists an* $n \times m$-*matrix* Q_0 *such that*

$$\det(\sigma I - A - BQ_0^*) = \pi_1(\sigma) \tag{6}$$

(in words: the characteristic polynomial of the matrix $(A + BQ_0^*)$ *is equal to the given polynomial).*

. . . .

12°. *There does not exist a non-zero vector* q *such that*

$$q^*(\sigma I - A)^{-1} B \equiv 0. \tag{7}$$

13°. *For every set of distinct complex numbers* σ_j *different from the eigenvalues of the matrix* A, *the matrix with* n *rows and* nm *columns*

$$((\sigma_1 I - A)^{-1}B \quad (\sigma_2 I - A)^{-1}B \ \ldots \ (\sigma_n I - A)^{-1}B) \tag{8}$$

is of rank n.

14°. *There exist* n *complex numbers* σ_j *such that the rank of the matrix in (8) equals* n.

15°. *The identity* $q^* e^{At} B \equiv 0$ *in an interval* $t_1 < t < t_2$ *is possible only for* $q = 0$.

16°. *For every pair of numbers* t_1 *and* $t_2 > t_1$ *the matrix*

$$W(t_1, t_2) = \int_{t_1}^{t_2} e^{At} B B^* e^{A^*t} \, dt \tag{9}$$

is positive definite.

2. Proof of Theorem 1

The greatest part of the proof is based on ideas already used in the proof of Theorem 1, § 31. We examine in detail only those parts of the proof which contain new elements.

The proof follows the scheme of implications

$$(\text{Def}) \to 1° \to 2° \to 3° \to 4° \to 7° \to 8° \to 9° \to 12° \to 13° \to \\ \to 14° \to 15° \to 16° \to (\text{Def}) \tag{10}$$

The symbol (Def) and the numbers $k°$ refer to Definition 1 and Theorem 1. The first four implications of scheme (10) are proved in exactly the same way as the corresponding implications of scheme (22, § 31). Therefore we shall start with the proof of the implication 4° → 7°. We shall show that if Property

4° is satisfied, then one can find the integers $0 = p_0 < p_1 < p_2 < \ldots < p_l = n (l \leqslant m)$ and the sequence of linearly independent vectors $e_1, e_2, \ldots, e_n$ which satisfy

$$\left.\begin{aligned} e_{p_j} &= b_j \\ e_{p_j-1} &= (A - \gamma_{j,n_j} I) b_j \\ e_{p_j-2} &= (A^2 - \gamma_{j,n_j} A - \gamma^{j,\, n_j-1} I)\, b_j, \\ &\ldots\ldots\ldots\ldots\ldots\ldots\ldots\ldots \\ e_{p_j-(n_j-1)} &= A^{n_j-1} - \gamma_{j,n_j} A^{n_j-2} - \ldots - \gamma_{j,2} E) b_j, \end{aligned}\right\} \begin{aligned} & j = 1, 2, \ldots, l \\ & (n_j = p_j - p_{j-1}), \end{aligned} \tag{11}$$

$$(A^{n_j} - \gamma_{j,n_j} A^{n_j-1} - \gamma_{j,n_j-1} A^{n_j-2} - \ldots - \gamma_{j,1} E)\, b_j = \sum_{k=1}^{p_{j-1}} \beta_{j,k}\, e_k \tag{12}$$

The vectors $b_1, b_2, \ldots, b_l$ from (11) are the corresponding columns of the matrix B, possibly after a suitable permutation to be described below. $\gamma_{j,k}$ and $\beta_{j,k}$ are constants. The relations (11) are written in the decreasing order of the indices of the vectors e_k so as to point out the rule by which these vectors are formed. In the case $l = 1$, Relations (11) have been used by S. Lefschetz in [3] in order to bring single-input systems to the form of Property 7°, Theorem 1, § 31.

The fact that we can find a set of vectors e_k satisfying Conditions (11) and (12) will be proved by induction. We assume that Relations (11) and (12) are satisfied for any integer $j \leqslant i-1$ and that inequality $p_{i-1} < n$ is also satisfied; we define the integer $p_i > p_{i-1}$ and the vectors $e_{p_{i-1}+1}$, $e_{p_{i-1}+2}, \ldots, e_{p_i}$ so that Relations (11) and (12) be satisfied for $j=1$. The reasonings will be also valid for the first step of the induction.

We remark first that the inequality $i - 1 < m$ must be satisfied and that among the last $m - i + 1$ columns of the matrix B there must exist at least one column which — considered as a vector — is not a linear combination of the vectors e_k, $k = 1, 2, \ldots, p_{i-1}$ determined in the previous steps [1]). Indeed, if all the columns of the matrix B were linear combinations of the vectors e_k, $k = 1, 2, \ldots, p_{i-1}$, then, by (11) and (12) written for $j = 1, 2, \ldots, i - 1$ all the columns of the matrices $A^k B$ — for any positive integer k — would be linear combina-

[1]) On the first step of the induction process (for $i = 1$) this amounts to the assertion: "there exists a column of the matrix B which contains at least one non-zero element".

tions of the vectors e_k, $k = 1, 2, \ldots, p_{i-1}$ whose number is smaller than n owing to the inequality $p_{i-1} < n$. Hence, the rank of the matrix C, (3) would be smaller than n, contrary to Property 4°. By permuting, if necessary, the order of the last $m - i + 1$ columns of the matrix B, we may obtain that the vector b_i formed by the column i of the matrix B is not a linear combination of the vectors e_k, $k = 1, 2, \ldots, p_{i-1}$. Considering the sequence of vectors b_i, $Ab_i, \ldots, A^k b_i$, one can find an integer n_i such that the vectors

$$e_1, e_2, \ldots, e_{p_{i-1}}, b_i, Ab_i, \ldots, A^{n_i - 1} b_i \tag{13}$$

are linearly independent and the vector $A^{n_i} b_i$ is a linear combination of the vectors in (13), of the form (12) for $j = i$. Using the coefficients γ of (12), thus obtained, one can then define the vectors e_k, $k = p_{i-1} + 1, p_{i-1} + 2, \ldots, p_i$ (where $p_i = p_{i-1} + n_i$) using (11) for $j = i$. The vectors e_k, $k = 1, 2, \ldots, p_i$ thus determined are linearly independent since if we could find a linear combination of these vectors equal to zero and having at least one non-zero coefficient, then, replacing the vectors e_k, $k = p_{i-1} + 1, p_{i-1} + 2, \ldots, p_i$ by their expressions taken from (11) for $j = i$, we would find a linear combination of the vectors (13) equal to zero and having at least one non-zero coefficient thus contradicting the fact that the vectors (13) are linearly independent.

By successively applying the above procedure, the number of the vectors e_k increases at each step by at least one unit. After a finite number of steps we shall obtain the greatest possible number of linearly independent vectors e_k after which the method can no longer be applied. The final step l is characterized by the equality $p_l = n$.

From (11) and (12) we derive the following equalities

$$e_{p_j} = b_j, \tag{14}$$

$$\left.\begin{aligned} Ae_{p_j} &= e_{p_j - 1} + \gamma_{j, n_j} e_{p_j}, \\ Ae_{p_j - 1} &= e_{p_j - 2} + \gamma_{j, n_j - 1} e_{p_j}, \\ &\ldots\ldots\ldots\ldots\ldots\ldots \\ Ae_{p_j - (n_j - 1)} &= \gamma_{j,1} e_{p_j} + \sum_{k=1}^{p_{j-1}} \beta_{j,k} e_k, \end{aligned}\right\} \tag{15}$$

for $j = 1, 2, \ldots, l$.

From (14) and (15) it follows that the matrices $\tilde{A}$ and $\tilde{B}$ written in the basis $e_1, e_2, \ldots, e_n$, have the form described in Property 7°. The matrix R which occurs in Property 7° has the expression

$$R^{-1} = (e_1 \;\; e_2 \; \ldots \; e_n). \tag{16}$$

Proof of implication $7° \to 8°$. We shall prove the chain of implications $7° \to 3° \to 8°$. First we prove that $7° \to 3°$. Indeed, if 3° is not satisfied there exists a vector x, such that

$$x^* \tilde{A} = \alpha\, x^*, \qquad x^* \tilde{B} = 0, \qquad x \neq 0. \tag{17}$$

The vector x can be also written in the form

$$x^* = (x_1^* \; x_2^* \; \ldots \; x_l^*), \tag{18}$$

where the vectors x_j have the same number of components as the corresponding vectors b_j^0 which occur in the expression of matrix $\tilde{B}$, (4). Since the components of the vectors b_j^0 are equal to zero except the last component which is equal to unity (see the statement of Property 7°), the equality $x^* \tilde{B} = 0$ of (17) may occur only if each of the vectors x_j has its last component equal to zero. Using this conclusion and taking into account the fact that the matrices A_{jj}^0 of (4) have the form (9, § 31) (see the statement of Property 7°), we find that the equality $x^* \tilde{A} = \alpha x^*$ may occur only if $x = 0$, which contradicts the last of Relations (17) and proves that $7° \to 3°$. It remains to prove that $3° \to 8°$. We bring the matrices A and B to the form (5) and we consider p Jordanean cells from expression (5) of $\tilde{A}$ having the same eigenvalue λ_0. Permuting if necessary the order of the cells of matrix $\tilde{A}$, we may assume that the cells $A_1, A_2, \ldots, A_p$ have the same eigenvalue. We denote by $f_1^*, f_2^*, \ldots, f_p^*$ the row vectors given by the last lines of the matrices $B_1, B_2, \ldots, B_p$. If Property 8° is not satisfied, then, after a convenient permutation of the cells of the matrix $\tilde{A}$ one can choose an eigenvalue λ_0 and the corresponding cells $A_1, \ldots, A_p$, as above, such that the rank of the matrix $(f_1 \; f_2 \ldots \ldots, f_p)$ is smaller than p. Hence, there exist some numbers $\alpha_1, \alpha_2, \ldots, \alpha_p$, not all zero, such that

$$(\alpha_1 \; \alpha_2 \ldots, \alpha_p)\,(f_1 \; f_2 \ldots f_p)^* = 0. \tag{19}$$

Let n_j be the number of rows and columns of matrix A_{jj}, $j = 1, 2, \ldots, p$. Using the formulas (5) and (11, § 31) one finds that the vector

$$x^* = (\overbrace{0\,0 \,\ldots\, \alpha_1}^{n_1} \; \overbrace{0\,0 \ldots \alpha_2}^{n_2} \ldots \overbrace{0\,0 \,\ldots\, \alpha_p}^{n_p} \; \overbrace{0\,0 \,\ldots\ldots\, 0}^{n - \sum_{j=1}^{p} n_j}),$$

$$(\max_j (|\alpha_j|) \neq 0) \tag{20}$$

satisfies the equality

$$x^*A = \lambda_0 x^*, \tag{21}$$

λ_0 being the common eigenvalue of the first p cells. The vector x^*B reduces to the expression in the left-hand member of Eq. (19) and hence

$$x^*B = 0. \tag{22}$$

Since $x \neq 0$, (21) and (22) show that Property 3° is not satisfied.

Proof of implication 8° → 9°. We prove the chain of implications 8° → 3° → 7° → 9° of which only the implications 8° → 3° and 7° → 9° have not been proved. The implication 8° → 3° is proved by remarking that if the first p Jordanean cells of the matrix $\tilde{A}$, (5), have the same eigenvalue λ_0 and if there does not exist another cell with the same eigenvalue, then any eigenvector of the matrix $\tilde{A}$, corresponding to the eigenvalue λ_0, has the form (20). We write the product $x^*\tilde{B}$ in the form (19) and since, by virtue of Property 8° the matrix $(f_1 f_2 \ldots f_p)$ is of rank p, it follows that $x^*\tilde{B} \neq 0$. By conveniently permuting the order of the cells this conclusion can be extended to any group of cells with the same eigenvalue regardless of their position in the matrix $\tilde{A}$. Hence, any eigenvector x of the matrix $\tilde{A}^*$ satisfies the inequality $x^*\tilde{B} \neq 0$. Using relations $\tilde{A} = RAR^{-1}$ and $\tilde{B} = RB$, we immediately find out that any eigenvector x of matrix A^* verifies the inequality $x^*B \neq 0$.

This proves that 8° → 3°. In order to prove that 7° → 9° consider an $n \times m$-matrix $\tilde{Q}$ of the form

$$\tilde{Q} = \begin{pmatrix} q_1 & 0 & \ldots & 0 & 0 \\ 0 & q_2 & \ldots & 0 & 0 \\ \cdot & \cdot & \cdots & \cdot & \cdot \\ 0 & 0 & \ldots & q_l & 0 \end{pmatrix}, \tag{23}$$

where q_j has the same number of dimensions as b_j^0, $j = 1, 2, \ldots, l$, and the last column of (23) (containing only zero terms) has the same number of dimensions as the last column of the expression of $\tilde{B}$, (4). Using Relations (4) one sees that the characteristic polynomial of the matrix $\tilde{A} + \tilde{B}\tilde{Q}^*$ is equal to the product of the characteristic polynomials of the matrices $A_{jj}^0 + b_j^0 q_j^*$. The pairs (A_{jj}^0, b_j^0) have the form of Property 7°, Theorem 1, § 31. Applying the equivalent Property 9° of Theorem 1, § 31, we deduce that the vectors q_j can be chosen so that the characteristic polynomial of the matrix $A^0_{jj} + b^0_j q^*_j$ be equal to any fixed beforehand polynomial of the form (14, § 31). Therefore all the roots of the characteristic equation of the matrix $\tilde{A} + \tilde{B}\tilde{Q}^*$ can be fixed beforehand. Using the relations $\tilde{A} = RAR^{-1}$ and $\tilde{B} = RB$ we find that the characteristic polynomial of the matrix $A + BQ^*$ becomes equal to an arbitrary polynomial of the form (14, § 31) if we choose the matrix Q (equal to $R^*\tilde{Q}$) adequately.

Proof of implication 9° → 12°. We prove the chain of implications 9° → 3° → 7° → 12° of which only the implications 9° → 3° and 7° → 12° are not yet proved. First we prove that 9° → 3°. If Property 3° is not satisfied, then there exists a vector x with the Properties (17). From (17) if follows that $x^*(A + BQ^*) = x^*A = \alpha x^*$ for every matrix Q. Hence, regardless of the value of matrix Q, the characteristic polynomial of the matrix $A + BQ^*$ has the root α, which shows that Property 9° cannot take place. It remains to prove that 7° → 12°. If Property 12° is not satisfied, then there exists a non-zero vector q which satisfies (7). Since Property 7° takes place, we can bring the pair (A, B) to the form $(\tilde{A}, \tilde{B})$ of Property 7° using a nonsingular matrix R. From (7) it follows directly that the vector

$$\tilde{q} = (R^{-1})^* \, q \neq 0 \tag{24}$$

satisfies the identity

$$\tilde{q}^* \, (\sigma I - \tilde{A})^{-1} \, \tilde{B} \equiv 0. \tag{25}$$

Writing the vector $\tilde{q}$ in the form

$$\tilde{q}^* = (q^*_1 \; q^*_2 \; \ldots \; q^*_l), \tag{26}$$

where q_j has the same number of dimensions as b^0_j, one obtains from (4) and (25) the identities $q^*_1(\sigma I_1 - A^0_{11})b^0_1 \equiv 0$ and :

$$\text{“if } q_s = 0 \text{ for } s < j, \text{ then } q^*_j(\sigma I_j - A^0_{jj})^{-1} b^0_j \equiv 0, j = 2, \ldots, l\text{”}, \tag{27}$$

where I_j is the identity matrix with the same dimensions as A^0_{jj}. But the pairs (A^0_{jj}, b^0_j) satisfy the Property 7° of Theorem 1, § 31 and hence they also satisfy the equivalent Property 12° of the same theorem. It follows that $q_1 = 0$ and — step by step — all the vectors q_j are zero and hence $\tilde{q} = 0$, which contradicts (24).

The last five implications of the scheme (10) are proved in exactly the same manner as the corresponding implications of Theorem 1, § 31.

3. Other properties of completely controllable multi-input systems

The following proposition is an analogue of the results given in § 31, Section 4:

Proposition 1. *If the pair* (A_0, B_0) *is completely controllable, then so is also any other pair* (A_1, B_1) *of the form*

$$A_1 = R_0(A_0 + B_0 Q_0^*) R_0^{-1}, \quad B_1 = R_0 B_0 K_0, \qquad (28)$$

where R_0 *is a nonsingular* $n \times n$*-matrix,* K_0 *is a non-singular* $m \times m$*-matrix and* Q_0 *is an* $n \times m$*-matrix.*

Indeed, for any $n \times m$-matrix Q one obtains from (28) the equality

$$A_1 + B_1 Q^* = R_0 (A_0 + B_0 Q_0^* + B_0 K_0 Q^* R_0) R_0^{-1},$$

and hence

$$\det (\sigma I - A_1 - B_1 Q^*) = \det \ (\sigma I - A_0 - B_0 (Q_0^* + K_0 Q^* R_0)).$$

Thus if Property 9° of Theorem 1 is satisfied for the pair (A_0, B_0), then it is also satisfied for the pair (A_1, B_1).

Note. 1°. It can be shown that if the pair (A_0, B_0) is not completely controllable, then any pair (A_1, B_1) of the form (28) has the same property.

2.° In contradistinction to the case of the single-input systems, given two completely controllable pairs (A_0, B_0) and (A_1, B_1) with the same dimensions, it cannot be asserted that the two pairs are connected by relations of the form (28). This becomes obvious if one remarks that the rank of matrix B_0 is equal to the rank of matrix B_1 given by (28), while the property of complete controllability does not set any condition upon the rank of matrix B.

We shall further prove a result which allows one to reduce the property of complete controllability of a multi-input systems to the complete controllability of a single system.

Proposition 2. *The $n \times n$-matrix A and the $n \times m$-matrix B constitute a completely controllable pair if and only if there exists an $n \times m$-matrix Q and an n-vector v such that the pair $(A + BQ^*, Bv)$ is completely controllable.*

Proof. If there exists a matrix Q and a vector v such that the pair $(A + BQ^*, Bv)$ is completely controllable, then the pair (A, B) is also completely controllable. For otherwise, by virtue of Property 1° of Theorem 1, the pair (A, B) can be brought to the form (2). Then any pair of the form $(\tilde{A} + \tilde{B}\tilde{Q}^*, \tilde{B}\tilde{v})$ has the form (2, § 31). By Property 1° of Theorem 1, § 31, no pair of this form is completely controllable. Since the pair $(\tilde{A} + \tilde{B}\tilde{Q}^*, \tilde{B}\tilde{v})$ can also be written in the form $(R(A + BQ^*)R^{-1}, RB\tilde{v})$ where $Q^* = \tilde{Q}^* R$, it follows from Theorem 2, § 31, that no pair of the form $(A + BQ^*, B\tilde{v})$ is completely controllable contrary to the hypothesis.

Conversely, if the pair (A, B) is completely controllable, then, by virtue of Property 9° of Theorem 1, one can find a matrix Q such that the roots of the characteristic equation of matrix $A_0 = A + BQ^*$ are all distinct. By Proposition 1 the pair (A_0, B) is also completely controllable. We bring the pair (A_0, B) to the form $(\tilde{A}_0, \tilde{B})$ from Property 8°, Theorem 1. The Jordan form $\tilde{A}_0$ of matrix A_0 will be a diagonal matrix whose diagonal entries are all distinct. By Property 8° of Theorem 1 each row of matrix $\tilde{B}$ has at least one non-zero entry. It is easy to see that under these conditions there exists a vector v such that all the components of the vector $\tilde{B}v$ are different from zero. Therefore the pair $(\tilde{A}_0, \tilde{B}v)$ is completely controllable (see Property 8° of Theorem 1, § 31). Using the relations

$$\tilde{A}_0 = R(A + BQ^*)R^{-1}, \quad \tilde{B}v = RBv,$$

we deduce by Theorem 2, § 31 that the pair $(A + BQ^*, Bv)$ is also completely controllable; this concludes the proof of Proposition 2.

The propositions which follow are proved by the same reasonings as in § 31, Section 3.

Proposition 3. *Suppose that the pair (A, B) is not completely controllable and that $B \neq 0$. Then there exists a non-singular matrix R such that the matrices $\tilde{A} = RAR^{-1}$ and $\tilde{B} = RB$ have the form (2) and the pair (A_{11}, B_1) is completely controllable. Furthermore, if one applies the transformation $\tilde{x} = Rx$ and one writes the vector $\tilde{x}$, in agreement with (2) and (3), as*

$$\tilde{x} = \begin{pmatrix} y \\ z \end{pmatrix}, \tag{29}$$

then Eq. (1) can be written as

$$\frac{dy}{dt} = A_{11}y + A_{12}\,z + B_1\,u(t),\ (A_{11},\ B_1) = \tag{30}$$

$$= \textit{completely controllable pair}$$

$$\frac{dz}{dt} = A_{22}\,z. \tag{31}$$

Proposition 4. *If $B \neq 0$, then for every m-dimensional vector a with the property $Ba \neq 0$ there exists a continuous function u such that the corresponding solution of equation (1) for $x(0) = 0$ satisfies the condition $x(1) = \det(\sigma I - A)^{-1}\,Ba$.*

In fact, Proposition 4 is a direct consequence of Proposition 2, § 31.

§ 35. Completely observable multi-output systems

By a direct extension of Definition 1, § 32 we obtain the following definition:

Definition 1. *The system*

$$dx/dt = Ax + Bu(t), \quad v = Q^*x \tag{1}$$

is said to be completely observable (or the pair (Q^*, A) is called completely observable) if the identity $Q^*e^{At}x_0 \equiv 0$, $(t > 0)$, is possible only for $x_0 = 0$.

Using Property 15° of Theorem 1, § 34 we obtain again the result that the property "the pair (Q^*, A) is completely observable" is equivalent to the property "the pair (A^*, Q) is completely controllable", in accordance with the duality relation of Kalman. Taking this into account, one obtains from Theorem 1, § 34 the following result:

Consequence 1. *The property: "the pair (Q^*, A) is completely observable" is equivalent to any of the properties obtained by replacing in the properties of Theorem 1, § 34, A by A^* and B by Q. In particular, this property is equivalent to the following properties:*

1°. *There does not exist a nonsingular matrix* R *for which the matrices* $\widetilde{A} = RAR^{-1}$ *and* $\widetilde{Q} = (R^{-1})^* Q$ *have the form*

$$A = \begin{pmatrix} A_{11} & 0 \\ A_{21} & A_{22} \end{pmatrix}, \tag{2}$$

$$\widetilde{Q}^* = (Q_1^* \;\; 0) = Q^* R^{-1} \tag{3}$$

where A_{11} *is a* $k \times k$*-matrix* $(k < n)$, *(possibly* $k = 0$*) and* Q_1 *is a* $k \times m$*-matrix.*

. . . .

3°. *There exists no eigenvector* x *of the matrix* A *satisfying the condition* $Q^*x = 0$.

4°. *The* $n \times nm$*-matrix*

$$\Theta = (Q \quad A^*Q \quad (A^*)^2 Q \; \ldots (A^*)^{n-1}Q)$$

is of rank n.

. . . .

12°. *There does not exist any vector* $b \neq 0$ *such that the expression* $Q^*(\sigma I - A)^{-1}b$ *is identically zero.*

. . . .

16°. *For every pair of numbers* t_1 *and* $t_2 > t_1$ *the matrix*

$$W(t_1, t_2) = \int_{t_1}^{t_2} e^{A^*t} Q Q^* e^{At} \, dt \tag{4}$$

is positive definite.

The proposition which follows corresponds to Proposition (3, § 34) :

Proposition 1. *Suppose that the pair* (Q^*, A) *is not completely observable and that* $Q \neq 0$. *Then there exists a non-singular matrix* R *such that the matrices* $\widetilde{A} = RAR^{-1}$ *and* $\widetilde{Q} = (R^{-1})^*Q$ *have the form (2) and the pair* (Q_1^*, A_{11}) *is completely controllable. Furthermore if one applies the transformation* $\widetilde{x} = Rx$ *and one writes the vector* $\widetilde{x}$, *in agreement with formulas (2) and (3), as*

$$x = \begin{pmatrix} y \\ z \end{pmatrix} \tag{5}$$

then Eqs. (1) can be written as

$$\frac{dy}{dt} = A_{11}\, y + B_1\, u\,(t) \tag{6}$$

$$\frac{dz}{dt} = A_{21} y + A_{22} z + B_2\, u\,(t) \tag{7}$$

$$v = Q_1 y,\ [(Q_1^*, A_{11}) = \text{completely observable pair.}] \tag{8}$$

§ 36. Special forms for multi-input blocks

If we were to follow the same order of exposition as in the case of single-input systems we should treat now the properties of the nondegenerate multi-input and multi-output blocks (completely controllable and completely observable). However, since throughout this book no use is made of these properties we describe instead a procedure of bringing general multi-input blocks to a special form (see Theorem 1 below) used several times in Chapters 3 and 4.

Consider a multi-input multi-output block of the form

$$\frac{dx}{dt} = Ax + Bu \tag{1}$$

$$v = C^*x + Du \tag{2}$$

where x is an n-vector, u is an m_2-vector and v is an m_1-vector. (It follows that the matrices from these equations have the following dimensions: $A : n \times n$; $B : n \times m_2$; $C : n \times m_1$; $D : m_1 \times m_2$.)

Consider a family of transformations of the following form: The variables u, x and v are replaced by the new variables $\tilde{u}$, $\tilde{x}$, $\tilde{v}$ by the relations

$$\tilde{u} = Pu - PQ^*x, \tag{3}$$

$$\tilde{x} = R\, x, \tag{4}$$

$$\tilde{v} = U\, v, \tag{5}$$

where the matrices introduced have the following dimensions : $P : m_2 \times m_2$; $Q : n \times m_2$; $R : n \times n$; $U : m_1 \times m_1$. Here P and R are nonsingular matrices and U is a unitary matrix, i.e. :

$$\det P \neq 0, \qquad \det R \neq 0, \qquad U^* = U^{-1}. \tag{6}$$

After applying the above transformation our block (1)—(2) becomes

$$\frac{\mathrm{d}\tilde{x}}{\mathrm{d}t} = \tilde{A}\,\tilde{x} + \tilde{B}\,\tilde{u} \tag{7}$$

$$\tilde{v} = \tilde{C}^*\,\tilde{x} + \tilde{D}\,\tilde{u} \tag{8}$$

where

$$\tilde{A} = R\,(A + B\,Q^*)\,R^{-1} \tag{9}$$

$$\tilde{B} = R\,B\,P^{-1} \tag{10}$$

$$\tilde{C}^* = U\,(C^* + D\,Q^*)\,R^{-1} \tag{11}$$

$$D = U\,D\,P^{-1}. \tag{12}$$

Moreover the transfer functions of the blocks (1)—(2) and (7)—(8), defined by the expressions

$$G\,(\sigma) = D + C^*\,(\sigma\,I - A)^{-1}\;B \tag{13}$$

$$\tilde{G}\,(\sigma) = \tilde{D} + \tilde{C}^*\,(\sigma\,I - \tilde{A})^{-1}\,\tilde{B}, \tag{14}$$

satisfy the relation

$$G\,(\sigma) = U^{-1}\,\tilde{G}\,(\sigma)\,P\,(I_m - Q^*\,(\sigma\,I - A)^{-1}\,B), \tag{15}$$

for every σ for which $\det(\sigma I - A) \neq 0$ and $\det(\sigma I - \tilde{A}) \neq 0$. (In (15) we denoted by I_m the $m_2 \times m_2$-identity-matrix). To establish (15), consider an arbitrary number σ satisfying the conditions $\det(\sigma I - A) \neq 0$ and $\det(\sigma I - \tilde{A}) \neq 0$ and note that (1) and (2) are satisfied by

$$u(t) = u_0\,\mathrm{e}^{\sigma t}, \quad x(t) = (\sigma\,I - A)^{-1}\,B\,u_0\,\mathrm{e}^{\sigma t}, \quad v(t) = G\,(\sigma)\,u_0\,\mathrm{e}^{\sigma t}, \tag{16}$$

for an arbitrary constant m_2-vector u_0. Using (3)—(5) one can find the corresponding functions which satisfy the transformed system (7)—(8). In particular from (3) one obtains

$$\widetilde{u}(t) = K\, u_0\, e^{\sigma t}, \quad K = P\,(I_m - Q^*\,(\sigma I - A)^{-1} B). \tag{17}$$

From (4) and (16) it follows that $\widetilde{x}(t)$ must have the form $Hu_0e^{\sigma t}$ where H is a constant matrix. But the only solution of (7) which has this form — for $\widetilde{u}(t)$ given by (17) — is $\widetilde{x}(t) = (\sigma I - \widetilde{A})^{-1}\widetilde{B}\, Ku_0e^{\sigma t}$. Therefore (see (8))

$$\widetilde{v}\,(t) = \widetilde{G}\,(\sigma)\, K\, u_0\, e^{\sigma t}. \tag{18}$$

Writing that the functions v and $\widetilde{v}$ given by (16) and (18) satisfy (5) and using the expression of K (see (17)) one obtains (15) (since u_0 is arbitrary).

If $m_1 = m_2$, $G(\sigma)$ and $\widetilde{G}(\sigma)$ are square matrices and one can compute their determinants, using (15). Remarking that det $U = 1$ (see (6)) and

$$\det\,(I_m - Q^*(\sigma I - A)^{-1}B) = \det\,(\sigma I - A - BQ^*)/\det\,(\sigma I - A)$$

(see (47, § 5) [1]) one obtains the relation

$$\det G(\sigma)\det(\sigma I - A) = \det\widetilde{G}(\sigma)\det(\sigma I - \widetilde{A})\det P, \text{ (if } m_1 = m_2) \tag{19}$$

which is true for every σ (as one can see, by continuity). We shall repeatedly make use of the following result

Lemma 1. *Let M be an $m_1 \times m_2$-matrix of rank $r_0 > 0$. Then there exist two unitary matrices U_0, $m_1 \times m_1$ and V_0, $m_2 \times m_2$, such that*

$$M = U_0\, D_0\, V_0 \tag{20}$$

where D_0 is a matrix of the form

$$D_0 = \begin{pmatrix} \overbrace{D_1}^{r_0} & \overbrace{0}^{m_2 - r_0} \\ 0 & 0 \end{pmatrix} \begin{matrix} \}\, r_0 \\ \}\, m_1 - r_0 \end{matrix}, \tag{21}$$

and D_1 is a nonsingular diagonal $r_0 \times r_0$-matrix whose diagonal entries are real and strictly positive.

[1]) For the reader who has not already followed § 5, we remark that (47, §5) directly follows from the matrix identity (46, § 5) and the latter can be easily verified.

Proof. We proceed as in F. R. Gantmacher [1] (see the proof of Theorem 8 of § 12, Chapter IX of the quoted book). Consider the $m_2 \times m_2$-matrix M^*M which is clearly hermitian and positive semidefinite. Hence one can find a complete orthonormal system of eigenvectors x_k such that

$$M^*M\, x_k = \rho_k^2 x_k, \quad x_k^* x_l = \delta_{kl}, \quad k, l = 1, 2, \ldots, m_2, \tag{22}$$

where δ_{kl} is the Kronecker delta. The eigenvalues ρ_k^2 are real and not all zero (since $r_0 \neq 0$). Changing, if necessary, the order of the eigenvectors x_k, one can find an integer $j \leqslant m_2$ such that $\rho_k^2 > 0$ for $k = 1, 2, \ldots, j$ and $\rho_k = 0$ for $j < k \leqslant m_2$. (If $j = m_2$ all the numbers ρ_k^2 are strictly positive.) Hence for $k > j$ one has $M^*Mx_k = 0$, and also $x_k^*\, M^*Mx_k = 0$, whence

$$Mx_k = 0, \quad \text{for } k = j+1, \ldots, m_2. \tag{23}$$

Define now the m_1-vectors z_k, $k = 1, 2, \ldots, j$ with the relations

$$Mx_k = \rho_k z_k, \qquad k = 1, 2, \ldots, j. \tag{24}$$

It is easy to see, that $z_k^* z_l = \delta_{kl}$, $k, l = 1, 2, \ldots, j$. If $j < m_1$ one can complete the system of vectors with other vectors $z_{j+1}, \ldots, z_{m_1}$ chosen so as to obtain an orthonormal system of vectors with the properties [1])

$$z_k^* z_l = \delta_{kl}, \qquad k, l = 1, 2, \ldots, m_1. \tag{25}$$

Introduce now the matrices

$$X = (x_1\ x_2 \ldots x_{m_2}), \quad U_0 = (z_1\ z_2 \ldots z_{m_1}), \tag{26}$$

with the dimensions $m_2 \times m_2$ and $m_1 \times m_1$ respectively.

From (22) and (25) one obtains $X^*X = I_2$ and $U_0^* U_0 = I_1$, where I_1 and I_2 are the identity-matrices of dimensions $m_1 \times m_1$ and $m_2 \times m_2$ respectively. Thus X and U_0 are unitary matrices. Now the Relations (23) and (24) can be written as

$$M\, X = U_0\, D_0 \tag{27}$$

[1]) It is sufficient to choose the vectors z_k, for $k > j$, as an orthonormal system in the subspace of the vectors z which satisfy the condition $M^*z = 0$. This subspace is clearly orthogonal to all the vectors z_k, $k = 1, 2, \ldots, j$. (see (24)).

where D_0 has the form (21) and $r_0 = j$ = rank of matrix M. The proof of Lemma 1 concludes with the remark that from (27) one obtains (20) for $V_0 = X^{-1}$ (the matrix V_0 is, like X, a unitary matrix).

Using Lemma 1 we prove

Proposition 1. *If the rank r_0 of the matrix D is different from zero, then there exists a transformation (3)—(5) which brings (1)—(2) to the form*

$$\frac{d\tilde{x}}{dt} = \tilde{A}\,\tilde{x} + B_0\,u_0 + B_1\,u_1 \qquad (28)$$

$$\tilde{v} = \begin{pmatrix} u_0 \\ C_1^* x \end{pmatrix} \begin{matrix} \} \, r_0 \\ \} \, m_1 - r_0 \end{matrix} \quad , \quad \tilde{u} = \begin{pmatrix} u_0 \\ u_1 \end{pmatrix} \begin{matrix} \} \, r_0 \\ \} \, m_2 - r_0 \end{matrix} \qquad (29)$$

(It follows that the matrices from these relations have the following dimensions: B_0: $n \times r_0$; B_1: $n \times m_2 - r_0$; C_1: $n \times m_1 - r_0$). If $r_0 = m_2$, then the component u_1 is absent. If $r_0 = m_1$ then the component $C_1^* x$ of (29) is absent [1]).

Proof. All we have to show is that there exists a transformed system (7)—(8) in which $\tilde{D}$ and $\tilde{C}$ have the form:

$$\tilde{D} = \begin{pmatrix} I_0 & 0 \\ 0 & 0 \end{pmatrix} \begin{matrix} \} \, r_0 \\ \} \, m_1 - r_0 \end{matrix} \quad ; \quad \tilde{C}^* = \begin{pmatrix} 0 \\ C_1^* \end{pmatrix} \begin{matrix} \} \, r_0 \\ \} \, m_1 - r_0 \end{matrix} \quad , \qquad (30)$$

where I_0 is the $r_0 \times r_0$-identity-matrix. Then if the vector $\tilde{u}$ and the matrix $\tilde{B}$ are partitioned as

$$\tilde{u} = \begin{pmatrix} u_0 \\ u_1 \end{pmatrix} \begin{matrix} \} \, r_0 \\ \} \, m_2 - r_0 \end{matrix} \quad , \quad \tilde{B} = (\overbrace{B_0}^{r_0} \quad \overbrace{B_1}^{m_2 - r_0})$$

one immediately obtains from (7)—(8) Eqs. (28) and (29).

The formulas (30) can be obtained by applying Lemma 1 for $M = D$. It follows that there exist two unitary matrices U_1 and V_1 such that [2])

$$U_1\,D\,V_1 = \begin{pmatrix} \overbrace{D_1}^{r_0} & \overbrace{0}^{m_2 - r_0} \\ 0 & 0 \end{pmatrix} \begin{matrix} \} \, r_0 \\ \} \, m_1 - r_0 \end{matrix} \quad , \qquad (31)$$

---

[1]) Obviously, $r_0 \leqslant \min(m_1, m_2)$.

[2]) We multiply Relation (20) on the left and — respectively—on the right by the unitary matrices $U_1 = U_0^{-1}$ and $V_1 = V_0^{-1}$ respectively.

where D_1 is a nonsingular diagonal $r_0 \times r_0$ matrix. Using (12) one finds that the first of Relations (30) is satisfied if one chooses the matrices U and P such that

$$U = U_1, \; V_1^{-1} P^{-1} = \begin{pmatrix} D_1^{-1} & 0 \\ 0 & I_2 \end{pmatrix} \begin{matrix} \} \, r_0 \\ \} \, m_2 - r_0 \end{matrix},$$

where I_2 is the identity-matrix with $m_2 - r_0$ rows and columns. Then, taking $R = I$, we can write (11) as

$$\widetilde{C}^* = U_1 \, C^* + U_1 \, D \, V_1 \, V_1^{-1} \, Q^*. \tag{32}$$

Using (31) and writing the matrices $\widetilde{C}^*$, $U_1 C^*$ and $V_1^{-1} Q^*$ in the form

$$\widetilde{C}^* = \begin{pmatrix} \widetilde{C}_0^* \\ \widetilde{C}_1^* \end{pmatrix}; \; U_1 C^* = \begin{pmatrix} C_0^* \\ C_1^* \end{pmatrix} \begin{matrix} \} \, r_0 \\ \} \, m_1 - r_0 \end{matrix};$$

$$V_1^{-1} Q^* = \begin{pmatrix} Q_0^* \\ Q_1^* \end{pmatrix} \begin{matrix} \} \, r_0 \\ \} \, m_2 - r_0 \end{matrix},$$

one splits (32) into two equalities : $\widetilde{C}_0^* = C_0^* + D_1 Q_0^*$ and $\widetilde{C}_1^* = C_1^*$. Since D_1 is nonsingular, one can choose Q_0 such that $Q_0^* = - D_1^{-1} C_0^*$; then $\widetilde{C}_0^* = 0$. Hence $\widetilde{C}$ takes the form from (30) and Proposition 1 is proved.

We consider now a block of the form obtained from (28)—(29) by retaining only the last $m_1 - r_0$ components of the output vector v, that is (simplifying the notations)

$$\frac{\mathrm{d}x}{\mathrm{d}t} = A \, x + B_0 \, w_0 + B \, u \tag{33}$$

$$v = C^* x \tag{34}$$

where w_0 is an r_0-vector, B_0 and $n \times r_0$-matrix and the other symbols have the same meaning as in (1)—(2).

Proposition 2. *If the block (33)—(34) has the property that* $G(\sigma) = C^*(\sigma I - A)^{-1} B$ *is not identically zero, then there exist two integers* r_1 *and* p *in the intervals* $0 < r_1 \leqslant \min(m_1, m_2)$, $0 < p \leqslant n/r_1$ *and there exists a transformation of the form (3)—*

(5) such that the transformed block can be split into a block of the form

$$\left.\begin{aligned} \frac{dy_1}{dt} &= y_2 + T_1 w_0 \qquad (\textit{if } p > 1) \\ \frac{dy_k}{dt} &= y_{k+1} + T_k w_0, \quad \textit{for any integer } k \textit{ from the interval } 0 < k < p \end{aligned}\right\} \tag{35}$$

$$\frac{dy_p}{dt} = T_p w_0 + u_1 \tag{36}$$

$$v_1 = y_1 \tag{37}$$

and a block of the form

$$\frac{dz}{dt} = F_r y_1 + A_r z + T_r w_0 + B_2 u_2, \quad (\textit{if } n - pr_1 > 0) \tag{38}$$

$$v_2 = C_0^* z. \tag{39}$$

Furthermore $\widetilde{u}$ *and* $\widetilde{v}$ *are connected to* u_1, u_2, v_1 *and* v_2 *by the relations*

$$\widetilde{v} = \begin{pmatrix} v_1 \\ v_2 \end{pmatrix} = \begin{pmatrix} y_1 \\ C_0^* z \end{pmatrix} \begin{matrix} \} \ r_1 \\ \} \ m_1 - r_1 \end{matrix} \tag{40}$$

$$\widetilde{u} = \begin{pmatrix} u_1 \\ u_2 \end{pmatrix} \begin{matrix} \} \ r_1 \\ \} \ m_2 - r_1 \end{matrix} \tag{41}$$

where u_1 *and* y_k $(k = 1, 2, \ldots, p)$ *are* r_1*-vectors,* u_2 *is an* m_2*-*r_1*-vector and* z *is an* n*-*pr_1*-vector. It follows that the matrices occurring in this system have the following dimensions:* T_k *(for* $k = 1, 2, \ldots, p$*):* $r_1 \times r_0$*;* T_r*:* $(n - pr_1) \times r_0$*;* A_r*:* $(n - pr_1) \times (n - pr_1)$*;* F_r*:* $(n - pr_1) \times r_1$*;* B_2*:* $(n - pr_1) \times (m_2 - r_1)$*;* C_0*:* $(n - pr_1) \times (m_1 - r_1)$ [1]*.*

---

[1]) Here and throughout the paragraph we agree that whenever the dimension of a vector or one of the dimensions of a matrix is zero, the respective vector or matrix is absent. In particular, Eq. (35) (or (38)) is absent if $p = 1$ (or respectively $n - pr_1 = 0$).

Proof. It is easy to see that if Eqs. (33)—(34) can be brought to the form (35)—(41) (by a transformation (4)—(6)) under the additional assumption $w_0 = 0$, then the same is true even if $w_0 \neq 0$. Therefore we take $w_0 = 0$. Then the block (33)—(34) takes the form (1)—(2) with $D = 0$. Moreover Eqs. (35), (36), and (38) take the simpler form

$$\frac{dy_k}{dt} = y_{k+1}, \text{ for } 0 < k < p \quad (\text{if } p > 1) \tag{35'}$$

$$\frac{dy_p}{dt} = u_1 \tag{36'}$$

$$\frac{dz}{dt} = F_r y_1 + A_r z + B_2 u_2, \ (\text{if } n - r_1 p > 0). \tag{38'}$$

We prove now that there exists at least one integer k from the interval $0 \leqslant k \leqslant n-1$ such that $C^* A^k B \neq 0$. For otherwise we would have $C^* A^k B = 0$, $k = 0, 1, 2, \ldots, n-1$. But we have $B \neq 0$ since $C^*(\sigma I - A)^{-1} B \not\equiv 0$ (see the assumptions in Proposition 2). Suppose that (A, B) is not completely controllable (if (A, B) is completely controllable the proof is simpler). Then, by Proposition 3, § 34, (A, B) can be written as

$$\tilde{A} = RAR^{-1} = \begin{pmatrix} A_{11} & A_{12} \\ 0 & A_{22} \end{pmatrix}, \tilde{B} = RB = \begin{pmatrix} B_1 \\ 0 \end{pmatrix}$$

where (A_{11}, B_1) is completely controllable. Introduce also the matrix $\tilde{C}^* = C^* R^{-1}$. We can write $\tilde{C}^* = C^* R^{-1} = (C_1^* \, C_2^*)$ such that the equalities $C^* A^k B = 0$ imply $C_1^* A_{11}^k B_1 = 0 (k = = 0, 1, \ldots, n-1)$ i.e. $C_1^* S = 0$ where $S = (B_1 \, A_{11} B_1 \ldots A_{11}^{n-1} B_1)$. Since (A_{11}, B_1) is completely controllable, the rank of S equals the number of rows of B_1 (see Property 4° of Theorem 1, § 34). Therefore the condition $C_1^* S = 0$ obtained above implies $C_1 = 0$. Then it is easy to see that $\tilde{C}^*(\sigma I - \tilde{A})^{-1} \tilde{B} \equiv 0$. Since this matrix is equal to $C^*(\sigma I - A)^{-1} B$ (as follows directly from the relations $\tilde{A} = RAR^{-1}$, $\tilde{B} = RB$ and $\tilde{C}^* = C^* R^{-1}$) one obtains a contradiction which proves that indeed there exists an integer in the interval $0 \leqslant k \leqslant n-1$ such that $C^* A^k B \neq 0$.

Let p be the smallest positive integer for which $C^*A^{p-1}B \neq 0$. Therefore one has

$$C^*A^kB = 0, \text{ for } 0 \leqslant k < p - 1 (\text{if } p > 1). \tag{42}$$

By Lemma 1 there exist two unitary matrices U_1 and V_1 such that

$$U_1C^*A^{p-1}BV_1 = \begin{pmatrix} \overbrace{D_1}^{r_1} & \overbrace{0}^{m_2-r_1} \\ 0 & 0 \end{pmatrix} \begin{matrix} \}r_1 \\ \}m_1-r_1 \end{matrix}, \tag{43}$$

where D_1 is a nonsingular diagonal matrix of rank r_1 (equal to the rank of the matrix $C^*A^{p-1}B$). By writing the matrix U_1C^* as

$$U_1C^* = \begin{pmatrix} C_1^* \\ C_2^* \end{pmatrix} \begin{matrix} \}r_1 \\ \}m_1-r_1 \end{matrix}, \tag{44}$$

we split (43) into two relations

$$C_1^* A^{p-1} B V_1 = (D_1 \quad 0) \tag{45}$$

$$C_2^* A^{p-1} B V_1 = 0. \tag{46}$$

Introduce the r_1-vector

$$y_p = C_1^* A^{p-1} x. \tag{47}$$

Differentiating and using (1) we find that

$$\frac{dy_p}{dt} = C_1^* A^p x + C_1^* A^{p-1}Bu.$$

Hence, using (3)

$$\frac{dy_p}{dt} = C_1^* A^p x + C_1^* A^{p-1} B (P^{-1} \tilde{u} + Q^* x). \tag{48}$$

This equation can be written in the form (36′). Indeed, by (45), the term in x of (48) vanishes if one chooses Q such that

$$V_1^{-1} Q^* = \begin{pmatrix} -D_1^{-1} C_1^* A^p \\ \\ Q_2^* \end{pmatrix} \begin{matrix} \} r_1 \\ \\ \} m_1 - r_1 \end{matrix} , \tag{49}$$

(where Q_2 is arbitrary). Moreover the term in $\widetilde{u}$ becomes equal to the r_1-vector u_1 (see (41) and (45)) if one chooses P such that

$$V_1^{-1} P^{-1} = \begin{pmatrix} D_1^{-1} & 0 \\ \\ 0 & I_2 \end{pmatrix} \begin{matrix} \} r_1 \\ \\ \} m_2 - r_1 \end{matrix} . \tag{50}$$

If $p > 1$ we introduce the r_1-vectors

$$y_k = C_1^* A^{k-1} x , \quad 1 \leqslant k \leqslant p - 1. \tag{51}$$

Differentiating and using (1), (42), (44) and (51), one sees that these vectors satisfy (35′). We now write (47) and (51) as

$$y = R_1 x \tag{52}$$

where y is an r_1, p-vector and R_1 is an r_1, $p \times n$-matrix given by

$$y = \begin{bmatrix} y_1 \\ y_2 \\ \cdot \\ \cdot \\ y_p \end{bmatrix} , \quad R_1 = \begin{bmatrix} C_1^* \\ C_1^* A \\ \cdot \\ \cdot \\ C_1^* A^{p-1} \end{bmatrix} . \tag{53}$$

We show that the rank of R_1 is $r_1 p$. Indeed, otherwise there exist the r_1-vectors v_k, for $k = 1, 2, \ldots, p$, such that

$$\sum_{k=1}^{p} v_k^* \, C_1^* \, A^{k-1} = 0, \quad (\max \, (\| v_k \|) \neq 0). \tag{54}$$

Multiplying the above relation on the right by $x(t)$ (where $t \mapsto x(t)$ is an arbitrary solution of Eq. (1)) and using (47) and (51), one obtains the identity

$$\sum_{k=1}^{p} v_k^* \, y_k \, (t) \equiv 0. \tag{55}$$

This identity holds in particular for the solution $t \mapsto x(t)$ which corresponds to the initial condition $x(0) = 0$ and to the control function u given by (3) and (41), where $u_2 = 0$ and u_1 is equal to an arbitrary r_1-vector u_{10}. Then, differentiating (55) at $t = 0$ and taking into account the initial conditions $y_k(0) = 0$ and (35′) and (36′), one obtains $v_p^* u_{10} = 0$. Thus $v_p = 0$ (since u_{10} is arbitrary). Introducing this equality in (55) and differentiating as many times as necessary, we obtain in the same way the equalities $v_k = 0$ for $k = 1, 2, \ldots, p$; this contradicts the last condition of (54) and proves that the rank of R_1 equals $r_1 p$.

If $r_1 p = n$, R_1 is nonsingular. Then (52) has the form (4) and the reasonings above show that (1) can be written in the form (35′)–(36′) (Eq. (38′) is absent). If $r_1 p < n$, we can find an $n - r_1 p \times n$ matrix R_2 such that the matrix

$$R = \begin{pmatrix} R_1 \\ R_2 \end{pmatrix} \begin{matrix} \}r_1 p \\ \}n - r_1 p \end{matrix} \tag{56}$$

is nonsingular. We perform the transformation (4) where R has the form (56). If the vector x is written as

$$\tilde{x} = \begin{pmatrix} y \\ \check{z} \end{pmatrix} \begin{matrix} \}r_1 p \\ \}n - r_1 p \end{matrix} \tag{57}$$

then transformation (4) splits into $y = R_1 x$ and $\check{z} = R_2 x$. We proved above that the component y satisfies (35′) and (36′). For the component $\check{z}$ one obtains by differentiating and by using (1),

$$\frac{d\check{z}}{dt} = R_2 A x + R_2 B u. \tag{58}$$

If in the right-hand member one expresses u by means of (41) and (3) where P and Q are chosen as shown above (such that (49) and (50) are satisfied) and then if one expresses x as $x = R^{-1}\tilde{x}$ (see (4)), where $\tilde{x}$ is given by (57) and y has the expression of (53), one can write (58) in the form

$$\frac{d\check{z}}{dt} = \sum_{k=1}^{p} F_k y_k + A_r \check{z} + B_1 u_1 + B_2 u_2 \tag{59}$$

where the matrices have the following dimensions : $F_k : (n - r_1 p) \times r_1$; $A_r : (n - r_1 p) \times (n - r_1 p)$; $B_1 : (n - r_1 p) \times r_1$ and $B_2 : (n - r_1 p) \times (m_2 - r_1)$.

To bring (59) to the form (38′) we eliminate the terms in u_1 and in y_k, for $k > 1$ by performing a new transformation of the form (4), chosen as follows : The vectors y_k remain unchanged while $\check{z}$ is replaced by the new vector

$$z = \check{z} - \sum_{k=1}^{p} L_k y_k , \tag{60}$$

where the matrices L_k have the dimensions $(n - r_1 p) \times r_1$. (One readily sees that the transformation thus defined is invertible, as required.) Using (35′), (36′) and (59) and differentiating (60), one obtains

$$\frac{dz}{dt} = (F_1 + A_r L_1) y_1 + \sum_{k=2}^{p} (F_k + A_r L_k - L_{k-1}) y_k + A_r z + \\ + (B_1 - L_p) u_1 + B_2 u_2 . \tag{61}$$

Choosing $L_p = B_1$, the term in u_1 vanishes. We then choose step by step the matrices L_k for $k = p - 1, p - 2, \ldots, 1$ (if $p > 1$), such that the terms in y_k vanish for $k = 2, 3, \ldots, p$. Then (61) takes the form (38′). Thus we proved that, under our assumptions, Eq. (1) can be brought to the form (35′), (36′), (38′).

We now show that v can be brought to the form (40). Since we chose $U = U_1$, we obtain from (44), (5) and (34)

$$\tilde{v} = \begin{pmatrix} C_1^* x \\ C_2^* x \end{pmatrix} = \begin{pmatrix} y_1 \\ C_2^* x \end{pmatrix} \tag{62}$$

(see (51), for $k = 1$). It remains to show that $C_2^* x$ can be written as $C_0^* z$. Notice that by inverting the transformation (4) which relates x to the vector $\tilde{x}$ (of components y_k and z) one can express x in terms of y_k and z, thus bringing $C_2^* x$ to the form

$$C_2^* x = \sum_{k=1}^{p} T_k y_k + C_0^* z, \tag{63}$$

where T_k are $(m_1 - r_1) \times r_1$ matrices and C_0 is an $(n - r_1 p) \times (m_1 - r_1)$-matrix. Consider now Relation (63) taking x, y_k and z as the solutions of the system (1) (or of the transformed system (35′), (36′), (38′)) corresponding to the initial condition $x(0) = 0$ (or $y_k(0) = 0$, $z(0) = 0$) and to the control function for which we have (in (35′), (36′), (38′)) $u_2 = 0$ and $u_1 = u_{10}$, where u_{10} is a constant vector. Then, differentiating the left-hand member of (63) and using (42), (46) and $x(0) = 0$, one obtains

$$\left(\frac{d^j C_2^* x(t)}{dt^j}\right)_{t=0} = C_2^* A^j x(0), \quad j = 1, 2, \ldots, p.$$

Similarly, using Eqs. (35′), (36′) and (38′) and the conditions $y_k(0) = 0$, $z(0) = 0$ and $u_2 = 0$, we obtain

$$\left(\frac{d^j C_0^* z(t)}{dt^j}\right)_{t=0} = 0, \quad j = 1, 2, \ldots, p.$$

Therefore, by differentiating (63) one obtains

$$\left(\frac{d^j \sum_{k=1}^{p} T_k y_k(t)}{dt^j}\right)_{t=0} = 0, \; j = 1, 2, \ldots, p. \tag{64}$$

Using (35′), (36′) and $y_k(0) = 0$ one sees that for $j = 1$ Relation (64) becomes $T_p u_{10} = 0$, whence $T_p = 0$ (since u_{10} is arbitrary). Introducing this equality into (64), one obtains, for $j = 2$, the equality $T_{p-1} = 0$. Similarly one obtains $T_k = 0$, $k = 1, 2, \ldots, p$. Now from (63) one sees that (62) has the form (40). Thus Proposition 2 is proved in the case $w_0 = 0$. As shown at the beginning of the proof, it follows that the proposition also holds for $w_0 \neq 0$.

If $G(\sigma)$ does not satisfy the condition required in Proposition 2, one can use

Proposition 3. *If the block (33), (34) (where $m_1 \geqslant 0$ and $m_2 \geqslant 0$*[1]*) has the property that $C^*(\sigma I - A)^{-1}B$ is identically zero, then there exists an integer n_1 from the interval $0 \leqslant n_1 \leqslant n$ and a transformation (3)—(5) such that the transformed block can be written as*

$$\frac{dz}{dt} = A_{11} z + T_1 w_0, \quad (\text{if } n_1 > 0) \tag{65}$$

$$\frac{dw}{dt} = A_{21} z + A_{22} w + T_2 w_0 + B_1 u, \quad (\text{if } n - n_1 > 0) \tag{66}$$

$$\tilde{v} = C_1^* z. \tag{67}$$

[1]) In Proposition 3 we also include the cases $m_1 = 0$ or $m_2 = 0$ or $m_1 = m_2 = 0$.

Furthermore, if $n_1 > 0$ *the pair* (C_1^*, A_{11}) *is completely observable. In (65)—(67)* z *is an* n_1*-vector,* w *is an* n-n_1*-vector and the matrices which occur have the following dimensions:* $A_{11}: n_1 \times n_1$; $T_1: n_1 \times r_0$; $A_{21}: (n - n_1) \times n_1$; $A_{22}: (n - - n_1) \times (n - n_1)$; $T_2: (n - n_1) \times r_0$ *and* $B_1: (n - n_1) \times m_2$.[1])

Proof. It is again sufficient to prove the proposition for $w_0 = 0$. If $C = 0$ or if $m_1 = 0$ the block (33)—(34) has from the beginning the form (65)—(67), for $n_1 = 0$. (If $m_1 = 0$ Eq. (67) is absent). If $C \neq 0$, applying Proposition 1, § 35, we see that there exists a transformation $\tilde{x} = Rx$ (det $R \neq 0$) such that, for $w_0 = 0$, Eqs. (33)—(34) can be written as a system consisting of the equation

$$\frac{dz}{dt} = A_{11} z + B_0 u \tag{68}$$

and Eqs. (66) and (67), with $w_0 = 0$. Furthermore the pair (C_1^*, A_{11}) is completely observable. The obtained block has the (identically zero) transfer function (see the assumption of Proposition 3) $\sigma \mapsto G(\sigma) = C_1^*(\sigma I_1 - A_{11})^{-1} B_0 = 0$ (I_1 is the $n_1 \times \times n_1$-identity-matrix. Since (C_1^*, A_{11}) is completely observable the obtained identity implies $B_0 = 0$ (see Property 12° of Consequence 1, § 35). Thus (68) takes the form (65) for $w_0 = 0$; this proves Proposition 3 for $w_0 = 0$. As remarked above, it follows that Proposition 3 holds also for $w_0 \neq 0$.

By applying successively the propositions established above we prove the following theorem:

Theorem 1. *For every block of the form (1)—(2) there exists :1° a list of nonnegative integers (*s, p_j, r_j $(j = 1, 2, \ldots, s)$, r_0, n_1 *and* n_2*) satisfying the conditions*

$$n_1 + n_2 + \sum_{j=1}^{s} p_j r_j = n \text{ and } r_0 + \sum_{j=1}^{s} r_j \leqslant \min(m_1, m_2),$$

and 2° an invertible transformation (3)—(6) which brings the block to the form

$$\left.\begin{aligned} \frac{dy_{kj}}{dt} &= \sum_{i=1}^{j-1} T_{kji} y_{1i} + y_{k+1,j} + B_{kj} u_0, \\ &\qquad \text{for } 0 < k < p_j (\text{if } p_j > 1) \\ \frac{dy_{p_j j}}{dt} &= \sum_{i=1}^{j-1} T_{p_j ji} y_{1i} + B_{p_j j} u_0 + u_j \end{aligned}\right| \begin{aligned} &\text{for} \\ &0 < j \leqslant s \\ &(\text{if } s > 0) \end{aligned} \tag{69}$$

$$\frac{dz}{dt} = \sum_{i=1}^{s} F_{1i} y_{1i} + A_{11} z + B_{10} u_0, \quad (\text{if } n_1 > 0) \tag{70}$$

---

[1]) We make use of the convention specified in footnote 1, page 338. Thus, if $n_1 = 0$, Eq. (65) is absent and Eq. (67) becomes $\tilde{v} = 0$. If $n = n_1$ Eq. (66) is absent.

$$\frac{dw}{dt} = \sum_{i=1}^{s} F_{21} y_{1i} + A_{21} z + A_{22} w + B_{20} u_0 + B_{2r} u_r, \text{ (if } n_2 > 0) \quad (71)$$

$$v = \begin{pmatrix} u_0 \\ y_{11} \\ y_{12} \\ \cdot \\ y_{1s} \\ C_1^* z \end{pmatrix} \begin{matrix} \} r_0 \\ \} r_1 \\ \} r_2 \\ \\ \} r_s \\ \} m_1 - r_0 - \sum_1^s r_j \end{matrix} \qquad \tilde{u} = \begin{pmatrix} u_0 \\ u_1 \\ u_2 \\ \cdot \\ u_s \\ u_r \end{pmatrix} \begin{matrix} \} r_0 \\ \} r_1 \\ \} r_2 \\ \\ \} r_s \\ \} m_2 - r_0 - \sum_1^s r_j \end{matrix} \quad (72)$$

Furthermore, if $n_1 > 0$ *and* $m_1 - r_0 - \sum_1^s r_j > 0$*, the pair* (C_1^*, A_{11}) *is completely observable. In (69)–(72),* u_j *and* y_{kj} $(k = 1, 2, \ldots, p_j)$ *are* r_j*-vectors and the other vectors have the following dimensions:* $u_0: r_0; z: n_1; w: n_2; u_r: m_2 - r_0 - \sum_1^s r_j$ *(this also determines the dimensions of the matrices occurring in the above formulas).*

B. *If* $m_1 = m_2$ *and if* $\det G(\sigma) \not\equiv 0$*, one may take* $n_1 = 0$ *and bring the system (69)–(72) to the form:*

$$\left.\begin{aligned} \frac{dy_{kj}}{dt} &= \sum_{i=1}^{j-1} T_{kji} y_{1i} + y_{k+1,j} + B_{kj} u_0, \\ &\qquad \text{for } 0 < k < p_j (\text{if } p_j > 1) \\ \frac{dy_{p_j j}}{dt} &= \sum_{i=1}^{j-1} T_{p_j j} y_{1i} + B_{p_j j} u_0 + u_j \end{aligned}\right\} \begin{matrix} \text{for} \\ 0 < j \leqslant s \\ (\text{if } s > 0) \end{matrix} \quad (69')$$

$$\frac{dw}{dt} = \sum_{i=1}^{s} F_{2i} y_{1i} + A_{22} w + B_{20} u_0, \text{ if } n_2 > 0 \quad (71')$$

$$\tilde{v} = \begin{pmatrix} u_0 \\ y_{11} \\ \cdot \\ y_{1s} \end{pmatrix}, \quad \tilde{u} = \begin{pmatrix} u_0 \\ u_1 \\ \cdot \\ u_s \end{pmatrix} \quad (72')$$

Furthermore

$$\det(\sigma I_2 - A_{22}) = \varkappa \det(\sigma I - A) \det G(\sigma) \tag{73}$$

where $\varkappa$ *is a constant and* I_2 *is the identity matrix* $n_2 \times n_2$.

Proof. If D has the rank $r_0 \neq 0$, we apply Proposition 1 and bring the block (1)—(2) to the form (28)—(29). (If the rank of D is zero, we pass to the next step.) If $r_0 = m_1$ the block (28)—(29) has the form (69) — (72) for $s = 0, n_1 = 0, n_2 = n, \widetilde{x} = w^1$).[1]

If $m_1 > r_0$ we consider the block consisting of Eq. (28) and the output $v_1 = C_1^* x$ (given by the last component of $\widetilde{v}$ in (29)). This block is of the form (33)—(34), for $w_0 = u_0$. If the transfer function of the block is not identically zero, we apply Proposition 2 (otherwise we pass to the next step). Thus, the block is split into the block (35)—(37) and the block (38)—(39). The latter is also of the form (33)—(34) where $B_0 w_0$ is now a linear combination of the form $F_r y_1 + T_r u_0$. If its transfer function is not identically zero we apply again Proposition 2 (otherwise we pass to the final step). Proceeding in this way as many times as possible we apply successively Proposition 2 to blocks of the form (33)—(34) with diminishing dimensions. The term $B_0 w_0$ is, each time, a linear combination containing the vectors u_0 and y_{1i} for all the blocks of the form (33)—(34) obtained in the preceding steps. After applying Proposition 2 a finite number of times the proposition cannot be applied any longer either because the equation in z, (38), is absent or because the transfer function of the block (38)—(39) (at the considered stage) becomes identically zero. If the equation in z is absent our block takes the desired form (69)—(72) for $n_1 = n_2 = 0$. If the transfer function of the last block (38)—(39) is identically zero, then, as a final step, we apply Proposition 3 and bring the last block (38)—(39) into the form (65)—(67). Rearranging all the blocks of the form (28)—(29), (35)—(41) and (65)—(67) into which the initial block has been split, we obtain the form (69)—(72) of Theorem 1. Implicitly we have also proved the existence of the integers specified in Theorem 1, thus concluding the proof of part A of the theorem.

We now prove part B. We shall consider in the sequel that Eqs. (7)—(8) and (69)—(72) represent the same block expressed in two different forms. The transfer function of this block can be calculated by observing that if σ_0 satisfies the condition $\det(\sigma_0 I - \widetilde{A}) \neq 0$ and if we take $\widetilde{u}(t) = k e^{\sigma_0 t}$ (k=constant), then the only solution of Eq. (7) of the form $\widetilde{x}(t) = x_0 e^{\sigma_0 t}$ has the expression $\widetilde{x}(t) = (\sigma_0 I - \widetilde{A})^{-1} \widetilde{B} k e^{\sigma_0 t}$, whence, using (8)

[1]) See the convention in footnote 1, page 338.

and (14) we obtain $\tilde{v}(t) = \tilde{C}^* x(t) + \tilde{D}u(t) = \tilde{G}(\sigma_0)ke^{\sigma_0 t}$. Hence, by computing the indicated solution we can determine $\tilde{G}(\sigma_0)k$. Since k is arbitrary, we thus determine $\tilde{G}(\sigma_0)$. From the form of Eqs. (69)—(72) it follows that $\tilde{v}(t)$ is independent of $u_r(t)$ which only occurs in (71)). Hence if $m_2 - r_2 - \sum_1^s r_j > 0$ the last $m_2 - r_0 - \sum_1^s r_j$ columns of the transfer function of the block (69)—(72) (i.e. the columns which correspond to u_r) are identically zero. If $m_1 = m_2$, we obtain the identity det $\tilde{G}(\sigma) \equiv 0$ and also, using (19), det $G(\sigma) \equiv 0$. Hence, the condition det $G(\sigma) \not\equiv 0$, required in part β of Theorem 1 can be satisfied only if $m_2 = r_0 + \sum_1^s r_j$. Therefore, if $m_1 = m_2$ the last component of the vector $\tilde{v}$ of (72) (i.e. the vector $C_1^* z$) is absent and Eqs. (69)—(72) can be brought to the form (69′)—(72′).

In order to calculate the transfer function of the block (69′) —(72′) by the method explained above, we put

$$u_0 (t) = u_{0c}\, e^{\sigma_0 t},\ u_j (t) = u_{jc}\, e^{\sigma_0 t},\ j = 1, 2, \ldots, s\,, \tag{74}$$

where u_{0c} and u_{jc} are constant vectors with r_0 and r_j components respectively while the scalar σ_0 satisfies the conditions $\sigma_0 \neq 0$ and det $(\sigma_0 I_2 - A_{22}) \neq 0$. Writing Eqs. (69′) and (71′) in the form (7) we find the expression of matrix $\tilde{A}$ and we easily obtain

$$\det (\sigma I - \tilde{A}) = \sigma^{\sum_1^s p_j r_j} \det (\sigma I_2 - A_{22}). \tag{75}$$

Therefore det$(\sigma_0 I - A) \neq 0$. Now we determine the solution of equations (69′), (74) which contains $e^{\sigma_0 i}$ as a factor. For the computation of the transfer function we need only the components which occur in $\tilde{v}(t)$, i.e. only the $v_{1j}(t)$ (see (72′)). Their expressions are

$$y_{1j} (t) = \sum_{k=0}^{j-1} G_{jk} (\sigma_0)\, u_{kc}\, e^{\sigma_0 t} + \frac{I_{rj}}{\sigma^{p_s}} u_{jc}\, e^{\sigma_0 t},\ j = 1., 2, \ldots, s\,,$$

where the I_{rj} are the identity-matrices $r_j \times r_j$ ($j = 1, 2, \ldots$ $\ldots, s$), while the $G_{jk}(\sigma_0)$ are some $r_j \times r_k$-matrices whose expressions do not bear on the formula we want to establish.

Considering the output $\tilde{v}(t)$ given by (72′), we deduce, as shown above, that the transfer function is given by [1]:

$$\tilde{G}(\sigma) = \begin{pmatrix} \overbrace{I_{r0}}^{r_0} & \overbrace{0}^{r_1} & \overbrace{0}^{r_2} & \ldots \overbrace{0}^{r_s} \\ 0 & \dfrac{I_{r1}}{\sigma^{p_1}} & 0 & \ldots 0 \\ 0 & G_{21}(\sigma) & \dfrac{I_{r2}}{\sigma^{p_1}} & \ldots 0 \\ \cdot & \cdots\cdots & \cdots\cdots & \cdots\cdots \\ 0 & G_{s1}(\sigma) & G_{s2}(\sigma) & \dfrac{I_{rs}}{\sigma^{p_s}} \end{pmatrix} \begin{matrix} \}\, r_0 \\ \}\, r_1 \\ \}\, r_2 \\ \\ \}\, r_s \end{matrix}$$

where I_{r0} is the identity matrix $r_0 \times r_0$. The relation

$$\det \tilde{G}(\sigma) = \frac{1}{\sigma^{\sum_1^s p_j r_j}}, \quad (\sigma \neq 0). \tag{76}$$

follows immediately. Using (75) and (76) one easily sees that the right-hand member of (19) is equal to $\det(\sigma I_2 - A_{22})$ multiplied by a non-zero constant. By multiplying (19) with the inverse of this constant, we obtain (73), and this concludes the proof of Theorem 1.

---

[1]) The method used here allows the computation of $\tilde{G}(\sigma)$ for every σ, except a finite number of values for which $\det(\sigma I - A) = 0$. However, in this way the transfer function is completely determined, since each of its entries is a quotient of two polynomials and these polynomials are completely determined (except for a common factor) if one knows the values of the transfer function at a finite number of distinct points.

FACTORIZATION OF POLYNOMIAL MATRICES

This appendix plays an essential part in the proof of the theorem on positiveness of multi-input systems (see Chapter 3). The main goal of the appendix is to prove a theorem on factorization on the unit circle (Theorem 1, § 38). The first paragraph of the appendix contains some auxiliary propositions used in the second paragraph for proving Theorem 1, § 38. A sufficiently general proof of Theorem 1, § 38 is found in the first nine sections of § 38, if one is willing to accept a few additional conditions which are nearly always satisfied. Sections 10—11 serve to complete the proof of the theorem in its purer form, free from unnecessary restrictions. The theorem of factorization on the imaginary axis (Theorem 1, § 39) is obtained from Theorem 1, § 38 by a change of variable.

The origins of the general problem of factorization may be found in classical studies of D. Hilbert and N. Wiener. The problem is amply treated in the work [1] of I. T. Gohberg and M. G. Krein from which one can derive results very close to the theorems established below. However since our assumptions are quite different, an independent treatment of the problem is preferred.

In the proof, we also use an idea of N. I. Mushelishvili, presented in the work [1] of N. P. Vekua. The following exposition is more general than the treatment of the same problem in the author's work [16] (see Appendix D of the work quoted). Under certain additional conditions, a different solution of the factorization problem was obtained by D.C. Youla [1].

§ 37. Auxiliary propositions

1. In the present appendix any symbol of the form $A(\lambda)$ will represent a matrix whose entries are polynomials of the complex variable λ or quotients of such polynomials. If $[A(\lambda)]_{ij}$ denotes the entry of the matrix $A(\lambda)$ situated on the i-th

row and on the j-th column, the general expression of these entries is

$$[A(\lambda)]_{ij} = \frac{\sum_{k=1}^{n_1} \alpha_{ij,k} \lambda^k}{\sum_{k=1}^{n_2} \beta_{ij,k} \lambda^k}. \tag{1}$$

As in the rest of the work, the notations $\bar{A}(\lambda)$ and $\bar{A}'(\lambda)$ used in connection with a matrix $A(\lambda)$ of the form (1) have the following meaning

$$\bar{A}(\lambda) = \overline{A(\bar{\lambda})}\,; \bar{A}'(\lambda) = (A(\bar{\lambda}))^* \tag{2}$$

If the entries of the matrix $A(\lambda)$ have the expression (1), the entries $\bar{A}(\lambda)$ and $\bar{A}'(\lambda)$ are given by

$$[\bar{A}(\lambda)]_{ij} = \frac{\sum_{k=1}^{n_2} \bar{\alpha}_{ij,k} \lambda^k}{\sum_{k=1}^{n_1} \bar{\beta}_{ij,k} \lambda^k}; \quad [\bar{A}'(\lambda)]_{ij} = \frac{\sum_{k=1}^{n_1} \bar{\alpha}_{ji,k} \lambda^k}{\sum_{k=1}^{n_2} \bar{\beta}_{ji,k} \lambda^k}. \tag{3}$$

In other words, if the sign of complex conjugation (—) is applied only to the symbol A then it does not affect the variable λ. The matrices $\bar{A}(\lambda)$ and $\bar{A}'(\lambda)$ must not be confused with $\overline{A(\lambda)}$ and $(A(\lambda))^*$ whose entries differ from the expressions (3) by the fact that λ is replaced by $\bar{\lambda}$. According to the above convention, if $\rho(\lambda)$ is a polynomial in λ, we denote by $\bar{\rho}(\lambda)$ another polynomial in λ defined as

$$\bar{\rho}(\lambda) = \overline{\rho(\bar{\lambda})}. \tag{4}$$

Thus the polynomial $\bar{\rho}(\lambda)$ is obtained by replacing each coefficient of the polynomial $\rho(\lambda)$ by its complex conjugate.

2. *Consider the expressions of the form*

$$\sigma(\lambda) = \sum_{j=-n}^{n} \alpha_j \lambda^j, \quad (\max |\alpha_j| \neq 0), \tag{5}$$

where λ is a complex variable, α_j are complex coefficients, not all zero, and n is a positive integer.

Proposition 1. *If $\sigma(\lambda)$ has the form (5) and is real and nonnegative on the unit circle ($|\lambda| = 1$), then the following properties are satisfied :*

1°. *The coefficients α_j satisfy the relations*

$$\alpha_j = \bar{\alpha}_{-j}\,, \; j = 0, 1, \ldots, n\,. \tag{6}$$

2°. *The number n' of the non-zero roots of the polynomial $\lambda \mapsto \lambda^n \sigma(\lambda)$ is even, and these roots can be grouped in pairs of the form*

$$(\lambda_1, \lambda_{-1}), (\lambda_2, \lambda_{-2}) \ldots, (\lambda_{n_0}, \lambda_{-n_0}), \qquad \left(n_0 = \frac{1}{2} n' \leqslant n\right), \tag{7}$$

$$\lambda_{-j} = \frac{1}{\bar{\lambda}_j} \quad j = 1, 2, \ldots, n_0. \tag{8}$$

3°. $\sigma(\lambda)$ *admits at least one factorization of the form* *)

$$\sigma(\lambda) = \bar{\rho}\left(\frac{1}{\lambda}\right) \rho(\lambda), \tag{9}$$

where ρ *is a polynomial of the form*

$$\rho(\lambda) = \varkappa (\lambda - \lambda_{i_1})(\lambda - \lambda_{i_2}) \ldots (\lambda - \lambda_{i_{n_0}}), \tag{10}$$

in the above expressions the numbers λ_{i_j}, $j = 1, 2, \ldots, n_0$ are arbitrarily chosen, one from each pair from (7) (the index i_j may take one of the values $+j$ and $-j$), and $\varkappa$ is a complex constant (depending on the choice of the numbers λ_{ij}). The argument of the constant $\varkappa$ may be arbitrarily chosen.

4°. *Any polynomial $\tilde{\rho}$ which satisfies (9) can be written in the form*

$$\tilde{\rho}(\lambda) = \lambda^p \rho(\lambda), \tag{11}$$

where ρ is one of the polynomials described in the statement of Property 3° and p is a nonnegative integer.

5°. *If all the coefficients α_j are real, there exists at least one factorization of the form (9)—(10) for which the polynomial ρ has real coefficients.*

Proof. On the unit circle we have $\bar{\lambda} = 1/\lambda$ and hence the assumption that $\sigma(\lambda)$ is real for $|\lambda| = 1$ implies that

$$\sigma(\lambda) = \bar{\sigma}\left(\frac{1}{\lambda}\right). \tag{12}$$

Substituting (5) in (12) we obtain

$$\sum_{j=-n}^{n} (\alpha_j - \bar{\alpha}_{-j}) \lambda^j = 0, \quad (|\lambda| = 1). \tag{13}$$

---

*) See the convention of notation given by (4).

Multiplying by λ^n the left-hand member of (13) gives a polynomial in λ whose degree is at most $2n$. By (13), this polynomial vanishes at any point on the unit circle, and therefore is identically zero, whence (6) follows. It also follows that (12) holds for every $\lambda \neq 0$, a conclusion which will be immediately used.

If there exists $\lambda_0 \neq 0$ satisfying the equality $\lambda_0^n \delta(\lambda_0) = 0$, then $\sigma(\lambda_0) = 0$ and also $\sigma(1/\bar{\lambda}_0) = 0$, according to (12) and the above remark. If $|\lambda_0| = 1$ then we have $\pi(\lambda_0) = 0$, where $\pi(\lambda) = \lambda^n \sigma(\lambda)/(\lambda - \lambda_0)$. Indeed, since $|\lambda_0| = 1$, the points on the unit circle can be written as $\lambda = \lambda_0 e^{i\varphi}$, where φ is real. We now substitute this value of λ in the equality $\sigma(\lambda) = (\lambda - \lambda_0)\pi(\lambda)/\lambda^n$ (deduced from the foregoing expression of $\pi(\lambda)$); we then expand the obtained expression in a Taylor series with respect to φ, in the neighbourhood of the point $\varphi = 0$, and we obtain the expression $\sigma(\lambda_0 e^{i\varphi}) = i\varphi\pi(\lambda_0)/\lambda_0^{n-1} + \mathcal{O}(\varphi^2)$, where $\lim_{\varphi\to 0} \mathcal{O}(\varphi^2)/\varphi = 0$; but the expression $\sigma(\lambda_0 e^{i\varphi})$ is, by hypothesis, real and nonnegative for any real φ, whence it follows that the term $i\varphi\pi(\lambda_0)/\lambda_0^{n-1}$ must be real and nonnegative for every real φ. Therefore $\pi(\lambda_0) = 0$. Hence, since $|\lambda_0| = 1$, we also obtain $\pi(1/\bar{\lambda}_0) = 0$.

From the conclusions of the preceding paragraph, it follows that the expression

$$\tilde{\sigma}(\lambda) = \frac{\sigma(\lambda)}{(\lambda - \lambda_0)\left(\frac{1}{\lambda} - \bar{\lambda}_0\right)} \tag{14}$$

is of the form (5) (where instead of n, we have $n - 1$). Furthermore, $\tilde{\sigma}(\lambda)$ is clearly real and positive on the unit circle and therefore the above reasoning may be repeated. By repeating the argument as many times as possible, we obtain Property 2°.

Knowing that all the non-zero solutions of the equation $\sigma(\lambda) = 0$ are included in the pairs (7) let us rearrange these roots in the form

$$\lambda_{i_1}, \lambda_{i_2}, \ldots, \lambda_{i_{n_0}},$$
$$\lambda_{-i_1}, \lambda_{-i_2}, \ldots, \lambda_{-i_{n_0}},$$

the numbers λ_{i_j} being chosen as specified in the statement of Property 3°. Assuming that the polynomial $\lambda \mapsto \lambda^n \sigma(\lambda)$ has p roots equal to zero, we have

$$\lambda^n \sigma(\lambda) = \mu_1 \lambda^p \prod_{j=1}^{n_0} (\lambda - \lambda_{i_j}) \prod_{j=1}^{n_0} (\lambda - \lambda_{-i_j}), \ \mu_1 = \text{constant } 0. \tag{15}$$

From (8) we obtain

$$(\lambda - \lambda_{i_j}) = - \frac{\lambda}{\lambda_{i_j}} \left(\frac{1}{\lambda} - \bar{\lambda}_{i_j} \right),$$

and hence (15) gives

$$\sigma(\lambda) = \mu(\lambda)\, \rho_0(\lambda)\, \bar{\rho}_0\left(\frac{1}{\lambda}\right), \tag{16}$$

where

$$\mu(\lambda) = (-1)^{n_0} \lambda^{p+n_0-n} \frac{\mu_1}{\bar{\lambda}_{i_1} \bar{\lambda}_{i_2} \cdots \bar{\lambda}_{i_{n_0}}}, \tag{17}$$

$$\rho_0(\lambda) = \prod_{j=1}^{n_0} (\lambda - \lambda_{i_j}). \tag{18}$$

The equality $1/\lambda = \bar{\lambda}$ is satisfied on the unit circle and hence the product $\rho_0(\lambda)\bar{\rho}_0(\lambda^{-1})$ is real and nonnegative (it may vanish at some distinct points). From (16) it follows that $\sigma(\lambda)$ is real and positive on the unit circle only if $\mu(\lambda)$ is real and nonnegative for $|\lambda| = 1$. From (17) we see that this condition is satisfied only if $\mu(\lambda)$ is independent of λ (hence $p = n - n_0$) being equal to a real and nonnegative constant.

$$\mu(\lambda) = \mu_0 = \frac{(-1)^{n_0} \mu_1}{\bar{\lambda}_{i_1} \bar{\lambda}_{i_2} \cdots \bar{\lambda}_{i_{n_0}}}, \qquad \operatorname{Im} \mu_0 = 0. \tag{19}$$

From (16) and (19) we immediately obtain (9) and (10) if we denote

$$\varkappa = \sqrt{|\mu_0|}\, e^{i\varphi}, \tag{20}$$

where φ is an arbitrary, real constant. The polynomial (10) is related to the polynomial (18) by $\rho(\lambda) = \varkappa \rho_0(\lambda)$. It is obvious that besides ρ, any polynomial of the form (11) satisfies the relation of factorization (9). Conversely, if a polynomial $\tilde{\rho}$ satisfies (9), we can number the non-zero roots of the polynomial such that every pair of roots $(\lambda_j, \lambda_{-j})$ from (7) contains one root of the polynomial ρ. The polynomial $\tilde{\rho}$ may also have an arbitrary number of roots equal to zero, whence Assertion 4°.

If all the coefficients α_j are real, the polynomial $\lambda \mapsto \lambda^n \sigma(\lambda)$ has real coefficients and hence the non-real roots occur in pairs of complex conjugate roots. However, it is easy to see that no pair of roots of the form $(\lambda_j, \lambda_{-j}) = (\lambda_j, 1/\bar{\lambda}_j)$ (see (7) and (8)) may contain two non-real, complex conjugate roots. Therefore, all the pairs $(\lambda_j, \lambda_{-j})$ of (7) which contain non-real roots can be arranged in pairs of the form

$$(\lambda_k , \lambda_{-k}), (\lambda_{\tilde{k}} , \lambda_{-\tilde{k}}), \quad \text{where } \lambda_k = \bar{\lambda}_{\tilde{k}}.$$

Using this property, the roots of the polynomial (10) can be chosen such that for every non-real root the complex conjugate root occurs also in (10). The constant $\varkappa$ given by (20) can be chosen real. Then the polynomial ρ has real coefficients. This concludes the proof of Proposition 1.

3. We examine now the expressions of the form

$$\psi(\lambda) = \frac{\pi_1(\lambda)}{\pi_2(\lambda)}, \tag{21}$$

where

$$\pi_1(\lambda) = \sum_{j=0}^{n_1} \alpha_j \lambda^j, \ (\max |\alpha_j| \neq 0) \tag{22}$$

$$\pi_2(\lambda) = \sum_{j=0}^{n_2} \beta_j \lambda^j, \ \beta_{n_2} \neq 0. \tag{23}$$

Proposition 2. *Assume that $\psi(\lambda)$ has the form (21) and is real and nonnegative on the unit circle ($|\lambda| = 1$). Assume also that the polynomial π_2 has the following properties : 1° $\pi_2(\lambda) \neq 0$ for $|\lambda| = 1$ and for $\lambda = 0$; 2° there exists no pair of roots (λ_1, λ_2) such that $\lambda_1 \bar{\lambda}_2 = 1$. Then $\psi(\lambda)$ is equal to a real, positive constant.*

Proof. Since, by the assumptions of the proposition, the expression

$$\pi_1(\lambda)\overline{\pi_2(\lambda)} = \psi(\lambda)|\pi^2(\lambda)|^2 \tag{24}$$

is real and nonnegative for $|\lambda| = 1$, so is also the expression

$$\pi_1(\lambda)\overline{\pi_2(1/\bar{\lambda})} = \pi_1(\lambda)\bar{\pi}_2(1/\lambda), \tag{25}$$

obtained by replacing $\bar{\lambda}$ by $1/\lambda$ in $\overline{\pi_2(\lambda)}$ of (24). Indeed, for $|\lambda| = 1$ we have $\bar{\lambda} = 1/\lambda$ and hence (24) and (25) are equal on the unit circle. Expression (25) has the form and the properties from Proposition 1 and therefore all non-zero numbers for which (25) vanishes, can be arranged in pairs of the form

$$\left(\lambda_j, \frac{1}{\bar{\lambda}_j}\right), \ j = 1, 2, \ldots, n'. \tag{26}$$

Let $\lambda_1', \lambda_2', \ldots, \lambda_{n_2}'$ be the roots of the polynomial $\pi_2(\lambda)$. By Condition 1° of Proposition 2 they are all different from zero. Since the expression (25) vanishes for $\lambda = 1/\bar{\lambda}_k'$, $k = 1, 2, \ldots, n_2$, the numbers $1/\bar{\lambda}_k'$ belong to the sequence (26) while all the other numbers belonging to (26) are roots of the polynomial $\pi_1(\lambda)$. There cannot exist two numbers $1/\bar{\lambda}_j'$ and $1/\bar{\lambda}_k'$ belonging to the same pair (26) since otherwise we would have the equality $1/\bar{\lambda}_j' = \lambda_k'$ which contradicts Condition 2° of Proposition 2. Therefore we can number the roots (26) such that

$$\frac{1}{\bar{\lambda}_k} = \frac{1}{\bar{\lambda}_k'} \text{ i.e. } \lambda_k = \lambda_k', \qquad k = 1, 2, \ldots, n_2. \tag{27}$$

Then it is clear that any root λ_j of the polynomial $\pi_2(\lambda)$ belongs to the numbers (26) and is a root of the polynomial $\pi_1(\lambda)$. Hence $\pi_1(\lambda)$ is divisible by $\pi_2(\lambda)$. Therefore the expression $\psi(\lambda)$ (21) reduces to a polynomial. Moreover this polynomial satisfies the conditions of Proposition 1. But any polynomial which satisfies these conditions reduces to a constant since we have the equalities $\alpha_j = 0$ for $j < 0$ and Relations (6) of Proposition 1. Thus $\psi(\lambda)$ is equal to a constant. According to the assumptions of Proposition 2, this constant must be real and positive, as stated.

§ 38. Theorem of factorization on the unit circle

1. Statement of the theorem

Let $\lambda \mapsto X(\lambda)$ be a matrix-valued function which assigns to every complex number λ the $m \times m$-matrix $X(\lambda)$, with complex entries. We assume that:

a) There exists a positive integer n such that the entries of the matrix-valued function $\lambda \mapsto \lambda^n X(\lambda)$ are polynomials whose degrees do not exceed $2n$.

b) For every $\lambda \neq 0$, $X(\lambda)$ satisfies the equality [1])

$$X(\lambda) = \bar{X}'(1/\lambda). \tag{1}$$

c) $X(\lambda)$ is positive semi-definite on the unit circle. In other words, the inequality

$$v^* X(\lambda)\, v \geqslant 0, \; (|\lambda| = 1). \tag{2}$$

[1]) See the conventions in Section 1 of § 37. If the matrix $\lambda^n X(\lambda)$ has real coefficients, the expression $X'(1/\lambda)$ is equal to $\bar{X}'(1/\lambda)$. From (1) it follows that the matrix $X(\lambda)$ is hermitian on the unit circle since $|\lambda| = 1$ entails $1/\lambda = \bar{\lambda}$.

is satisfied for every m-vector v and every complex number λ such that $|\lambda| = 1$.

Under these conditions we have the following property of factorization:

Theorem 1. *If $X(\lambda)$ has the properties specified above, there exists at least one polynomial ρ satisfying the equality*

$$\det X(\lambda) = \bar{\rho}\left(\frac{1}{\lambda}\right) \rho(\lambda), \quad (\lambda \neq 0) \tag{3}$$

and for every polynomial ρ with this property, there exists a matrix polynomial D such that

$$X(\lambda) = \bar{D}'(1/\lambda)\, D(\lambda), \quad (\lambda \neq 0) \tag{4}$$

$$\det D(\lambda) = \rho(\lambda). \tag{5}$$

If the polynomial ρ satisfies the condition $\rho(0) \neq 0$, the degree of the matrix polynomial D cannot exceed n. If $\lambda^n X(\lambda)$ and $\rho(\lambda)$ have real coefficients, then $D(\lambda)$ can be chosen such that its coefficients are real [1])

2. Preliminary remarks

The fact that, if $\rho(0) \neq 0$, the degree of the matrix polynomial D cannot exceed n is proved by simply writing the matrix polynomial D in the form

$$D(\lambda) = (D_0 + \lambda D_1 + \ldots + \lambda^q D_q), \quad D_q \neq 0,$$

where $D_0, D_1, \ldots, D_q$ are constant matrices and q is a positive integer (equal to the degree of D). The condition $\rho(0) \neq 0$ entails $\det D_0 \neq 0$ (see (5)). The hypothesis $q > n$ leads to a contradiction. Indeed by introducing the above expression of $D(\lambda)$ in (4) we find that the term of the function $\lambda \mapsto X(\lambda)$ with λ at the lowest power is $\lambda^{-q} \bar{D}_q' D_0$ and is different from zero, since the matrix D_0 is nonsingular. Hence the inequality $q > n$ implies that the function $\lambda \mapsto \lambda^n X(\lambda)$ contains negative powers of λ. This contradicts (a) and proves that $q \leqslant n$.

---

[1]) A matrix polynomial is said to have real coefficients if all the polynomials which constitute the entries of the matrix have real coefficients. The highest of the degrees of these polynomials is, by definition, the degree of the matrix polynomial.

If all the entries of $X(\lambda)$ are zero, the proof of Theorem 1 is immediate. Eq. (3) is satisfied by $\rho(\lambda) \equiv 0$. Any polynomial ρ which satisfies (3), that is $\bar{\rho}(1/\lambda)\rho(\lambda) \equiv 0$, satisfies for $|\lambda| = 1$ the equality $|\rho(\lambda)|^2 \equiv 0$ (since for $|\lambda| = 1$ we have $\bar{\lambda} = 1/\lambda$), which implies $\rho(\lambda) \equiv 0$. Relations (4) and (5) are satisfied by $D(\lambda) \equiv 0$ and Theorem 1 is trivial. We assume in the sequel that $X(\lambda)$ is not identically zero.

We now prove that there exists at least one polynomial ρ which satisfies (3). If det $X(\lambda) \equiv 0$. Then (3) is satisfied by $\rho(\lambda) \equiv 0$. If det $X(\lambda) \not\equiv 0$ then, by expressing Assumption (c) with the help of Sylvester's condition, we find that det $X(\lambda)$ is real and nonnegative whenever $|\lambda| = 1$. Hence det $X(\lambda)$ has the form and the properties from Proposition 1, § 37 and therefore can be factorized according to (9, § 37). This gives (3). As shown in the proof of Proposition 1, § 37, there exist in general several distinct polynomials which satisfy (3).

We finally show that it is sufficient to prove Theorem 1 under the assumption that det $X(\lambda) \not\equiv 0$. Indeed, since $\lambda \mapsto \lambda^n X(\lambda)$ is a matrix polynomial, it follows; as is well known (e.g. see F. R. Gantmacher [1], Chapter VI, Theorem 3 and Definition 2′), that there exist three matrix polynomials P, Q and R such that

$$\lambda^n \mathrm{X}(\lambda) = P(\lambda)\, Q(\lambda)\, R(\lambda) \tag{6}$$

$$\det P(\lambda) = \varkappa_1, \quad \det R(\lambda) = \varkappa_2 \tag{7}$$

$$Q(\lambda) = \mathrm{diag}\,(\eta_1(\lambda),\ \eta_2(\lambda),\ \ldots,\ \eta_m(\lambda)), \tag{8}$$

where $\varkappa_1$ and $\varkappa_2$ are non-zero constants, η_j, $j = 1, 2, \ldots, m$ are polynomials and η_j divides η_{j+1} for $j = 1, 2, \ldots, m - 1$. If det $X(\lambda) \equiv 0$ then $Q(\lambda)$ can be written as

$$Q(\lambda) = \begin{pmatrix} \overbrace{Q_{11}(\lambda)}^{m_1} & \overbrace{0}^{m_2} \\ 0 & 0 \end{pmatrix} \begin{matrix} \}\, m_1 \\ \}\, m_2 \end{matrix}, \quad m_1 + m_2 = m,$$

where (in the case $X(\lambda) \not\equiv 0$) the expression det $Q_{11}(\lambda)$ is not identically zero. (The trivial case $X(\lambda) \equiv 0$ has been considered before). Let us introduce the matrix

$$Z(\lambda) = \bar{T}'\left(\frac{1}{\lambda}\right) \mathrm{X}(\lambda)\, T(\lambda), \quad (\lambda \neq 0),$$

where $T(\lambda) = R^{-1}(\lambda)$. We note that T is a matrix polynomial. Since $X(\lambda)$ satisfies Conditions (a), (b) and (c), $Z(\lambda)$ satisfies the same conditions. Furthermore, from (6) and from the form of $Q(\lambda)$ it follows that $Z(\lambda)$ can be written as

$$Z(\lambda) = \begin{pmatrix} \overbrace{Z_{11}(\lambda)}^{m_1} & \overbrace{0}^{m_2} \\ Z_{21}(\lambda) & 0 \end{pmatrix} \begin{matrix} \} \; m_1 \\ \} \; m_2 \end{matrix}$$

Using the property $Z'(1/\lambda) = \overline{Z}(\lambda)$ (which follows as we remarked, from Condition (b)), we find that $Z_{21}(\lambda)$ must be identically zero. Since $Z(\lambda)$ satisfies Conditions (a), (b) and (c), $Z_{11}(\lambda)$ satisfies the same conditions. Clearly, if we can factorize $Z_{11}(\lambda)$ as required in the theorem, then $X(\lambda)$ admits also a factorization with the properties required in Theorem 1. Thus if the conditions $X(\lambda) \not\equiv 0$ and det $X(\lambda) \equiv 0$ are satisfied, our problem reduces to a similar problem of factorization for a non-zero matrix of lower dimensions.

By applying successively this procedure, if necessary, after a finite number of steps we reduce the problem to the factorization of a matrix whose determinant is not identically zero. Hence, it is sufficient to prove the theorem only for the case det $X(\lambda) \not\equiv 0$.

3. Some additional assumptions

In order to simplify the proof we shall temporarily introduce a few additional assumptions. We replace Assumption (c) by

C) $X(\lambda)$ *is positive definite on the unit circle* and we assume that the polynomial ρ which satisfies (3) has also the properties

d) $\rho(0) \neq 0$

e) *If the number* $\lambda_0 \neq 0$ *satisfies the equation* $\rho(\lambda_0) = 0$, *then*

$$\rho(1/\overline{\lambda}_0) \neq 0.$$

Conditions (C) and (e) will be eliminated in the course of the proof. As for Condition (d) we show right now that if Theorem 1 is proved assuming that $\rho(0) \neq 0$ is satisfied, then the theorem is also valid in the case $\rho(0) = 0$. Assume that (3) is satisfied by a polynomial ρ which has the property $\rho(0) = 0$. As we have seen in Section 2, we can assume without loss of generality, that the condition det $X(\lambda) \not\equiv 0$ is satisfied. Then, from (3) it follows that $\rho(\lambda)$ is not identically zero. Hence the condition $\rho(0) = 0$ can be satisfied only if the polynomial has the form

$$\rho(\lambda) = \lambda^p \tilde{\rho}(\lambda),$$

where p is a positive integer and $\tilde{\rho}(\lambda)$ is a new polynomial satisfying the relation of factorization (3) and the condition $\tilde{\rho}(0) \neq 0$. Assuming that Theorem 1 is proved for polynomials satisfying the condition $\rho(0) \neq 0$, we find that there exists a matrix polynomial $\tilde{D}$ such that

$$X(\lambda) = \overline{\tilde{D}}'\left(\frac{1}{\lambda}\right)\tilde{D}(\lambda), \qquad \det \tilde{D}(\lambda) = \tilde{\rho}(\lambda).$$

Let us write in an arbitrary manner the polynomial λ^p in the form of a product: $\lambda^p = \lambda^{j_1}\lambda^{j_2}\ldots\lambda^{j_m}$ where each one of the numbers j_k may take one of the values 0, 1, 2,..., p. We define the matrix $\Delta(\lambda)=\operatorname{diag}(\lambda^{j_1}, \lambda^{j_2},\ldots, \lambda^{j_m})$ which obviously satisfies the relations

$$\overline{\Delta}'\left(\frac{1}{\lambda}\right)\Delta(\lambda) = I, \qquad \det \Delta(\lambda) = \lambda^p$$

(where I is the $m \times m$-identity-matrix). Then $D(\lambda) = \Delta(\lambda)\tilde{D}(\lambda)$ satisfies (4) and (5). Therefore, in the proof of Theorem 1, we can accept Assumption (d) without loss of generality.

4. An asymmetrical factorization of the matrix $\lambda^n X(\lambda)$

We consider again Relations (6)—(8). From (6), (7) and (3) one obtains

$$\det Q(\lambda) = \varkappa_3 \lambda^{nm}\overline{\rho}\left(\frac{1}{\lambda}\right)\rho(\lambda), \quad \varkappa_3 = \frac{1}{\varkappa_1\varkappa_2} \tag{9}$$

Expressing $\det Q(\lambda)$ with the help of (8), we can write (9) as

$$\eta_1(\lambda)\,\eta_2(\lambda)\ldots\eta_m(\lambda) = \varkappa_3\lambda^{nm}\,\overline{\rho}\left(\frac{1}{\lambda}\right)\rho(\lambda). \tag{10}$$

If we write the polynomials of (10) as products of irreducible factors (in the field of complex numbers) it is obvious that the left-hand member of (10) contains the same factors as the right-hand member — arranged perhaps in a different order. Notice also that the relations $\rho(\lambda_0) = 0$ and $\lambda_0^{mn}\overline{\rho}(1/\lambda_0) = 0$ cannot be simultaneously satisfied, because the first equality implies that $\lambda_0 \neq 0$ (see Condition (d)) and the second equality gives $\rho(1/\overline{\lambda}_0) = 0$, which contradicts Assumption (e) of Section 3. Therefore, in the expression of each polynomial η_j from (10), we can determine the factors which are also factors of the polynomial ρ. Let $\eta_{j,2}$ be the product of all these factors multiplied by a convenient constant such that

$$\rho(\lambda) = \prod_{j=1}^{m} \eta_{j,2}(\lambda). \tag{11}$$

Let $\eta_{j,1}$ be the product of all the other factors of η_j, multiplied by a convenient constant such that

$$\eta_j(\lambda) = \eta_{j,1}(\lambda)\, \eta_{j,2}(\lambda). \tag{12}$$

The polynomials $\eta_{j,1}$ have no common factor with the polynomials ρ and $\eta_{j,2}$. Using (12) we can factorize $Q(\lambda)$, (8) in the form

$$Q(\lambda) = Q_1(\lambda)\, Q_2(\lambda), \tag{13}$$

where

$$Q_1(\lambda) = \operatorname{diag}(\eta_{1,1}(\lambda),\ \eta_{2,1}(\lambda),\ \ldots,\ \eta_{m,1}(\lambda)), \tag{14}$$

$$Q_2(\lambda) = \operatorname{diag}(\eta_{1,2}(\lambda),\ \eta_{2,2}(\lambda),\ \ldots,\ \eta_{m,2}(\lambda)). \tag{15}$$

Using (11) we can write

$$\det Q_2(\lambda) = \rho(\lambda). \tag{16}$$

From (9), (13) and (16) we deduce

$$\det Q_1(\lambda) = \varkappa_3 \lambda^{nm} \bar{\rho}(1/\lambda). \tag{17}$$

Replacing in (6) $Q(\lambda)$ by (13) we obtain

$$\lambda^n X(\lambda) = A_0(\lambda)\, B_0(\lambda) \tag{18}$$

where

$$A_0(\lambda) = P(\lambda)\, Q_1(\lambda), \qquad B_0(\lambda) = Q_2(\lambda)\, R(\lambda). \tag{19}$$

5. A family of factorization relations

From (18) and (19) we can obtain a sequence of relations of factorization

$$\lambda^n X(\lambda) = A_j(\lambda)\, B_j(\lambda), \qquad j = 1, 2, \ldots, \tag{20}$$

where the matrices $A_j(\lambda)$ and $B_j(\lambda)$ have the expressions

$$A_j(\lambda) = P(\lambda)\, Q_1(\lambda)\, K_1(\lambda)\, K_2(\lambda) \ldots K_j(\lambda), \tag{21}$$

$$B_j(\lambda) = K_j^{-1}(\lambda)\, K_{j-1}^{-1}(\lambda) \ldots K_1^{-1}(\lambda)\, Q_2(\lambda)\, R(\lambda), \tag{22}$$

and $K_1 \ldots K_j$ are matrix polynomials whose determinants are constant and different from zero. (Substituting (21) and (22) in (20) one easily sees that one obtains again (18), (19)).

Since the determinants of the matrix polynomials P, R and K_i are constant and different from zero, we obtain from (21) and (22)

$$\det A_j(\lambda) = \varkappa_{j,1} \det Q_1(\lambda) = \varkappa_{j,2} \lambda^{nm} \overline{\rho}(1/\lambda), \tag{23}$$

$$\det B_j(\lambda) = \varkappa_{j,3} \det Q_2(\lambda) = \varkappa_{j,3}\, \rho(\lambda) \tag{24}$$

where we have used (16) and (17) and introduced the constants $\varkappa_{j,1}$, $\varkappa_{j,2}$ and $\varkappa_{j,3}$.

6. A special way of writing polynomial matrices

Any matrix polynomial $\lambda \mapsto A(\lambda)$ can be written as

$$A(\lambda) = \left(A_0 + \frac{1}{\lambda} A_1 + \frac{1}{\lambda^2} A_2 + \ldots + \frac{1}{\lambda^p} A_p\right) \operatorname{diag}(\lambda^{n_1}, \lambda^{n_2}, \ldots, \lambda^{n_m}), \tag{25}$$

where $A_0, A_1, \ldots, A_p$ are constant matrices and $n_1, n_2 \ldots, n_m$ are positive integers. Indeed, the entry $\alpha_{jk}(\lambda)$ — situated on the j-th row and on the k-th column of $A(\lambda)$ — can be written as

$$\alpha_{jk}(\lambda) = \alpha_{jk,0} \lambda^{n_{jk}} + \alpha_{jk,1} \lambda^{n_{jk}-1} + \ldots + \alpha_{jk,n_{jk}} \tag{26}$$

for some positive integer n_{jk}. Defining n_k as

$$n_k = \max\,(n_{1k}, n_{2k}, \ldots, n_{mk}),$$

all the entries $A(\lambda)$ belonging to the k-th column can be written as

$$\alpha_{ik}(\lambda) = \tilde{\alpha}_{jk,0}\, \lambda^{n_k} + \tilde{\alpha}_{jk,1} \lambda^{n_k-1} + \\ + \ldots + \tilde{\alpha}_{jk,n_k}, \qquad j = 1, 2, \ldots, m. \tag{27}$$

The new coefficients $\tilde{\alpha}_{jk,i}$ are obtained by identifying the polynomials (26) and (27). Let p be the highest of the indices i for which there exists at least one coefficient $\tilde{\alpha}_{jk,i} \neq 0$. We introduce the matrices A_i, $i = 0, 1, \ldots, p$ whose entries situated on the j-th row and the k-th column are equal, respectively, to the coefficients $\tilde{\alpha}_{jk,i}$. With these notations, (25) is only another way of writing (27).

7. A nonsingular factorization

By an adequate choice of $K_1(\lambda)$, $K_2(\lambda), \ldots, K_j(\lambda)$ of (21) we obtain a "nonsingular" factorization with the property that, if $A_j(\lambda)$ is written in the form (25) then the corresponding term A_0 is a nonsingular matrix. This property can be ensured by a procedure indicated by N. I. Mushelishvili (see N. P. Vekua [1]).

Consider a factorization of the form (20) where $A_j(\lambda)$ and $B_j(\lambda)$ are expressed by (21) and (22) for a given index j. Assume that $A_j(\lambda)$, written as in (25), takes the form

$$A_j(\lambda) = \left(A_0 + \frac{1}{\lambda} A_1 + \ldots + \frac{1}{\lambda^p} A_p \right) \operatorname{diag}(\lambda^{n_1}, \lambda^{n_2}, \ldots, \lambda^{n_m}) \quad (28)$$

where A_0 is singular. Then there exists an m-vector $v \neq 0$ such that $A_0 v = 0$. Let v_k be the components of v. One can always find a nonnegative integer $q \leqslant m$ such that $n_q \geqslant n_k$, for every k such that $v_k \neq 0$. Consider a new factorization of the form (20) — (22) (with j replaced by $j + 1$) by introducing the new matrix

$$K_{j+1}(\lambda) = \begin{pmatrix} 1 & 0 & \ldots & v_1 \lambda^{n_q - n_1} & \ldots & 0 \\ 0 & 1 & \ldots & v_2 \lambda^{n_q - n_2} & \ldots & 0 \\ \cdot & \cdot & \ldots & \cdot & \ldots & \cdot \\ 0 & 0 & \ldots & v_m \lambda^{n_q - n_m} & \ldots & 1 \end{pmatrix} \quad (29)$$

$$\underbrace{}_{q} \quad \underbrace{}_{m-q}$$

(This matrix differs from the $m \times m$-identity-matrix only by the column q whose elements have the expression $v_j \lambda^{n_q - n_j}$, $j = 1, 2, \ldots, m$). Moreover K_{j+1} is a matrix polynomial (since we have $n_q - n_k \geqslant 0$ whenever $v_k \neq 0$) and its determinant equals v_q, hence is constant and different from zero (owing to our choice of the index q). Therefore K_{j+1} satisfies all the conditions required to K_i in (21), (22). Clearly we also have

$$\operatorname{diag}(\lambda^{n_1}, \lambda^{n_2}, \ldots, \lambda^{n_m})\, K_{j+1}(\lambda) = C \operatorname{diag}(\lambda^{n_1}, \ldots, \lambda^{n_m}), \quad (30)$$

where

$$C = \underbrace{\begin{pmatrix} 1 & 0 & \cdot\cdot & v_1 \\ 0 & 1 & \cdot\cdot & v_2 \\ \cdot & \cdot & \cdot\cdot & \cdot \\ 0 & 0 & \cdot\cdot & v_m \end{pmatrix}}_{q}\underbrace{\begin{matrix} \cdot & 0 \\ \cdot & 0 \\ \cdot & \cdot \\ \cdot & 1 \end{matrix}}_{m-q} \quad (31)$$

(this matrix differs from the identity matrix only by the column q which is equal to the vector v). Using (30) and multiplying on the right $A_j(\lambda)$ (28) by $K_{j+1}(\lambda)$ we obtain

$$A_{j+1}(\lambda) = A_j(\lambda)\, K_{j+1}(\lambda) =$$

$$= \left(\widetilde{A}_0 + \frac{1}{\lambda}\widetilde{A}_1 + \ldots + \frac{1}{\lambda^p}\widetilde{A}_p\right) \operatorname{diag}(\lambda^{n_1}, \lambda^{n_2}, \ldots, \lambda^{n_m}), \quad (32)$$

where

$$\widetilde{A}_k = A_k C, \qquad k = 0, 1, 2, \ldots, p. \quad (33)$$

From (31) and from the condition $A_0 v = 0$ it follows that the matrix $\widetilde{A}_0$, given by (33) for $k = 0$, has the column k equal to zero. Therefore the right-hand member of (32) does not change if the number n_q is replaced by $n_q - 1$ and if the column q of the matrices A_k is replaced by the column q of the matrices A_{k+1}, an operation to be performed successively for $k = 0, 1, 2, \ldots$ $\ldots, p - 1$, after which the column q of the matrix A_p is replaced by zero.

In general the operations described above can be repeated and each time one number n_q is replaced by $n_q - 1$. Since $n_j \geqslant 0$ the procedure can be applied at most a finite number of times. Then the procedure cannot be further applied either because the determinant of the matrix A_0 in (28) written for the respective step is different from zero, or because all the numbers n_j vanish. In the latter case $A_j(\lambda)$ reduces to the constant matrix A_0. Then from (20) and from Condition (C) of Section 3 it follows that A_0 is nonsingular. Therefore by applying the procedure described above the highest possible number of times we always obtain a nonsingular factorization.

8. Properties of the nonsingular factorizations

Using the procedure described in the preceding section we obtain a nonsingular factorization of the form (see (20))

$$\lambda^n X(\lambda) = A(\lambda)\, B(\lambda), \tag{34}$$

where (see (21) and (22))

$$A(\lambda) = P(\lambda)\, Q_1(\lambda)\, K(\lambda), \quad B(\lambda) = K^{-1}(\lambda)\, Q_2(\lambda)\, R(\lambda), \tag{35}$$

and where K is a matrix polynomial with the property $\det K(\lambda) =$ $=$ constant $\neq 0$. Hence, we also have using (16) and (17)

$$\det A(\lambda) = \varkappa_4 \lambda^{nm} \bar{\rho}(1/\lambda), \quad \det B(\lambda) = \varkappa_4' \rho(\lambda), \tag{36}$$

where $\varkappa_4$ and $\varkappa'_4$ are some constants different from zero. Furthermore $A(\lambda)$ can be written in the form (25) (where A_0 is nonsingular). Hence

$$\bar{A}'\left(\frac{1}{\lambda}\right) = \operatorname{diag}(\lambda^{-n_1}, \lambda^{-n_2}, \ldots, \lambda^{-n_m})(A_0^* + \lambda A_1^* + \ldots + \lambda^p A_p^*), \tag{37}$$

where

$$\det A_0 \neq 0. \tag{38}$$

From (34) we obtain

$$X(\lambda) = \lambda^{-n} A(\lambda)\, S(\lambda) \bar{A}'\left(\frac{1}{\lambda}\right) \lambda^n, \tag{39}$$

where

$$S(\lambda) = B(\lambda)\, \lambda^{-n} \left(\bar{A}'\left(\frac{1}{\lambda}\right)\right)^{-1}, \tag{40}$$

as one can easily see by substituting (40) into (39). The existence of $\left(A'\left(\frac{1}{\lambda}\right)\right)^{-1}$ (excepting the point $\lambda = 0$ and a finite number of values of λ equal to the roots of the polynomial ρ) results from (36). We now bring $S(\lambda)$ to a simpler form. Using (37) we can write

$$\lambda^{-n}\left(\bar{A}'\left(\frac{1}{\lambda}\right)\right)^{-1} =$$

$$=[A_0^*+\lambda A_1^*+\ldots+\lambda^p A_p^*]^{-1}\,\mathrm{diag}(\lambda^{n_1-n}, \lambda^{n_2-n}, \ldots, \lambda^{n_m-n}). \qquad (41)$$

We clearly have

$$[A_0^* + \lambda A_1^* + \ldots + \lambda^p A_p^*]^{-1} =$$

$$= \frac{F(\lambda)}{\det(A_0^* + \lambda A_1^* + \ldots + \lambda^p A_p^*)}, \qquad (42)$$

where F is a matrix polynomial. From (36) we deduce

$$\det \bar{A}'\left(\frac{1}{\lambda}\right) = \bar{\varkappa}_4 \lambda^{-nm}\rho(\lambda), \qquad (43)$$

whence, using (37), we obtain

$$\det(A_0^* + \lambda A_1^* + \ldots + \lambda^p A_p^*) = \bar{\varkappa}_4 \lambda^{n_0 - nm}\rho(\lambda), \qquad (44)$$

where

$$n_0 = n_1 + n_2 + \ldots + n_m. \qquad (45)$$

This shows that $n_0 - nm \leqslant 0$, since otherwise from (44) written for $\lambda = 0$ we would obtain $\det A_0^* = 0$ which contradicts (38). From (44) we also obtain $n_0 - nm \geqslant 0$ since otherwise, by multiplying (44) by the polynomial λ^{nm-n_0} and writing the resulting equality for $\lambda = 0$, we would obtain $\rho(0) = 0$ which contradicts Condition (d) from Section 3. Thus

$$n_0 = nm = n_1 + n_2 + \ldots + n_m. \qquad (46)$$

From (40), (41), (42), (44), and (46) we obtain

$$S(\lambda) = \frac{B(\lambda)\,F(\lambda)}{\bar{\varkappa}_4 \rho(\lambda)}\,\mathrm{diag}(\lambda^{n_1-n}, \lambda^{n_2-n}, \ldots, \lambda^{n_m-n}). \qquad (47)$$

and from (39) we deduce that

$$S(\lambda) = A^{-1}(\lambda)\,X(\lambda)\left(\bar{A}'\left(\frac{1}{\lambda}\right)\right)^{-1}. \qquad (48)$$

From Condition (C) (Section 3) and (34) it follows that $A(\lambda)$ is nonsingular for $|\lambda| = 1$. Since $X(\lambda)$ is hermitian and positive definite on the unit circle (see Condition (b) of Section 1 and Condition (C) of Section 3), we derive from (48) the conclusion that $S(\lambda)$ is also hermitian and positive definite on the unit circle. Hence for every vector $g \neq 0$ the expression $g^* S(\lambda) g$ is real on the unit circle and satisfies the inequality

$$g^* S(\lambda)\, g > 0, \qquad (|\,\lambda\,| = 1,\ g \neq 0). \tag{49}$$

Using these results we show that all the numbers n_j are equal to n. We introduce the vectors e_j whose j component is equal to the unity while all the other components are zero. From (47) we deduce

$$e_j^* S(\lambda)\, e_j = \frac{\pi_j(\lambda)}{\rho(\lambda)}\, \lambda^{n_j - n}, \tag{50}$$

where π_j is a polynomial in λ whose expression is $\pi_j(\lambda) = e_j^* B(\lambda) F(\lambda) e_j / \varkappa_4$. By Condition (49) the expression (50) is real and positive on the unit circle. By virtue of the Hypotheses (C), (d) and (e) of Section 3, the polynomial ρ satisfies the conditions required to the polynomial π_2 in Proposition 2, § 37. If $n_j - n \geqslant 0$, then the conditions of Proposition 2, § 37 are satisfied and therefore expression (50) is equal to a real and positive constant. If $n_j - n > 0$, then by writing that the right-hand member of Relation (50) is equal to a constant, multiplying the relation thus obtained by $\rho(\lambda)$ and considering the particular value $\lambda = 0$, we obtain the relation $\rho(0) = 0$, which contradicts condition $\rho(0) \neq 0$. Hence, the numbers n_j must satisfy the inequalities $n_j - n \leqslant 0$, $j = 1, 2, \ldots, m$. Taking also into account (46) we obtain

$$n_j = n, \qquad j = 1, 2, \ldots, m. \tag{51}$$

9. Bringing the nonsingular factorization to the form required in Theorem 1

Using (51) we obtain from (47)

$$g^* S(\lambda)\, g = \frac{\pi_g(\lambda)}{\rho(\lambda)}, \tag{52}$$

where $\pi_g(\lambda)$ is a polynomial, whose expression is $\pi_g(\lambda) = g^* B(\lambda) F(\lambda) g/\bar{\varkappa}_4$. As before, (52) satisfies the conditions of Proposition 2, § 37 and therefore is equal to a constant. Therefore the left-hand member of (52) is also equal to $g^* S(1) g$ whence one obtains the equality

$$g^* S g = 0, \quad (\text{where } S = S(\lambda) - S(1)), \tag{53}$$

which is valid for every vector g. From (53) and from the obvious relation

$$x^* S y = \frac{1}{4}((x + y)^* S(x + y) - (x - y)^* S(x - y) -$$

$$- \mathrm{i}(x + \mathrm{i}y)^* S(x + \mathrm{i}y) + \mathrm{i}(x - \mathrm{i}y)^* S(x - \mathrm{i}y))$$

one obtains the equality $x^* S y = 0$ for every x and y, which implies $S = 0$, i.e.

$$S(\lambda) = S(1). \tag{54}$$

Since the matrix $S(\lambda)$ is Hermitian and positive definite on the unit circle, the matrix $S(1)$ is also Hermitian and positive definite. Hence there exists a nonsingular constant matrix M_0 such that (see F. R. Gantmacher [1], Chapter IX, § 11):

$$S(\lambda) = S(1) = M_0^* M_0. \tag{55}$$

Replacing in (39) the matrix $S(\lambda)$ by the expression (55) and introducing the notation

$$\tilde{D}(\lambda) = M_0 \bar{A}'\left(\frac{1}{\lambda}\right) \lambda^n, \tag{56}$$

we obtain the equality

$$X(\lambda) = \bar{\tilde{D}}'\left(\frac{1}{\lambda}\right) \tilde{D}(\lambda), \tag{57}$$

which is of the same form (4). From (56) and (43), we obtain

$$\det \tilde{D}(\lambda) = \varkappa_5^m \rho(\lambda). \tag{58}$$

If we calculate the determinant of matrix $X(\lambda)$ taking into account (57), (58) and (3), we obtain the condition $|\varkappa_5|^2 = 1$. Then the matrix

$$D(\lambda) = \frac{1}{\varkappa_5} \tilde{D}(\lambda) = \frac{1}{\varkappa_5} \lambda^n M_0 \bar{A}'\left(\frac{1}{\lambda}\right) \tag{59}$$

satisfies Relation (4) as one can easily see by introducing in (57) the expression of matrix $\tilde{D}(\lambda)$ derived from (59) and by using the equality $|\varkappa_5|^2 = 1$. Furthermore from (58) and (59) it follows that (5) is satisfied. Substituting (37) into (59) and using (51) we obtain

$$D(\lambda) = \frac{1}{\varkappa_5} M_0(A_0^* + \lambda A_1^* + \ldots + \lambda^p A_p^*), \tag{60}$$

which shows that $D(\lambda)$ a matrix polynomial.

If $\lambda^n X(\lambda)$ and $\rho(\lambda)$ have real coefficients, then all the matrix polynomials which occur in the proof can be chosen such as to have real coefficients. Hence $D(\lambda)$ will also have real coefficients.

These remarks conclude the proof of Theorem 1 under the assumptions introduced in Section 3. The rest of the proof eliminates these restrictions.

10. More about Assumption (e).

We shall show that if $X(\lambda)$ has the Property

(e') "*All the nonvanishing roots of the polynomial* $\lambda \mapsto \lambda^{nm}$ *det* $X(\lambda)$ *are distinct*", then every polynomial ρ which satisfies (3) satisfies also Condition (e) of Section 3. Indeed, consider an arbitrary factorization (3) and assume that Property (e) is not satisfied. Hence there exists a pair of numbers λ_1 and λ_2 with the properties

$$\rho(\lambda_1) = 0, \quad \rho(\lambda_2) = 0, \quad \lambda_1\bar{\lambda}_2 = 1. \tag{61}$$

Therefore the polynomial ρ can be written as

$$\rho(\lambda) = (\lambda - \lambda_1)\left(\lambda - \frac{1}{\bar{\lambda}_1}\right)\rho_0(\lambda), \tag{62}$$

where ρ_0 is a new polynomial. By (62) we have

$$\lambda^{nm}\bar{\rho}\left(\frac{1}{\lambda}\right) = \lambda^{nm}\left(\frac{1}{\lambda} - \bar{\lambda}_1\right)\left(\frac{1}{\lambda} - \frac{1}{\lambda_1}\right)\bar{\rho}_0\left(\frac{1}{\lambda}\right) =$$

$$= \lambda^{nm-2}\frac{\bar{\lambda}_1}{\lambda_1}\left(\lambda - \frac{1}{\bar{\lambda}_1}\right)(\lambda - \lambda_1)\bar{\rho}_0\left(\frac{1}{\lambda}\right). \tag{63}$$

Substituting (62) and (63) in the equality

$$\lambda^{nm}\det X(\lambda) = \lambda^{nm}\bar{\rho}(1/\lambda)\rho(\lambda)$$

which is derived from (3), it is easy to see that λ_1 is a double root of the polynomial $\lambda \mapsto \lambda^{nm} \det X(\lambda)$; this contradicts condition (e').

11. Eliminating restrictions (C) and (e)

Using the discussion from Section 3 concerning Condition (d), we see that in order to finish the proof of Theorem 1, it only remains to eliminate restrictions (C) and (e). This will be achieved by a passage to the limit.

We introduce the matrix

$$J(\lambda) = \operatorname{diag}(\mu_1(\lambda), \mu_2(\lambda), \ldots, \mu_m(\lambda)), \tag{64}$$

where

$$\mu_k(\lambda) = (\lambda^n + k + 1)\left(\frac{1}{\lambda^n} + k + 1\right), \qquad k = 1, 2, \ldots, m. \tag{65}$$

It is easy to see, that the polynomial $\lambda \mapsto \lambda^{nm} \det J(\lambda)$ is of degree $2nm$ and has $2nm$ distinct roots. We show that the polynomial

$$\psi(\lambda; \varepsilon) = \lambda^{nm} \det (X(\lambda) + \varepsilon J(\lambda)) \tag{66}$$

has a similar property for every ε, except at most a finite number of values ε_k, $k = 1, 2, \ldots, N$. To this end we introduce the expression

$$\tilde{\psi}(\lambda; \nu) = \lambda^{nm} \det (\nu X(\lambda) + J(\lambda)). \tag{67}$$

Taking into account Condition (a) and (64), the above expression can be written as

$$\widetilde{\psi}(\lambda\,;\,\nu) = \sum_{j=0}^{2nm} a_j(\nu)\lambda^j, \tag{68}$$

where the functions $\nu \mapsto a_j(\nu)$, for $j = 0, 1, \ldots, 2nm$, are polynomials. From (67) we obtain

$$\widetilde{\psi}(\lambda\,;\,0) = \lambda^{nm} \det J(\lambda), \tag{69}$$

and hence $\lambda \mapsto \widetilde{\psi}(\lambda, 0)$ is a polynomial of degree $2nm$ with $2nm$ distinct roots. Hence $a_{2nm}(0) \neq 0$. Therefore, the equation $a_{2nm}(\nu) = 0$ can have only a finite number of solutions ν_k. Consider now, for an arbitrary number ν, the resultant of the polynomials $\lambda \mapsto \widetilde{\psi}(\lambda\,;\,\nu)$ and $\lambda \mapsto d\widetilde{\psi}(\lambda\,;\,\nu)/d\lambda$. Obviously the resultant of these polynomials is a polynomial in ν, which is different from zero for $\nu = 0$ (since, as we have seen, the polynomial in (69) has no multiple roots). Hence, the resultant of the polynomials considered is not identically zero. Therefore (since it is a polynomial in ν), it can vanish for at most a finite number of distinct values ν_k'. These values excepted, the polynomial $\lambda \mapsto \widetilde{\psi}(\lambda\,;\,\nu)$ has no multiple roots. Thus we have obtained the conclusion that, excepting a finite number of points ν_k'', the function $\lambda \mapsto \widetilde{\psi}(\lambda\,;\,\nu)$ is a polynomial of degree $2nm$, having $2nm$ distinct roots. Taking into account that (66) and (67) are connected by the relation

$$\psi(\lambda\,;\,\varepsilon) = \varepsilon^m \widetilde{\psi}\left(\lambda\,;\,\frac{1}{\varepsilon}\right).$$

we immediately see that there exists an interval $0 < \varepsilon < \varepsilon_0$ wherein the polynomial $\lambda \mapsto \psi(\lambda\,;\,\varepsilon)$ is of degree $2nm$ and has $2nm$ distinct roots. In other words, for the matrix $X(\lambda\,;\,\varepsilon) = X(\lambda) + \varepsilon J(\lambda)$, Condition (e′) of Section 10 is satisfied in an interval $0 < \varepsilon < \varepsilon_0$.

Furthermore, since $X(\lambda)$ satisfies Conditions (a), (b) and (c), $X(\lambda) + \varepsilon J(\lambda)$ satisfies the same conditions. In addition, the latter matrix satisfies even Condition (C) of Section 3, since from (64) and (65) one obtains the inequality $v^*J(\lambda)v > 0$, for every scalar λ and every vector v which satisfy the conditions $|\lambda| = 1$ and $\|v\| = 1$. Using also the conclusion of Section 10, it follows that $X(\lambda) + \varepsilon J(\lambda)$, $(0 < \varepsilon < \varepsilon_0)$ satisfies all the conditions under which Theorem 1 was proved in the previous sections. Therefore we can assert that for every polynomial $\lambda \mapsto \rho(\lambda;\ \varepsilon)$ which satisfies the equality

$$\det (X(\lambda) + \varepsilon J(\lambda)) = \bar{\rho}\left(\frac{1}{\lambda};\varepsilon\right) \rho(\lambda;\ \varepsilon), \qquad (0 < \varepsilon < \varepsilon_0), \tag{70}$$

there exists a matrix polynomial $\lambda \mapsto D(\lambda;\ \varepsilon)$, such that

$$X(\lambda) + \varepsilon J(\lambda) = \bar{D}'\left(\frac{1}{\lambda};\varepsilon\right) D(\lambda;\varepsilon) \tag{71}$$

$$\det D(\lambda;\ \varepsilon) = \rho(\lambda;\ \varepsilon). \tag{72}$$

We now show that, given an arbitrary polynomial $\lambda \mapsto \rho(\lambda)$ which satisfies (3), we can always find a polynomial $\lambda \to \rho(\lambda;\ \varepsilon)$ which satisfies (70) and

$$\lim_{\varepsilon \to 0} \rho(\lambda;\ \varepsilon) = \rho(\lambda). \tag{73}$$

Indeed the function $\lambda \mapsto \det (X(\lambda) + \varepsilon J(\lambda))$, $(0 < \varepsilon < \varepsilon_0)$ satisfies the conditions of Proposition 1, § 37 and hence the $2nm$ distinct roots of the polynomial $\lambda \mapsto \psi(\lambda;\ \varepsilon)$, (66), can be arranged in pairs of the form

$$(\lambda_j(\varepsilon), \lambda_{-j}(\varepsilon)), \quad \lambda_{-j}(\varepsilon) = \frac{1}{\bar{\lambda}_j(\varepsilon)}, \qquad j = 1, 2, \ldots, nm. \tag{74}$$

Similarly, for $\varepsilon = 0$, the roots of the polynomial (66) can be arranged in pairs $(\lambda_{j0}, \lambda_{-j0})$, where $\lambda_{-j0} = 1/\bar{\lambda}_{j0}$, for $j = 1, 2, \ldots$ $\ldots, n_0 \leqslant nm$. Since the coefficients of the polynomial (66) are continuous functions of ε, one can choose (for every ε) the order of the roots such that

$$\lim_{\varepsilon \to 0} |\lambda_j(\varepsilon)| = \infty, \qquad j = n_0 + 1, \ldots, nm, \tag{75}$$

$$\lim_{\varepsilon \to 0} \lambda_j(\varepsilon) = \lambda_{j0}, \qquad j = 1, 2, \ldots, n_0. \tag{76}$$

From Proposition 1, § 37 it also follows that the expression $\det(X(\lambda) + \varepsilon J(\lambda))$ in the interval $0 < \varepsilon < \varepsilon_0$ can be written in the form

$$\det(X(\lambda) + \varepsilon J(\lambda)) = \rho_0(\lambda;\varepsilon)\,\overline{\rho}_0\left(\frac{1}{\lambda};\varepsilon\right), \tag{77}$$

$$\rho_0(\lambda;\varepsilon) = |\varkappa(\varepsilon)|\,\rho_1(\lambda;\varepsilon)\,\rho_2(\lambda;\varepsilon), \tag{78}$$

$$\rho_1(\lambda;\varepsilon) = \prod_{j=1}^{n_0}(\lambda - \lambda_j(\varepsilon)) \tag{79}$$

$$\rho_2(\lambda;\varepsilon) = \prod_{j=n_0+1}^{nm}\left(\frac{\lambda}{\lambda_j(\varepsilon)} - 1\right).^{1)} \tag{80}$$

From (75) and (76) we obtain

$$\lim_{\varepsilon\to 0}\rho_1(\lambda;\varepsilon) = \prod_{j=1}^{n_0}(\lambda - \lambda_{j0}) \tag{81}$$

$$\lim_{\varepsilon\to 0}\rho_2(\lambda;\varepsilon) = (-1)^{nm-n_0-1}. \tag{82}$$

By (77) and (78) we have

$$\frac{\det(X(\lambda) + \varepsilon J(\lambda))}{\rho_1(\lambda;\varepsilon)\,\overline{\rho}_1\left(\frac{1}{\lambda};\varepsilon\right)\rho_2(\lambda;\varepsilon)\,\overline{\rho}_2\left(\frac{1}{\lambda};\varepsilon\right)} = |\varkappa(\varepsilon)|^2. \tag{83}$$

Taking into account (81) and (82) we derive from (83)

$$\lim_{\varepsilon\to 0}|\varkappa(\varepsilon)|^2 = \frac{\det X(\lambda)}{\prod_{j=1}^{n_0}(\lambda - \lambda_{j0})\prod_{j=1}^{n_0}\left(\frac{1}{\lambda} - \overline{\lambda}_{j0}\right)}. \tag{84}$$

Since the left-hand member of (84) does not depend on λ and since we may assume that $\det X(\lambda) \not\equiv 0$ (on the basis of the discussion of Section 2), it follows from (84) that

$$\lim_{\varepsilon\to 0}|\varkappa(\varepsilon)|^2 = |\varkappa_0|^2, \tag{85}$$

---

[1]) See (10, § 37). The factors $(\lambda - \lambda_j(\varepsilon))$, for $j = n_0 + 1, \ldots, nm$, have been written in the form $\lambda_j(\varepsilon)(\lambda/\lambda_j(\varepsilon) - 1)$ and the factors $\lambda_j(\varepsilon)$ have been included in the coefficient $\varkappa(\varepsilon)$.

where $\varkappa_0$ is a constant which may be taken real and strictly positive.

Consider now an arbitrary polynomial ρ which satisfies (3). Since the right-hand member of (84) is equal to a positive constant, each root of the polynomial $\rho(\lambda)$ must be equal to one of the numbers λ_{j0} and $1/\bar{\lambda}_{j0}$, $j = 1, 2, \ldots, n_0$. It also follows that n_0 is equal to the degree of the polynomial ρ and that one of the numbers of each pair $(\lambda_{j0}, 1/\bar{\lambda}_{j0})$, $j = 1, 2, \ldots, n_0$ must be a root of the polynomial ρ. We now consider each pair of numbers of (74) for $j = 1, 2, \ldots, n_0$ and taking into account (76), we permute (if necessary) the numbers $\lambda_j(\varepsilon)$ and $1/\bar{\lambda}_j(\varepsilon)$ of the respective pair, so that after this operation the first number of each pair tends to a root of the polynomial $\rho(\lambda)$ when ε approaches zero (for $j = n_0 + 1, \ldots, nm$, the pairs of (74) are chosen as specified in the course of the proof). We assume now that these permutations have already been effected in (74)—(85). Then, considering the way we choose the pairs of (74), we obtain the relation

$$\rho(\lambda) = \varkappa_1 \prod_{j=1}^{n_0} (\lambda - \lambda_{j0}), \tag{86}$$

where $\varkappa_1$ is a non-zero constant. Now, from (3), (84) and (85) it follows that

$$\lim_{\varepsilon \to 0} |\varkappa(\varepsilon)|^2 = |\varkappa_1|^2. \tag{87}$$

We write the number $\varkappa_1$ in the form $\varkappa_1 = |\varkappa_1| e^{i\vartheta}$ and consider the polynomial (in λ):

$$\rho(\lambda; \varepsilon) = (-1)^{nm-n_0-1} e^{i\vartheta} \rho_0(\lambda; \varepsilon), \tag{88}$$

which, as one can see from (77), satisfies (70). Using (78), (86), (87), (81) and (82) we find that (88) satisfies also (73).

We now show that from (70)—(73) it follows that there exists a matrix $D(\lambda)$ which satisfies (4) and (5). We first prove, for $D(\lambda)$, a property of boundedness. Introduce an arbitrary m-vector v satisfying the condition $\|v\| = 1$ and multiply on the right by v and on the left by v^* each member of (71). Writing the relations thus obtained for $|\lambda| = 1$ (whence $1/\lambda = \bar{\lambda}$), we obtain

$$v^*(X(\lambda) + \varepsilon J(\lambda))v = \|D(\lambda; \varepsilon)v\|^2.$$

Since all the entries of the matrices $X(\lambda)$ and $J(\lambda)$ are bounded on the unit circle, we see that there exists a positive constant such that

$$\|D(\lambda;\ \varepsilon)v\|^2 \leqslant \varkappa, \quad (|\lambda| = 1, \quad \|v\| = 1, \quad 0 < \varepsilon < \varepsilon_0). \tag{89}$$

Introducing another vector w, with $\|w\| = 1$, we derive from (89) the inequality

$$\begin{gathered} |w^* D(\lambda;\ \varepsilon)v| \leqslant \|w\|\, \|D(\lambda;\ \varepsilon)v\| \leqslant \sqrt{\varkappa}, \\ (|\lambda| = \|w\| = \|v\| = 1, \quad 0 < \varepsilon < \varepsilon_0). \end{gathered} \tag{90}$$

Choosing arbitrarily the numbers j and k from the numbers $1, 2, \ldots, m$ and defining the vectors w and v by the relations $w = e_j$, $v = e_k$, where the vectors e_i have the same meaning as in Section 8, we conclude from (90) that on the unit circle the absolute value of every entry of $D(\lambda;\ \varepsilon)$ is bounded by the constant $\sqrt{\varkappa}$.

Let λ be an arbitrary number satisfying the condition $|\lambda| = 1$ and consider a sequence of positive numbers ε converging to zero. Since the entries of $D(\lambda;\ \varepsilon)$ are bounded as shown above we can find a subsequence ε_k, $k = 1, 2, \ldots$, such that $D(\lambda;\ \varepsilon_k)$ is convergent. Let $D(\lambda)$ be the matrix

$$D(\lambda) = \lim_{k \to \infty} D(\lambda;\ \varepsilon_k), \quad (|\lambda| = 1). \tag{91}$$

As shown at the beginning of Section 2 the degrees of the polynomials representing the entries of the matrix polynomial $\lambda \mapsto D(\lambda;\ \varepsilon)$ do not exceed n. Relation (91) written for $n + 1$ distinct points λ_j, $j = 1, 2, \ldots, n + 1$ on the unit circle allows one to determine the (unique) matrix polynomial D which takes the values $D(\lambda_j)$ at the points λ_j.

The matrix polynomial D thus determined satisfies (4) and (5) of Theorem 1. Indeed, if (5) is not satisfied then, since both members of (5) are polynomials, we can find a number λ such that $|\lambda| = 1$ and $\det D(\lambda) \neq \rho(\lambda)$, contradicting (91), (72) and (73). If (4) is not satisfied, then by similar reasonings we obtain an inequality of the form $X(\lambda) \neq \bar{D}'\left(\frac{1}{\lambda}\right) D(\lambda)$, $(|\lambda| = 1)$, which contradicts (91) and (71).

To conclude the proof of Theorem 1 we remark that if $\lambda^n X(\lambda)$ and $\rho(\lambda)$ have real coefficients, we may secure the same property for $D(\lambda;\ \varepsilon_k)$ and then $D(\lambda)$ will also have real coefficients.

§ 39. The theorem of factorization on the imaginary axis

1. Statement of the theorem

Let $\sigma \mapsto Y(\sigma)$ be an $m \times m$-matrix polynomial having the following properties:

α) The equality

$$\bar{Y}'(-\sigma) = Y(\sigma) \tag{1}$$

is satisfied for every σ.

β) $Y(\sigma)$ is positive semidefinite on the imaginary axis. In other words, the inequality

$$v^* Y(i\omega)\, v \geqslant 0 \tag{2}$$

is satisfied for every real number ω and every m-vector v.

Under these assumptions $Y(\sigma)$ can be factorized on the imaginary axis as follows

Theorem 1. *If Y has the properties specified above, there exists at least one polynomial ψ satisfying the equality*

$$\det Y(\sigma) = \bar{\psi}(-\sigma)\, \psi(\sigma) \tag{3}$$

and for every polynomial ψ with this property, there exists a matrix polynomial C such that

$$Y(\sigma) = \bar{C}'(-\sigma)\, C(\sigma), \tag{4}$$

$$\det C(\sigma) = \psi(\sigma). \tag{5}$$

If Y and ψ have real coefficients, then C can be chosen such that its coefficients are real.

In the following sections, we derive Theorem 1 from Theorem 1, § 38 by performing the change of variable

$$\sigma = \frac{\lambda - 1}{\lambda + 1}. \tag{6}$$

Using the same reasonings as in Section 3, § 38 one sees that it is sufficient to prove Theorem 1 under the assumption $\det Y(\sigma) \not\equiv 0$. Moreover, we may even assume that there exists a real number α such that $\det Y(-\alpha) \neq 0$ (otherwise, since

$\sigma \mapsto \det Y(\sigma)$ is a polynomial, it would follow that $\det Y(\sigma) \equiv 0$). Effecting the change of variable $\sigma = \alpha\tilde{\sigma}$, one also sees that it is sufficient to prove Theorem 1 assuming that $\det Y(-1) \neq 0$. Then any polynomial ψ which satisfies (3), must also satisfy the condition $\psi(-1) \neq 0$. Therefore, in the proof which follows we shall treat only the case in which $\psi(-1) \neq 0$.

2. Definition of a matrix factorizable on the unit circle

Starting from an arbitrary matrix polynomial Y which satisfies the Conditions (α) and (β), we shall define a matrix polynomial X satisfying the Conditions (a), (b) and (c) of Section 1, § 38.

Replacing in $Y(\sigma)$ the variable σ by (6), we obtain the matrix $Y((\lambda - 1)/(\lambda + 1))$ for $\lambda \neq -1$. Let $2n$ be an even integer, higher or equal to the highest degree of the polynomials which constitute the entries of the matrix polynomial Y. Then $\lambda \mapsto Y((\lambda - 1)/(\lambda + 1))(\lambda + 1)^{2n}$ is a matrix polynomial which, by continuity, is also defined for $\lambda = -1$. Therefore if we define $X(\lambda)$ as

$$X(\lambda) = Y\left(\frac{\lambda - 1}{\lambda + 1}\right)\left(\frac{(\lambda + 1)^2}{\lambda}\right)^n, \quad (\lambda \neq -1) \tag{7}$$

$$X(-1) = \lim_{\lambda \to -1} X(\lambda). \tag{8}$$

then $\lambda \mapsto \lambda^n X(\lambda)$ is a matrix polynomial. Hence X satisfies Condition (a) of Section 1, § 38.

We now show that $X(\lambda)$ satisfies also the Conditions (b) and (c) of Section 1, § 38. From (7) and (1) we successively obtain

$$\overline{X}'\left(\frac{1}{\lambda}\right) = \overline{Y}'\left(\frac{1 - \lambda}{1 + \lambda}\right)\left(\frac{(\lambda + 1)^2}{\lambda}\right)^n = Y\left(\frac{\lambda - 1}{\lambda + 1}\right)\left(\frac{(\lambda + 1)^2}{\lambda}\right)^n = X(\lambda)$$

and hence Condition (b) is satisfied. For every number λ which satisfies the conditions $|\lambda| = 1$ and $\lambda + 1 \neq 0$, the corresponding variable σ given by (6) can be written as

$$\sigma = i\omega = \frac{\lambda - 1}{\lambda + 1},$$

where ω is a real number. By Condition (β), the matrix $Y(i\omega) = Y((\lambda - 1)/(\lambda + 1))$ is positive semidefinite. Hence, using (7), we see that the matrix $X(\lambda)(\lambda(\lambda + 1)^2)^n$ is positive semidefinite for every number λ which satisfies the conditions $|\lambda| = 1$ and $\lambda + 1 \neq 0$. It is easy to see that if λ satisfies these conditions then $\lambda/(\lambda + 1)^2$ is real and positive; hence, from the former conclusion it follows that $X(\lambda)$ is positive semidefinite for $|\lambda| = 1$ and $\lambda + 1 \neq 0$. By continuity it follows that the matrix $X(-1)$ is also positive semidefinite. Thus we proved that $X(\lambda)$ satisfies the conditions required in Theorem 1, § 38.

3. Relations between $\psi(\sigma)$ and $\rho(\lambda)$

Let $\sigma \mapsto \psi(\sigma)$ be a polynomial which satisfies (3) and $\lambda \mapsto \rho(\lambda)$, a polynomial which satisfies (3, § 38) for the matrix $X(\lambda)$ defined in the previous section. From (3) and (6) we obtain

$$\bar{\psi}\left(-\frac{\lambda - 1}{\lambda + 1}\right)\psi\left(\frac{\lambda - 1}{\lambda + 1}\right) = \det Y\left(\frac{\lambda - 1}{\lambda + 1}\right). \tag{9}$$

The right-hand member satisfies also the equality, derived from (7)

$$\left(\frac{(\lambda + 1)^2}{\lambda}\right)^{nm} \det Y\left(\frac{\lambda - 1}{\lambda + 1}\right) = \det X(\lambda), \tag{10}$$

and hence

$$\left(\frac{(\lambda + 1)^2}{\lambda}\right)^{nm} \bar{\psi}\left(-\frac{\lambda - 1}{\lambda + 1}\right)\psi\left(\frac{\lambda - 1}{\lambda + 1}\right) = \det X(\lambda). \tag{11}$$

If we define $\rho(\lambda)$ for $\lambda \neq -1$ as

$$\rho(\lambda) = \psi\left(\frac{\lambda - 1}{\lambda + 1}\right)(\lambda + 1)^{nm} \tag{12}$$

then

$$\bar{\rho}\left(\frac{1}{\lambda}\right) = \bar{\psi}\left(-\frac{\lambda - 1}{\lambda + 1}\right)\left(\frac{\lambda + 1}{\lambda}\right)^{nm}, \tag{13}$$

and using (11) and (12), we find that (3, § 38) is satisfied. By continuity, all these properties are extended to $\lambda = -1$. The fact that $\rho(\lambda)$ given by (12), represents a polynomial, follows from the definition of the number n and from (3) (which shows that the degree of the polynomial $\psi(\sigma)$ is at most nm).

Conversely, let ρ be a polynomial which satisfies (3, § 38), where the matrix $X(\lambda)$ is defined as in Section 2. Assume that $\rho(0) \neq 0$; (otherwise, since det $X(\lambda) \not\equiv 0$, this condition is satisfied after dividing $\rho(\lambda)$ by λ^p for an adequately chosen integer $p > 0$; then the polynomial $\lambda \mapsto \rho(\lambda)\lambda^{-p}$ will still satisfy (3, § 38). By Theorem 1, § 38 from $\rho(0) \neq 0$ it follows that the degree of the polynomial ρ cannot exceed nm [1]). Using (12) and (6) we define the polynomial $\sigma \mapsto \psi(\sigma)$ as

$$\psi(\sigma) = \rho\left(\frac{1+\sigma}{1-\sigma}\right)\left(\frac{1-\sigma}{2}\right)^{nm}, \qquad (1-\sigma \neq 0). \qquad (14)$$

Returning to the variable λ, we remark that from (3, § 38) and from (12) and (13) we obtain (11); the latter, considering (10) (which follows from (7)), leads to (9) which is only a different way of writing (3) .

These results — which have been proved for every λ and σ, except for $\lambda = -1$ and $\sigma = 1$, — are also extended to these values, by continuity.

4. Factorization on the imaginary axis

Let Y be a matrix polynomial with the Properties (α) and (β) and define $X(\lambda)$ by (7). Clearly there always exists a polynomial ψ which satisfies (3). Indeed, if det $Y(\sigma) \not\equiv 0$ (as we can always assume on the basis of the reasonings in Section 1), then (10) shows that det $X(\lambda) \not\equiv 0$. By Theorem 1, § 38 there exists a polynomial ρ which satisfies (3, § 38) and therefore is not identically zero. We can always take $\rho(0)\neq 0$ by dividing, if necessary $\rho(\lambda)$ by λ^p for a convenient integer $p > 0$. As shown in the previous section, the polynomial ψ defined by (14) satisfies (3).

Now let ψ be an arbitrary polynomial which satisfies (3) and the condition (see Section 1)

$$\psi(-1) \neq 0. \qquad (15)$$

As proved in the previous section, the corresponding polynomial ρ defined by (12) satisfies (3, § 38); it also satisfies the condition $\rho(0) \neq 0$ which results from (15) and (12). By Theorem 1, § 38 there exists D which satisfies (4, § 38) and (5, § 38). Theorem 1, § 38 also shows that from $\rho(0) \neq 0$ it follows that the degree of D is not greater than n.

---

[1]) See (5, § 38) and remark that, as mentioned in Theorem 1, § 38, the degree of $D(\lambda)$ is not higher than n.

Replacing in (4, § 38) the variable λ by $\lambda = (1+\sigma)/(1-\sigma)$ (see (6)) and using (7) we obtain

$$\overline{D}'\left(\frac{1-\sigma}{1+\sigma}\right) D\left(\frac{1+\sigma}{1-\sigma}\right) = X\left(\frac{1+\sigma}{1-\sigma}\right) = Y(\sigma)\left(\frac{2}{1+\sigma}\right)^n\left(\frac{2}{1-\sigma}\right)^n. \tag{16}$$

Therefore, the matrix

$$C(\sigma) = D\left(\frac{1+\sigma}{1-\sigma}\right)\left(\frac{1-\sigma}{2}\right)^n \tag{17}$$

satisfies (4). Since the degree of D is not greater than n, C is a matrix polynomial. From (17) and (5, § 38) one obtains

$$\det C(\sigma) = \left(\frac{1-\sigma}{2}\right)^{nm} \rho\left(\frac{1+\sigma}{1-\sigma}\right) \tag{18}$$

whence, using (14), one also obtains (5). The exceptional values $\lambda = -1$ and $\sigma = 1$ are covered as before, by continuity. From Theorem 1, § 38 and from the relations used for deducing the Theorem 1 from Theorem 1, § 38, it follows that if Y and ψ have real coefficients, then C has real coefficients. This concludes the proof of Theorem 1.

POSITIVE REAL FUNCTIONS

The main purpose of this appendix is to establish some properties which are used in § 15. For other properties of positive real functions see, for instance, Guillemin [1]. Consider the expressions of the form

$$\gamma(\sigma) = \frac{\rho(\sigma)}{\xi(\sigma)}, \tag{1}$$

where $\rho(\sigma)$ and $\xi(\sigma)$ are polynomials with complex coefficients and σ is a complex variable. We adopt the following definition :

Definition 1. *The function γ, (1), is said to be positive real if*

$$\operatorname{Re} \gamma(\sigma) \geqslant 0 \tag{2}$$

for every σ for which Re $\sigma > 0$ and for which Re $\gamma(\sigma)$ is defined.

It follows from (1) that γ is defined for every σ, except possibly the points for which $\xi(\sigma) = 0$. However, from Theorem 1, it will follow that $\xi(\sigma) \neq 0$, for every σ for which Re $\sigma > 0$.

Theorem 1. *The following properties are equivalent:*

(A): *"Function γ is positive real".*

(B): *"There exists no number σ with Re $\sigma > 0$, for which one of the numbers $\gamma(\sigma)\sigma$ and $\gamma(\sigma)/\sigma$ is finite, real and strictly negative".*

(C): *"The inequality Re $\gamma(i\omega) \geqslant 0$ is satisfied for every real number ω for which $\xi(i\omega) \neq 0$. Furthermore, the following conditions are satisfied:*

(a) *function γ has no poles in the half plane Re $\sigma > 0$.*

(b) *on the axis Re $\sigma = 0$, function γ has at most simple poles and the residues of the function at these poles are real and positive".*

The proof is based on the scheme of implications

$$(C) \to (A) \to (B) \to (C). \tag{3}$$

First we prove the implication (C) $\to$ (A). From (C) it follows that $\gamma(\sigma)$ can be also written

$$\gamma(\sigma) = \sum_{j=1}^{m} \frac{\delta_j}{(\sigma - i\omega_j)} + \tilde{\gamma}(\sigma), \qquad (\delta_j > 0) \tag{4}$$

where $i\omega_j$, $j = 1, 2, \ldots, m$ are *all* the poles situated on the axis $\operatorname{Re} \sigma = 0$ while $\tilde{\gamma}(\sigma)$ has no poles in an open set which contains the closed half plane $\operatorname{Re} \sigma \geqslant 0$. In (4) we might have $m = 0$; then the arguments which follow become simpler. Using the obvious equality

$$\operatorname{Re} \frac{\delta_j}{(i\omega - i\omega_j)} = 0$$

we deduce from $\operatorname{Re} \gamma(i\omega) \geqslant 0$ (see Property (C)) and from (4) the inequality $\operatorname{Re} \tilde{\gamma}(i\omega) \geqslant 0$. Hence, using the properties of analytic functions we obtain $\operatorname{Re} \tilde{\gamma}(\sigma) \geqslant 0$ for every σ with $\operatorname{Re} \sigma \geqslant 0$. Indeed, if there exists a point σ with $\operatorname{Re} \sigma \geqslant 0$ such that $\operatorname{Re} \tilde{\gamma}(\sigma) < 0$, then the function $\sigma \mapsto \operatorname{Re} \tilde{\gamma}(\sigma)$ has a local minimum in the right-hand half plane, which is impossible since the function $\tilde{\gamma}$ is analytic in an open set which contains the half plane $\operatorname{Re} \sigma \geqslant 0$ (see, for instance S. Stoilow [1], Vol. 1, Chapter 2, Section 27).

Using this result and remarking that $\operatorname{Re} \sigma > 0$ implies that

$$\operatorname{Re} \frac{\delta_j}{(\sigma - i\omega_j)} \geqslant 0$$

we deduce from (4), (2) for every σ such that $\operatorname{Re} \sigma > 0$, thus obtaining Property (A).

The implication (A) $\to$ (B) is nearly immediate: if Property (B) is not satisfied then there exists a number σ_0 satisfying the condition $\operatorname{Re} \sigma_0 > 0$ and such that $\sigma_0\gamma(\sigma_0) = -\delta_0$ (where δ_0 is real, finite and positive) or $\gamma(\sigma_0)/\sigma_0 = -\delta_0'$ (where δ_0' has the same properties as δ_0). In both cases we obtain the inequality $\operatorname{Re} \gamma(\sigma_0) < 0$ which contradicts Property (A) (since $\operatorname{Re} \sigma_0 > 0$).

We consider now the implication (B) $\to$ (C) which is the most difficult part of the proof. We first establish that part of Property (C) which refers to the poles of the function γ. Assume that γ has a pole of multiplicity k_0 at σ_0. Expanding in simple fractions and rearranging the terms we can write the equality

$$\gamma(\sigma) = \frac{\delta_1}{(\sigma - \sigma_0)^{k_0}}(1 + \mathcal{O}(\sigma, \sigma_0)), \qquad (k_0 \geqslant 1, \ \delta_1 \neq 0). \tag{5}$$

We shall examine separately the cases $\sigma_0 \neq 0$ and $\sigma_0 = 0$.

Case $\sigma_0 \neq 0$. Let us introduce the *real* numbers ρ, θ, ρ_1, θ_1, ρ_0 and θ_0 such that

$$\sigma - \sigma_0 = \rho e^{i\theta}, \qquad (\rho \geqslant 0) \tag{6}$$

$$\delta_1 = - \rho_1 e^{i\theta_1}, \qquad (|\theta_1| \leqslant \pi, \ \rho_1 > 0) \tag{7}$$

$$\sigma_0 = \rho_0 e^{i\theta_0}, \qquad (|\theta_0| \leqslant \pi, \ \rho_0 > 0) \tag{8}$$

Both $\gamma(\sigma)\sigma$ and $\gamma(\sigma)/\sigma$ can be written in the form $\dfrac{\gamma(\sigma)}{\sigma^{\varepsilon_0}}$, where ε_0 takes one of the values $+1$ and -1. We can write

$$\frac{\gamma(\sigma)}{\sigma^{\varepsilon_0}} = \frac{\delta_1}{\sigma_0^{\varepsilon_0}(\sigma - \sigma_0)^{k_0}}(1 + \widetilde{\mathcal{O}}(\sigma, \sigma_0 ; \varepsilon_0)),$$

where $\lim_{\sigma \to \sigma_0} \widetilde{\mathcal{O}}(\sigma, \sigma_0 ; \varepsilon_0) = 0$. Obviously we may further write

$$1 + \widetilde{\mathcal{O}}(\sigma, \sigma_0 ; \varepsilon_0) = \rho_2(\rho, \theta ; \varepsilon_0) e^{i\theta_2(\rho, \theta ; \varepsilon_0)},$$

where the functions $(\rho, \theta) \mapsto \rho_2(\rho, \theta ; \varepsilon_0)$ and $(\rho, \theta) \mapsto \theta_2(\rho, \theta ; \varepsilon_0)$ are real-valued and are continuous in ρ and θ for every real θ and every ρ in an interval $0 \leqslant \rho \leqslant \varkappa_0$ where $\varkappa_0$ is a positive constant. From $\lim_{\sigma \to \sigma_0} \widetilde{\mathcal{O}}(\sigma, \sigma_0 ; \varepsilon_0) = 0$ we obtain

$$\left.\begin{aligned} \lim_{\rho \to 0} \max_{|\theta| \leqslant \pi} \rho_2(\rho, \theta ; \varepsilon_0) &= 1, \\ \lim_{\rho \to 0} \max_{|\theta| \leqslant \pi} \theta_2(\rho, \theta ; \varepsilon_0) &= 0. \end{aligned}\right\} \tag{9}$$

Using the same notations, we may write

$$\frac{\gamma(\sigma)}{\sigma^{\varepsilon_0}} = - \frac{\rho_1 \rho_2(\rho, \theta ; \varepsilon_0)}{\rho_0^{\varepsilon_0} \rho^{k_0}} e^{i(\theta_1 - \varepsilon_0\theta_0 - k_0\theta + \theta_2(\rho, \theta ; \varepsilon_0))}.$$

From Property (B) it follows that there does not exist a set of numbers : ρ, θ, ε_0, $(\rho > 0, \ \varepsilon_0 = \pm 1)$ such that

$$\rho_2(\rho, \theta; \varepsilon_0) > 0, \qquad \operatorname{Re} \sigma = \rho \cos \theta + \operatorname{Re} \sigma_0 > 0$$

and

$$\theta_1 - \varepsilon_0 \theta_0 - k_0 \theta + \theta_2(\rho, \theta; \varepsilon_0) = 0. \tag{10}$$

(otherwise we obtain $\gamma(\sigma)/\sigma^{\varepsilon_0} < 0$, and $\operatorname{Re} \sigma > 0$ contradicting Property (B)). Instead of (10) we consider now the simpler equality

$$\theta_1 - \varepsilon_0 \theta_0 - k_0 \theta = 0, \tag{11}$$

and we claim that there does not exist a set of numbers θ, ε_0, $\varkappa_0$ ($\varepsilon_0 = \pm 1$, $\varkappa_0 > 0$) such that (11) holds and $\rho \cos \theta + \operatorname{Re} \sigma_0 > 0$ in the interval $0 < \rho < \varkappa_0$. Indeed, otherwise we can find an interval of the form $\theta_3 \leqslant \theta \leqslant \theta_4$ with the properties: (α) $\rho \cos\theta + \operatorname{Re} \sigma_0 > 0$ (for every θ in the interval $\theta_3 \leqslant \theta \leqslant \theta_4$ and every ρ in the interval $0 < \rho < \varkappa_0$); (β) $\theta_1 - \varepsilon_0 \theta_0 - k_0 \theta_4 = \varkappa_1 < 0$ and (γ) $\theta_1 - \varepsilon_0 \theta_0 - k_0 \theta_3 = \varkappa_2 > 0$. By virtue of Conditions (9) there exists a number $\varkappa_0' \leqslant \varkappa_0$ such that

$$|\theta_2(\rho, \theta; \varepsilon_0)| \leqslant \frac{1}{2} \min(|\varkappa_1|, |\varkappa_2|), \qquad (0 < \rho < \varkappa_0'; \ \theta_3 \leqslant \theta \leqslant \theta_4)$$

$$\rho_2(\rho, \theta; \varepsilon_0) > 0, \qquad (0 < \rho < \varkappa_0'; \ \theta_3 \leqslant \theta \leqslant \theta_4).$$

Therefore, for every given number ρ from the interval $0 < \rho < \varkappa'_0$ we have

$$\theta_1 - \varepsilon_0 \theta_0 - k_0 \theta_3 + \theta_2(\rho, \theta_3; \varepsilon_0) \geqslant \frac{1}{2} \varkappa_2 > 0$$

$$\theta_1 - \varepsilon_0 \theta_0 - k_0 \theta_4 + \theta_2(\rho, \theta_4; \varepsilon_0) \leqslant \frac{1}{2} \varkappa_1 < 0.$$

Hence, noticing also that the left-hand member of (10) defines a continuous function of θ we find that there exists a number $\theta \in [\theta_3, \theta_4]$ such that Relation (10) is satisfied; furthermore the conditions $\rho_2(\rho, \theta; \varepsilon_0) > 0$ and $\rho \cos\theta + \operatorname{Re} \sigma_0 > 0$ are also satisfied; but we have seen that these conditions cannot be satisfied simultaneously. The contradiction proves that indeed there exists no set of numbers θ, ε_0, $\varkappa_0$, ($\varepsilon_0 = \pm 1$, $\varkappa_0 > 0$) such that we have Relation (11) and $\rho \cos\theta + \operatorname{Re} \sigma_0 > 0$ in the interval $0 < \rho < \varkappa_0$.

Clearly, in the case $\operatorname{Re} \sigma_0 > 0$ there exists a set of numbers with the above properties. Hence the condition $\operatorname{Re} \sigma_0 > 0$ contradicts Property (B). Thus we proved that Condition (a) of Property (C) follows from Property (B).

In the case Re $\sigma_0 = 0$ and $\sigma_0 \neq 0$ we obtain from (8) the equality $|\theta_0| = \frac{\pi}{2}$. In the case $k_0 > 1$ we can always choose $\varepsilon_0 (= \pm 1)$ and θ so as to satisfy (11) and the condition $|\theta| < \frac{\pi}{2}$, (whence one obtains the inequality $\rho \cos\theta + \operatorname{Re}\sigma_0 > > 0$, for every $\rho > 0$). Thus we have proved a part of Condition (b) from Property (C), namely that on the axis Re $\sigma = 0$, $\sigma \neq 0$ function γ has at most simple poles.

Consider now the case Re $\sigma_0 = 0$, $\sigma_0 \neq 0$, $k_0 = 1$. As before, we have $|\theta_0| = \frac{\pi}{2}$. If $|\theta_1| \neq \pi$, then we can find ε_0 (equal to ± 1) and θ so as to satisfy (11) and the condition $|\theta| \leqslant \frac{\pi}{2}$, thus obtaining the same contradiction as before. Therefore we must have $|\theta_1| = \pi$ and hence (see (7)) we obtain another part of Condition (b) of Property (C), namely : in the case Re $\sigma_0 = 0$, $\sigma_0 \neq 0$, $k_0 = 1$ the residue of the function γ at the simple pole σ_0 is real and positive.

We now show that Condition (b) of Property (C) is also satisfied in the case of a pole at the origin $\sigma_0 = 0$.

The arguments are based on the same ideas and will be presented more concisely. From (5), for $\sigma_0 = 0$, we obtain

$$\frac{\gamma(\sigma)}{\sigma} = \frac{\delta_1}{\sigma^{k_0+1}}(1 + \mathcal{O}(\sigma, 0)), \qquad (k_0 \geqslant 1, \quad \delta_1 \neq 0).$$

We introduce the real numbers ρ, θ, ρ_1, θ_1 and the real-valued functions $(\rho, \theta) \mapsto \tilde{\rho}_2(\rho, \theta)$, $(\rho, \theta) \to \tilde{\theta}_2(\rho, \theta)$ by the relations $\sigma = \rho e^{i\theta}$, $(\rho \geqslant 0)$; $\delta_1 = -\rho_1 e^{i\theta_1}$, $(|\theta_1| \leqslant \pi, \ \rho_1 > 0)$; $1 + \mathcal{O}(\sigma, 0) = \tilde{\rho}_2(\rho, \theta) e^{i\tilde{\theta}_2(\rho, \theta)}$. We have the relations (similar to (9)) : $\lim_{\rho\to 0} \max_{|\theta|\leqslant\pi} \tilde{\rho}_2(\rho, \theta) = 1$ and $\lim_{\rho\to 0} \max_{|\theta|\leqslant\pi} \tilde{\theta}_2(\rho, \theta) = 0$. Using the above relations we have:

$$\frac{\gamma(\sigma)}{\sigma} = -\frac{\rho_1 \tilde{\rho}_2(\rho, \theta)}{\rho^{(k_0+1)}} e^{i(\theta_1 - (k_0+1)\theta + \tilde{\theta}_2(\rho_1 \theta))}.$$

As before, it can be shown that from Property (B) it follows that there exists no number θ such that $\theta_1 - (k_0 + 1)\theta = 0$ and $\cos\theta > 0$. But such a number exists if we have $k_0 > 1$ or if $k_0 = 1$ and at the same time $|\theta_1| < \pi$. Therefore, it follows from Property (B) that these cases cannot occur, and thus Condition (b) of Property (C) is also obtained in the case $\sigma_0 = 0$.

To conclude the proof of the implication (B) $\to$ (C) it only remains to show, that the condition $\operatorname{Re} \gamma(i\omega) \geqslant 0$ follows from Property (B). Assume that there exists a real number ω_0 such that

$$\operatorname{Re} \gamma(i\omega_0) < 0. \tag{12}$$

Assume that the number ω_0 is positive (an assumption which we shall eliminate in the sequel). Let us join by a straight segment in the plane of the variable σ, the point $\sigma_0 = i\omega_0$ and an arbitrary point σ_1 on the real axis and in the half plane $\operatorname{Re} \sigma > 0$. We can assume that this segment contains no zero and no pole of function γ. (Clearly, this property can be always secured by modifying, if necessary, the point σ_1). We shall denote the points on the segment by $\sigma(\tau)$ where the parameter τ is equal to the abscissa of the point σ. Hence we have $\sigma(0) = i\omega_0$ and $\sigma(\sigma_1) = \sigma_1$. Obviously the conditions

$$0 < \tau \leqslant \sigma_1 \tag{13}$$

$$0 \leqslant \arg \sigma(\tau) < \frac{\pi}{2} \tag{14*}$$

$$\operatorname{Re} \sigma(\tau) > 0. \tag{15}$$

are simultaneously satisfied on our segment. Since this segment contains no zeros and no poles of function γ we can define $\tau \mapsto \arg \gamma(\sigma(\tau))$ as a continuous function of τ, over the interval $0 \leqslant \tau \leqslant \sigma_1$; Moreover owing to the Condition (12) we can also secure the condition

$$\frac{\pi}{2} < \arg \gamma(i\omega_0) < \frac{3\pi}{2}. \tag{16}$$

Let τ_t be the largest number of the interval $[0, \sigma_1]$ such that

$$\frac{\pi}{2} \leqslant \arg \ \gamma(\sigma(\tau)) \leqslant \frac{3\pi}{2} \ \text{for every} \ \tau \in [0, \ \tau_t] \tag{17}$$

---

* Here and in the following we denote by $\arg z$ a real number satisfying the relation $z = |z| e^{i \arg z}$. Additional conditions will be always introduced to determine $\arg z$ uniquely.

(compare with (16)). Define also

$$\sigma_t = \sigma(\tau_t) \tag{18}$$

We might have $\tau_t = \sigma_1$. Then the number $\sigma_t = \sigma(\tau_t) = \sigma_1$ is real and strictly positive and hence we can have only one of the cases

$$\arg(\gamma(\sigma_t)\sigma_t) \leqslant \pi \tag{19}$$

$$\arg \frac{\gamma(\sigma_t)}{\sigma_t} > \pi \tag{20}$$

(depending on whether $\arg \gamma(\sigma_t) \leqslant \pi$ or $\arg \gamma(\sigma_t) > \pi$). If we have $\tau_t < \sigma_1$, then for $\tau = \tau_t$ the number $\arg \gamma(\sigma_t)$ is at the boundary of the interval (18) and hence one of the equalities $\arg \gamma(\sigma_t) = \dfrac{\pi}{2}$ or $\arg \gamma(\sigma_t) = \dfrac{3\pi}{2}$ is satisfied. In the first case, taking into account the condition $0 \leqslant \arg \sigma_t \leqslant \dfrac{\pi}{2}$ (see (14)), it follows that inequality (19) is satisfied, while in the second case inequality (20) is satisfied. Thus we can always consider that one of the cases (19) or (20) is taking place. Assume that inequality (19) is satisfied and consider the function $\sigma \mapsto \gamma(\sigma)\sigma$. The inequality

$$\arg \gamma(i\omega_0)\, i\omega_0 > \pi \tag{21}$$

(which results from (16)) is satisfied for $\sigma = \sigma(0) = i\omega_0$.

Since $\tau \mapsto \arg \gamma(\sigma(\tau))\sigma(\tau)$ is a continuous function of τ, defined on the interval $0 < \tau \leqslant \tau_t$ (where τ_t is the abscissa of the point σ_t), we obtain from (19) and (21) the conclusion that there exists a number τ_0 in the interval $0 < \sigma \leqslant \tau_t$ such that $\arg \gamma(\sigma(\tau_0))\sigma(\tau_0) = \pi$. Hence the number $\gamma(\sigma(\tau_0))\sigma(\tau_0)$ is real and strictly negative while at the same time the condition $\operatorname{Re} \sigma(\tau_0) > 0$ is satisfied (see (15)). Thus Property (B) is contradicted. In the case of (20) consider the function $\sigma \mapsto \gamma(\sigma)/\sigma$. From (16) we obtain for $\sigma = i\omega_0$,

$$\arg \frac{\gamma(i\omega_0)}{i\omega_0} < \pi. \tag{22}$$

Using also (20) we obtain by the same reasonings as before the conclusion that there exists a number τ'_0 in the interval $0 < \tau \leqslant \tau_t$ such that the number $\gamma(\sigma(\tau'_0))/\sigma(\tau'_0)$ is real and strictly negative and at the same time we have $\operatorname{Re} \sigma(\tau'_0) > 0$ which again contradicts Property (B).

It only remains to eliminate the assumption that the number ω_0 which occurs in the inequality (12) is strictly positive. If we have $\omega_0 = 0$ then, since the function $i\omega \mapsto \gamma(i\omega)$ is continuous in the neighbourhood of the point $i\omega_0$, inequality(12) is also satisfied for strictly positive and sufficiently small numbers ω_0 and we obtain a contradiction as shown above. If the number ω_0 is strictly negative, we introduce the new variable

$$\tilde{\sigma} = \bar{\sigma} \tag{23}$$

and define the function $\tilde{\sigma} \mapsto \tilde{\gamma}(\sigma)$ by the relation

$$\tilde{\gamma}(\tilde{\sigma}) = \overline{\gamma(\sigma)}. \tag{24}$$

From inequality (12) and Relations (23) and (24) we obtain the inequality

$$\operatorname{Re} \tilde{\gamma}(i\tilde{\omega}_0) < 0, \tag{24'}$$

where $\tilde{\omega}_0 = -\omega_0$. Hence the number $\tilde{\omega}_0$ is strictly positive. Applying the previous reasonings we see that there exists a number $\tilde{\sigma}_0$ which satisfies the condition $\operatorname{Re} \tilde{\sigma}_0 > 0$ and for which one of the numbers $\tilde{\gamma}(\tilde{\sigma}_0)\tilde{\sigma}_0$ or $\tilde{\gamma}(\tilde{\sigma}_0)/\tilde{\sigma}_0$ is real and strictly negative. Returning to the function γ we conclude with the help of (23) and (24) that there exists a number σ_0 (equal to $\bar{\tilde{\sigma}}_0$) such that we have $\operatorname{Re} \sigma_0 > 0$ and at the same time one of the numbers $\overline{\gamma(\sigma_0)}\bar{\sigma}_0$ or $\overline{\gamma(\sigma_0)}/\bar{\sigma}_0$ is real and strictly negative. Then, since the number $\overline{\gamma(\sigma_0)\sigma_0}$ (or $\overline{\gamma(\sigma_0)/\sigma_0}$) is real, it is also equal to its complex conjugate, and therefore the number $\gamma(\sigma_0)\sigma_0$ (or, respectively $\gamma(\sigma_0)/\sigma_0$) is also real and strictly negative. Thus we obtain again a contradiction of Property (B) and this concludes the proof of Theorem 1.

From Theorem 1 we derive

Consequence 1. *If the function γ is positive real, then*

$$\lim_{\sigma \to \infty} \frac{\gamma(\sigma)}{\sigma^2} = 0. \tag{25}$$

Moreover only one of the following four cases occurs:

(i) *The number β, defined by the relation*

$$\beta = \lim_{\sigma \to \infty} \frac{\gamma(\sigma)}{\sigma}, \tag{26}$$

is real and satisfies the inequality

$$\beta > 0 \tag{27}$$

(ii) *The number* β *satisfies the equality*

$$\beta = 0 \tag{28}$$

and the number α *defined by the relation*

$$\alpha = \lim_{\sigma \to \infty} \gamma(\sigma) \tag{29}$$

satisfies the inequality

$$\operatorname{Re} \alpha > 0. \tag{30}$$

(iii) *The numbers* β *and* α *satisfy the conditions*

$$\beta = \operatorname{Re} \alpha = 0 \tag{31}$$

and the number δ *defined by the relation*

$$\delta = \lim_{\sigma \to \infty} \sigma(\gamma(\sigma) - \alpha), \tag{32}$$

is real and satisfies the inequality

$$\delta > 0. \tag{33}$$

(iv) *Relations (31) take place and at the same time* $\gamma(\sigma) = \alpha$ *(for every* σ*).*

Indeed, if (25) is not satisfied, the degree of the numerator of the function γ is greater than the degree of the denominator by at least two units. Then the function $\sigma \mapsto \gamma(1/\sigma)$ has at least a double pole at the origin and hence is not real positive (see Property (C) of Theorem 1). But it is easy to see that if the function $\sigma \mapsto \gamma(\sigma)$ is positive real (as assumed in Consequence 1), then the function $\sigma \mapsto \gamma(1/\sigma)$ also is positive real. The contradiction obtained proves the necessity of Condition (25).

From (25) it also follows that there exists a number β as defined by (26). Assume first that $\beta \neq 0$. Clearly then the degree of the numerator of the function γ is higher than the denominator's degree by one unit. Therefore the function $\sigma \mapsto \gamma(1/\sigma)$ has a simple pole at the origin ; moreover the residue of the function at that pole is precisely β. Since, as we have remarked, the function $\sigma \mapsto \gamma(1/\sigma)$ must be positive real, the residue β must be positive as stated in (27) (case (i)).

It remains to examine the alternative $\beta = 0$. Then there exists a number α given by (29). If the inequality Re $\alpha < 0$ were satisfied, then Re $\gamma(\sigma)$ would be strictly negative for real, positive and sufficiently large numbers σ, and hence the function γ would not be positive real. Therefore we have either inequality (30) (case (ii)) or else the equality Re $\alpha = 0$ which will be examined below.

If $\beta = 0$ and Re $\alpha = 0$, we remark that the degree of the numerator of the expression $\gamma(\sigma) - \alpha$ is smaller than the denominator's degree by at least one unit. Therefore the number δ is well defined by (32). Since Re $\alpha = 0$, the function $\sigma \mapsto \tilde{\gamma}(\sigma) = \gamma(\sigma) - \alpha$ is positive real. Hence the function $\sigma \mapsto 1/\tilde{\gamma}(1/\sigma)$ is also positive real. Assume now that $\delta \neq 0$. Then the function $\delta \mapsto 1/\tilde{\gamma}(1/\sigma)$ has a pole at the origin and this pole is simple since the function is positive real. The residue of the function at that pole is equal to $1/\delta$, and must be positive, as shown by (33) (case (iii)).

It only remains to examine the last case $\beta = \text{Re}\ \alpha = \delta = 0$. Then the function $\sigma \mapsto \tilde{\gamma}(\sigma) = \gamma(\sigma) - \alpha$ either is identically zero or else has the property that the degree of its numerator is smaller than the degree of its denominator by at least two units. However the latter case cannot occur because then the function $\sigma \mapsto 1/\tilde{\gamma}(1/\gamma)$ (which, as we saw, must be positive real) would have a multiple pole at the origin, and this is again contradictory. Hence the only possibility left is the identity $\gamma(\sigma) - \alpha \equiv 0$. Thus we obtain the case (iv) concluding the proof of Consequence 1.

THE PRINCIPAL HYPERSTABLE BLOCKS

This appendix presents a catalogue of the various hyperstable blocks introduced in the work. For uniformity of notations the inputs are designated by μ_i, and the outputs by ν_i.

This collection of hyperstable blocks can be completed with new hyperstable blocks by combining, after the rules established in § 13 the hyperstable blocks listed below.

The catalogue of hyperstable blocks — completed as shown above — can be used to build up stable systems by applying the results of Section 9, § 13.

Table

N°.	A Type of block	B Equations describing the block	C Principal conditions under which the block is hyperstable	D The paragraph (section) in which the block is studied
1.	Linear block described by differential equations with constant coefficients	$\dfrac{dx}{dt} = Ax + b\mu_i$ $\nu_0 = a^* x + \alpha\mu_i$	Re $\gamma_i(i\omega) \geqslant 0$ (for every real ω for which det $(i\omega I - A) \neq 0$) where $\gamma_i(i\omega) = \alpha + a^*(i\omega I - A)^{-1} b$. (The block is assumed to be minimally stable)	§ 15
2.	Non-linear block with the characteristic in a sector	$\beta \dfrac{d\nu}{dt} + \left(\nu - \dfrac{1}{\lambda_0}\varphi(\nu)\right) = \mu_i$ $\nu_i = \varphi(\nu)$	$\beta > 0$, φ = continuous function satisfying the conditions $\varphi(0) = 0$ $\varphi(\nu)\left(\nu - \dfrac{1}{\lambda_0}\varphi(\nu)\right) > 0$ (for every $\nu \neq 0$	§ 25 (Section 8)

N°	A	B	C	D
3.	Block of the nuclear reactor type	$\frac{d\rho}{dt} = -\sum_{j=1}^{p} \delta_j (\rho - \eta_j) + (1+\rho)\mu_\iota$ $\frac{d\eta_j}{dt} = \lambda_j(\rho - \eta_j),\ j = 1, 2, \ldots, p$ $\nu_\iota = \rho$	$\delta_j > 0, \lambda_j > 0, 1 + \rho(0) > 0$ $1 + \eta_j\ (0) > 0,\ j = 1, \ldots, p$ (The inequalities $1 + \rho(t) > 0$ and $1 + \eta_i\ (t) > 0$ must be satisfied for every $t > 0$))	§ 28
4.	Non-linear block with monotonic characteristic	$\frac{d\xi}{dt} = -\rho_2\xi + \mu_\iota$ $\nu = \rho_1\xi + \mu_\iota,$ $\nu_\iota = \varphi(\nu)$	$\rho_1 > 0,\ \rho_2 \geqslant 0,$ φ = continuous function satisfying the conditions $\varphi\ (0) = 0$ and $\varphi\ (\nu_2) - \varphi\ (\nu_1) > 0$ (whenever $\nu_2 > \nu_1$).	§ 29 (Section 1)
5.	Ditto (a more general case)	$\beta\frac{d\nu}{dt} + \alpha\left(\nu - \frac{1}{\lambda_0}\varphi(\nu)\right) + \sum_{j=1}^{p} \delta_j(\nu - \rho\xi_j) = \mu_\iota$ $\frac{d\xi_j}{dt} = -(\rho_{1,j} + \rho_{2,j})\xi_j + \nu, j = 1, \ldots, p$ $\nu_\iota = \varphi(\nu)$	$\rho_{1,j} > 0,\ \rho_{2,j} \geqslant 0$ $\delta_j > 0,\ \beta > 0,\ \alpha \geqslant 0$ φ = continuous function satisfying the conditions $\varphi(0) = 0,\ \varphi(\nu)\left(\nu - \frac{1}{\lambda_0}\varphi(\nu)\right) > 0$ (for every ν_2) $\varphi(\nu_2) - \varphi(\nu_1) > 0$ (whenever $\nu_2 > \nu_1$).	§ 29 (Section 1)
6.	Block with a non-linearity depending on two variables	$\beta\frac{d\nu}{dt} + \alpha\left(\nu - \frac{1}{\lambda_0}\varphi_1(\nu - \varphi_2(\xi))\right) = \mu_\iota$ $\frac{d\xi}{dt} = \varphi_1(\nu - \varphi_2(\xi))$ $\nu_\iota = \varphi_1(\nu - \varphi_2(\xi))$	$\beta > 0,\ \alpha \geqslant 0$ φ_1 = continuous function satisfying the conditions: $\varphi_1\ (0) = 0$ and $\varphi_1(\nu)\left(\nu - \frac{1}{\lambda_0}\varphi_1(\nu)\right) > \varepsilon\nu^2$ (for every ν) φ_2 = differentiable function satisfying the conditions $\varphi_2(0) = 0$ and $\frac{d\varphi_2(\xi)}{d\xi} > 0$, (for every ξ)	§ 29 (Section 2)

NOTATIONS

The notations are those generally used in the literature. With a few exceptions indicated in the text, Greek letters are used for scalars, lower case letters for vectors and capital letters for matrices. The real and the imaginary parts of a complex number α are designated by $\mathrm{Re}\,\alpha$ and $\mathrm{Im}\,\alpha$ respectively; thus $\alpha = \mathrm{Re}\,\alpha + \mathrm{i}\,\mathrm{Im}\,\alpha$, $(\mathrm{i} = \sqrt{-1})$. The operation of complex conjugation is designated by an upper bar; e.g. $\overline{\alpha} = \mathrm{Re}\,\alpha - \mathrm{i}\,\mathrm{Im}\,\alpha$. The same symbols are used with vectors and matrices.

The transpose of a given matrix A is denoted by A' and the symbol* has the meaning $A^* = A'$. The same notations are used with vectors. If $A(\lambda)$ is a matrix polynomial, the symbol $A(\lambda)$ has the meaning $\overline{A}(\lambda) = A(\lambda)$, a convention used also with vectors and scalars.

The norm of a vector is defined by

$$\|v\| = \sqrt{v^*v}$$

and the norm of a matrix is

$$\|A\| = \sup_{\|v\|=1} \|Av\|$$

Besides the symbols mentioned above we sometimes use special symbols to denote particular operations (e.g. the sign $^{\Delta}$ introduced in Section 4, § 2). The identity matrix is designated by I. If necessary, we add indices to distinguish between identity matrices of different dimensions: e.g. I_1, I_2, etc. By diag $(\alpha_1, \alpha_2, \ldots, \alpha_n)$ we denote the diagonal $n \times n$-matrix with diagonal entries $\alpha_1, \alpha_2, \ldots, \alpha_n$. To denote functions, we use in general single letters (e.g. x, in § 2). When necessary we use the more precise notation $t \mapsto x(t)$. Unless otherwise specified in the context, the domain and the codomain of the functions are sets of complex numbers (or sets of complex vectors or complex matrices, as indicated by our convention of denoting vectors and matrices). If $x(t)$ for instance occurs in a relation which holds for every t we very often write simply x instead of $x(t)$ (see also footnote 1, § 25).

The matrix transfer functions are denoted by $\sigma \mapsto G(\sigma)$ or $i\omega \mapsto G(i\omega)$ (if we are interested only in the values of the transfer function for $\sigma = i\omega$, where ω is a real number). The scalar transfer functions are denoted by $\sigma \mapsto \gamma(\sigma)$ or $i\omega \mapsto \gamma(i\omega)$. (However, the letter γ alone or with an index added, might represent a constant.)

For the values of the characteristic function defined in § 5, we reserve throughout the work the symbol $H(\lambda, \sigma)$ while the symbol $\chi(\lambda, \sigma)$ is used in the scalar case (see § 3). It should be noted that $H(\lambda, \sigma)$ has different expressions in the continuous case (examined in § 5) and in the discrete case (examined in § 6). The characteristic polynomials introduced in the above paragraphs are denoted throughout the work by $(\lambda, \sigma) \mapsto \pi(\lambda, \sigma)$.

The letters N, P, Q and R (and in the scalar case N, ρ, q and R) have been generally reserved for the transformations introduced in §§ 5 and 2, respectively; the letters J, K, L, M (and in the scalar case J, $\varkappa$, l, M) have been reserved for the coefficients of the expressions under the summation (or integral) sign in formulas (2, § 2), (2, § 5), (2, § 6), etc. For writing in a compact form such expressions as (5, § 2), (4, § 5) or (4, § 6) we always use the letter D. However D has different meanings depending on the type of system we are investigating (see (6, § 2), (5, § 5) or (5, § 6)).

BIBLIOGRAPHY *

AIZERMAN, M. A., 1. *On a problem concerning the "overall" stability of dynamic systems*. Uspekhi Matemat. nauk, IV, 4, 1949 (in Russian).

AIZERMAN, M. A., GANTMACHER, F. R., 1. *Absolute stability of controlled systems*. Izd. Akad. Nauk SSSR, 1963 (in Russian).

BARBĂLAT, I., 1. *Systèmes d'équations différentielles d'oscillations non linéaires*. Revue de mathématiques pures et appliquées, Bucharest, IV, 2, 267—270, 1959.

BARBASHIN, E. A., KRASOVSKIY, N. N. 1. *On the overall stability of motion*. Dokl. A.N. SSSR, LXXXVI, 3, 1952 (in Russian).

BARBU VIOREL, 1. *Sur une équation intégrale non-linéaire*. Analele Ştiinţifice ale Universităţii "Al. I. Cuza", Iaşi, Secţiunea I a, Matematică, X, 1, 61—65, 1961.

BOCHNER, S., 1. *Vorlesungen über Fouriersche Integrale*. New York, Chelsea Publishing Company, 1918.

BOURBAKI, N., 1. *Eléments de mathématiques*. Livre II, Algèbre.

BROCKETT, R. W., WILLEMS, J. L. 1. *Frequency Domain Stability Criteria*. P.I. II. IEEE Trans. A.C., July, October, 1965.

BRUNOVSKI, P., 1. *On a problem concerning the linear system stability with a determinant class of inconsistencies actioning permanently*. Differentsjalnye uravnenija, II, 6, 1966 (in Russian).

CHANG, S. S. L., 1. *Synthesis of optimum control systems*. McGraw-Hill, 1961.

CORDUNEANU, C., 1. *Problèmes globaux dans la théorie des équations intégrales de Volterra*. Annali di matematica pura ed applicata, Bologna, IV, LXVII, 319—363, 1965.

2. *Sur une équation intégrale de la théorie du réglage automatique*. C.R. Acad. Sci., Paris, 256, 3564—3567.

3. *Sur une équation intégrale non-linéaire*. Analele Ştiinţifice ale universităţii "Al. I. Cuza" Iaşi, I. Matematică, Fizică, Chimie, IX, 2, 369—375, 1963.

DESOER C. A., 1. *Generalization of the Popov Criterion*. IEEE Trans. on Automatic Control. AC—10, 2, 182—185-April 1965.

GANTMACHER, F. R., 1. *Matrix theory*. Gost. izdat., Moskva, 1955 (in Russian).

* The present list is strictly confined to the references used in this work. This explains the omission of many works which are of great importance for stability theory. With the exception of very few papers and books — mentioned in the notes added at the revision of this translation — the list does not contain works which appeared after 1966 (the date of publication of the original of this book).

GELIG, A. H., 1. *The stability of controllable non-linear systems with distributed parameters in special cases.* Avtomatika i telemekhanika. XXVII, *4*, 1966 (in Russian).

2. *On the stability of controllable non-linear systems with an infinite number of degrees of freedom.* Prikladnaja Matematika i mekhanika, 30, *4*, 1966 (in Russian).

GOHBERG, I. T., KREIN, M. G., 1. *A system of integral equations on a semiaxis with kernels depending on different arguments.* Uspekhi matemat. nauk, 13, *2*, 1958 (80) (in Russian).

HALANAY, A., 1. *Qualitative theory of differential equations.* Editura Academiei, Bucureşti, 1963 (in Romanian).

2. *Absolute stability of some controlled non-linear systems,* Avtomatika i telemekhanika, XXV, 1964 (in Russian).

3. *Absolute stability of controlled non-linear systems with delayed argument.* Avtomatika i telemekhanika, XXV, *10*, 1393—98, 1964 (in Russian).

4. *Positive definite kernels and the stability of automatic systems.* Revue Roumaine de Math. pures et appliquées, IX, *8*, 751—765, 1964 (in Russian).

5. *Asymptotic behaviour of the solutions of certain non-linear integral equations.* Revue Roumaine de Math. pures et appliquées, IX, *8*, 765—777, 1965.

JURY, E. L., LEE, R. W., 1. *On the stability of a certain class of non-linear sampled-data systems. IEEE.* Transactions of automatic control AC—9, *1*, 51—62, Jan. 1964.

KALMAN, R. E., 1. *Contributions to the theory of optimal control.* Bot. Soc. Mat. Mexicana, 102—119, 1960.

2. *On the general theory of control systems.* Proc. 1st International Congress on Automatic Control, Moscow, 1960 ; published by Butterworths, *1*, 481—492, 1961.

3. *Canonical structure of linear dynamical systems.* Proc. Nat. Acad. Sci. USA, 48, 596—600, 1962.

4. *Mathematical description of linear dynamical systems.* SIAM J. Control, 1963.

5. *Lyapunov functions for the problem of Lur'e in automatic control.* Proc. Mat. Acad. Sci. USA, 49, 201—205, 1963.

6. *When is a linear control system optimal?.* J. Basic Engineering (Trans. ASME part D) *1*, 1964.

KALMAN, R. E., HO, Y. C., NARENDRA, K. S., 1. *Controllability of linear dynamical systems.* Contrib. Differential equations, *1*, 189—213, 1962.

KUDREVICH, I., 1. *A pewnym kryterium stabilnosti i jego zastosowaniach.* Warszawa — Pazdziernik, 1962.

2. *The stability of non-linear systems with inverse connection.* Avtomatika i telemekhanika, XXV, *3*, 1145—1155, 1964 (in Russian).

KURZWEIL, I., 1. *On the analytical construction of control systems.* Avtomatika i telemekhanika, 22, *6*, 1961 (in Russian).

LEFSCHETZ, S., 1. *Some mathematical considerations of non-linear automatic controls.* Contrib. Differential Equations, *1*, 1—28, 1963.

2. *Differential equations: Geometric theory.* Wiley (Interscience), New York, 1963.

3. *Stability of nonlinear control systems.* Academic Press, New York, London, 1965.

LETOV, A. M., 1. *The stability of controlled non-linear systems.* Gostkizdat, Moscow (2nd edition), 1962 (in Russian).
2. *The analytical construction of control systems.* Avtomatika i telemekhanika, 21, 1960 (in Russian).

LUCA, N., 1. *Sur quelques systèmes d'équations intégrales à noyau transitoire qui s'appliquent aux problèmes de réglage automatique.* Analele Ştiinţifice ale Universităţii "Al. I. Cuza" Iaşi, (I-a Matematică), X, *2*, 317—355, 1954.

LUR'E, A. I., 1. *Non-linear problems of the theory of automatic control.* Gostekizdat, Moscow, 1951 (in Russian).

LUR'E, A. I., POSTNIKOV, V. N., 1. *On the theory of the stability of controlled systems.* Prikladnaya Matematika i mekhanika, IX, *5*, 1945 (in Russian).

LIAPUNOV, A. M., 1. *The general problem of the stability of motion.* Gostkizdat, Moscow-Leningrad, 1950 (in Russian).

MAIGARIN, B. J., 1. *On the absolute stability of non-linear systems of third order automatic control.* Avtomatika i telemekhanika, XXIV, *6*, 1963 (in Russian).
2. *On the absolute stability of non-linear systems of automatic control with inverse tachometric connection.* Avtom. i telemekh., XXIV, *9*, 1963 (in Russian).

MASSERA, J. L., 1. *Contributions to stability theory.* Annals of Mathematics, July, 1956.

MOISIL, GR. C., 1. *Introduction in algebra I. Rings and ideals.* Ed. Academiei, Bucureşti, 1954 (in Russian).

MOROZAN, T., 1. *Remarques sur une note de Yakubovich.* C. R. Acad. Sci. (Paris), *254*, 1127—1129, 1962.

NAUMOV, B. N., 1. *A study of the absolute stability of the equilibrium position in non-linear systems of automatic control with the help of the logarithmic frequency characteristics.* Avtom. i telemekh. XXVI, *4*, 1965 (in Russian).

NAUMOV, B. N., TSYPKIN, I. Z., 1. *The frequency criterion of absolutely stable processes in the linear systems of automatic control.* Avtom. i telemekh., XXV, *6*, 1964 (in Russian).

ONICESCU, O., MIHOC, G., TULCEA, C. I. T., 1. *Probability calculus and applications.* Editura Academiei, Bucureşti, 1956 (in Romanian).

PEARSON, J. R., GIBSON, J. E., 1. *On the stability of a certain class of saturating sampled-data systems.* IEEE Trans. on Appl. and Ind., *70*, 1964.

PLISS, V. A., 1. *Some problems of the theory of the overall stability of motion.* Izd. L.G.U., 1958 (in Russian).

POPOV, V. M., 1. *The usage of the Bezoutienne in the theory of stability.* Studii şi cercetări de energetică, VIII, 1958, 87—103, in Romanian (Translated in French in Revue d'Electrotechnique et Energétique III, *1*, 1958).
2. *On the weakening of the sufficiency conditions of absolute stability.* Avtom. i telemekh., XIX, *1*, 1958 (in Russian).
3. *Stability criteria for non-linear systems of automatic control based on the utilization of the Laplace transform.* Studii şi cercetări de energetică, IX, *1*, 1959 (in Romanian).
4. *Sufficiency criteria of asymptotic stability for non-linear control systems with more executing parts.* Studii şi cercetări de energetică, IX, *4*, 1959 (in Romanian).

5. *New stability criteria for non-linear control systems.* Studii şi cercetări de energetică, IX, *1*, 1960, in Romanian (Translated in French in Revue d'El. et d'En., Bucharest, V, *1*, 1960).

6. *New graphical criteria for the stability of the steady state of non-linear control systems.* Studii şi cercetări de energetică, X, *3*, 1960, in Romanian (Translated in English in Rev. d'El. et d'En., V, *1*, 1961).

7. *Stability criteria for control systems with non-univocal elements.* Probleme de automatizare, *13*, Oct. 1960 (in Romanian).

8. *On the absolute stability of non-linear systems of automatic control.* Avtom. i telemekh., XXII, *8*, 1961 (in Russian).

9. *Stability criteria for some non-linear control systems. Automatica şi electronica*, 5, *3*, 1961 (in Romanian).

10. *On a critical case of absolute stability.* Avtom. i telemekh., XXIII, *2*, 1963 (in Russian).

11. *Solution of a new problem of the stability of controlled systems.* Avtom. i telemekh., XXIV, 1, 1963 (in Russian).

12. *Sur certaines inégalités intégrales concernant la théorie du réglage auto matique.* C. R. Acad. Sci. (Paris), *256*, 3568—3570, 22 April 1963.

13. *On a problem of the theory of the absolute stability of controlled systems.* Avtom. i telemekh., XXV, *9*, 1257—1262, 1964 (in Russian).

14. *Hyperstability of automatic control systems with several non-linear elements.* Revue Roumaine des Sciences techniques, *1*, 1961.

15. *A new criterion for the stability of systems containing nuclear reactors.* Revue d'El. et d'En. Série A, VIII, *1*, 1963.

16. *Hyperstability and optimality of automatic systems with several control functions.* Revue Roumaine des Sciences techniques, Série Electrotechnique et Energétique, 1964, *9* (Abridged version in Romanian, published in "Studii şi Cercetări de Energetică şi Electrotehnică", 14, *4*, 913—949, 1964).

17. *Hyperstability of control systems with variable coefficients.* Studii şi cercetări de energetică şi electrotehnică, 15, *2*, 1965, in Romanian (Translated in English in Revue Roumaine des Sciences Techniques. Série Electrotechnique et Energétique, 10, *2*, 1965).

Popov, V. M., Halanay, A., 1. *On the stability of the non-linear systems of the automatic control with delayed argument.* Avtom. i telemekh., XXIII, *7*, 1962 (in Russian).

Rekasius, Z. V., 1. *A stability criterion for feedback systems with non-linear element.* IEEE Transactions on Automatic Control, AC—9, *1*, 46—51, Jan. 1961.

Rozenwasser, E. N., 1. *On the absolute stability of non-linear systems.* Avtom. i telemekh., XXIV, *3*, 1963 (in Russian).

LaSalle, J. P., 1. *Complete stability of a non-linear control system.* Proc. D. Nat. Acad. Sci. USA, *48*, 600—603, 1962.

LaSalle, J. P., Lefschetz, S., 1. *Stability by Lyapunov's direct method with applications.* Academic Press, New York, 1961.

SANDBERG, I. W., 1. *A frequency-domain condition for the stability of feedback systems containing a single time varying non-linear element.* The Bell System Technical Journal, XLIII, *4*, Part. 2, 1601—1608, July 1964.

2. *On the L_2-boundedness of solutions of non-linear functional equations.* BSTJ, XLIII, *4*, 1581—1598, July 1964.

SMETS, H. B., 1. *On Welton's stability criterion for nuclear reactors*, J. Appl. Phys., *30*, 1623, 1958.

2. *Stability in the large of heterogeneous power reactors.* Bull. Acad. Roy. Belgique, Cl. Sci. *47*, 382—405, 1961.

3. *Problems in nuclear power reactor stability.* Brussels, Université Libre de Bruxelles, 1962.

STOILOW, S., 1. *The theory of functions of a complex variable.* Editura Academiei, București, 2 vol.; 2nd vol. in cooperation with Cabiria Andreian Cazacu (in Romanian).

SZEGÖ. G., 1. *On the absolute stability of sampled-data control systems.* Proc. Nat. Acad. Sci. USA, 50, *3*, 1963.

2. *Sur la stabilité absolue d'un système non linéaire discret.* C. R. Acad. Sci., Paris, 257, *11*, 1963.

SZEGÖ, G., KALMAN, R. E., 1. *Sur la stabilité absolue d'un système d'équations aux différences finies.* C. R. Acad. Sci. (Paris), *357*, 338—390, 1963.

TSYPKIN, I. Z., 1. *On the overall stability of non-linear automatic systems and impulses.* Dokl. A.N. SSSR, 145, *1*, 1962 (in Russian).

2. *Die absolute Stabilität nicht linearer Impulsregelsysteme.* Regelungstechnik, *4*, 1963.

3. *Frequency criteria of the absolute stability of non-linear impulse systems*, Avtom. i telemekh., XXV, *3*, 1964 (in Russian).

4. *On the absolute stability of a class of non-linear automatic impulse systems.* Avtom. i telemekh., XXV, *7*, 1030—1035, 1964 (in Russian).

5. *The basic theory of non-linear automatic impulse systems.* (Papers of the second IFCA Congress, Basel, 1963, in Russian).

6. *Einige Verallgemeinerungen des Kriteriums der absoluten Stabilität.* Regelungstechnik., H. 11, *10*, 1963.

VEKUA, N. P., 1. *Systems of singular integral equations.* Gostkizdat, Moskva — Leningrad, 1950 (in Russian).

WELTON, T. A., 1. *Kinetics of Stationary Reactor Systems.* Proc. First Int. Conf. Peaceful Uses of Atomic Energy, 1955, 5, 377—388.

YAKUBOVICH, V. A., 1. *On the systems of non-linear differential equations of automatic control with a single control unit.* Vestn. L.G.U., 7, *2*, 1960 (in Russian).

2. *Solutions of some matrix inequalities occurring in the theory of automatic control.* Dokl. A. N. SSSR, 143, *6*, 1962 (in Russian).

3. *The frequency conditions of the absolute stability of controlled systems with hysteresis non-linearities.* Dokl. A.N. SSSR, 149, *2*, 1963 (in Russian).

4. *Solution of special matrix inequalities occurring in the theory of non-linear control.* Dokl. A.N. SSSR, 156, *2*, 1964 (in Russian).

5. *The absolute stability of controlled non-linear systems in critical cases.* Avtomatika i telemekhanika, XXIV, 3, 1963 (1st part), *6*, 1963 (2nd part); XXV, *5*, 1964 (3rd part), (in Russian).

6. *The method of matrix inequalities in the theory of the stability of controlled non-linear systems.* Avtomatika i telemekhanika, XXV, *7*, 1964 (1st part), XXVI, *4*, 1965 (2nd part); XXVI, *5*, 1965 (3rd part) (in Russian).

7. *The frequency conditions of absolute stability and dissipativity of controlled systems with a single differentiable nonlinearity.* Dokl. A.N. SSSR, 160, *2*, 1965 (in Russian).

8. *The frequency conditions of absolute stability of controlled systems with several non-stationary, linear or non-linear units.* Avtomatika i telemekhanika, 1967 (in Russian).

YOULA, D. C., 1. *On the factorization of rational matrices.* IRE Trans. on Information Theory, 1, *3*, 7 July 1961.

ZAMES, G., 1. *On the input-output stability of time varying non-linear feedback systems.* IEEE Trans. A.C. AC−11, P I, II, April, July, 1966.